Applied Mathematical Sciences
Volume 172

Editors
S.S Antman
Department of Mathematics
and
Institute for Physical Science and Technology
University of Maryland
College Park, MD 20742-4015
USA
ssa@math.umd.edu

J.E. Marsden
Control and Dynamical Systems, 107-81
California Institute of Technology
Pasadena, CA 91125
USA
marsden@cds.caltech.edu

L. Sirovich
Laboratory of Applied Mathematics
Department of Biomathematical Sciences
Mount Sinai School of Medicine
New York, NY 10029-6574
lsirovich@rockefeller.edu

For other titles published in this series, go to
http://www.springer.com/series/34

Henk Broer · Floris Takens

Dynamical Systems
and Chaos

Springer

Henk Broer
University of Groningen
Johann Bernoulli Institute
  for Mathematics and Computer Science
The Netherlands
h.w.broer@rug.nl

Floris Takens
University of Groningen
Johann Bernoulli Institute
  for Mathematics and Computer Science
The Netherlands

An earlier version of this book was published by Epsilon Uitgaven (Parkstraat 11, 3581 PB te Utrecht, Netherlands) in 2009.

ISSN 0066-5452
ISBN 978-1-4614-2712-4   ISBN 978-1-4419-6870-8 (eBook)
DOI 10.1007/978-1-4419-6870-8
Springer New York Dordrecht Heidelberg London

Mathematics Subject Classification (2010): 34A26, 34A34, 34CXX, 34DXX, 37XX, 37DXX, 37EXX, 37GXX, 54C70, 58KXX

Printed on acid-free paper

Springer is part of Springer Science+Business Media (www.springer.com)

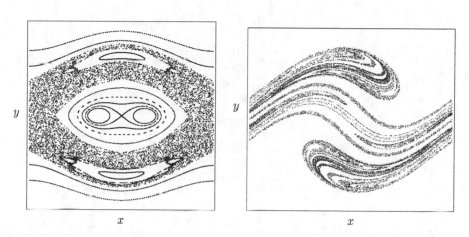

Iterations of a stroboscopic map of the swing displaying complicated, probably chaotic dynamics. In the bottom figure the swing is damped while in the top figure it is not. For details see Appendix C.

Preface

Everything should be made as simple as possible,
but not one bit simpler
Albert Einstein (1879 1955)

The discipline of *Dynamical Systems* provides the mathematical language describing the time dependence of deterministic systems. For the past four decades there has been ongoing theoretical development. This book starts from the phenomenological point of view, reviewing examples, some of which have guided the development of the theory. So we discuss oscillators, such as the pendulum in many variations, including damping and periodic forcing, the Van der Pol system, the Hénon and logistic families, the Newton algorithm seen as a dynamical system, and the Lorenz and Rössler systems. The doubling map on the circle and the Thom map (also known as the Arnold cat map) on the 2-dimensional torus are useful toy models to illustrate theoretical ideas such as symbolic dynamics. In the appendix the 1963 Lorenz model is derived from appropriate partial differential equations.

The phenomena concern equilibrium, periodic, multi- or quasi-periodic, and chaotic dynamics as these occur in all kinds of modelling and are met both in computer simulations and in experiments. The application area varies from celestial mechanics and economical evolutions to population dynamics and climate variability. One general motivating question is how one should produce intelligent interpretations of data that come from computer simulations or from experiments. For this thorough theoretical investigations are needed.

One key idea is that the dynamical systems used for modelling should be 'robust,' which means that relevant qualitative properties should be persistent under small perturbations. Here we can distinguish between variations of the initial state or of system parameters. In the latter case one may think of the coefficients in the equations of motion. One of the strongest forms of persistence, called structural stability, is discussed, but many weaker forms often are more relevant. Instead of individual evolutions one also has to consider invariant manifolds, such as stable and unstable manifolds of equilibria and periodic orbits, and invariant tori as these often mark the geometrical organisation of the state space. Here a notion like (normal) hyperbolicity comes into play. In fact, looking more globally at the structure of dynamical systems, we consider attractors and their basin boundaries in the examples of the solenoid and the horseshoe map.

The central concept of the theory is chaos, defined in terms of unpredictability. The prediction principle we use is coined *l'histoire se répète*, that is, in terms of and based on past observations. The unpredictability then is expressed in a dispersion exponent, a notion related to entropy and Lyapunov exponents. Structural stability turns out to be useful in proving that the chaoticity of the doubling and Thom maps is persistent under small perturbations. The ideas on predictability are also used in the development of reconstruction theory, where important dynamical invariants such as the box-counting and correlation dimensions of an attractor, which often are fractal, and related forms of entropy are reconstructed from time series based on a long segment of the past evolution. Also numerical estimation methods of the invariants are discussed; these are important for applications, for example, in early warning systems for chemical processes.

Another central subject is formed by multi- and quasi-periodic dynamics. Here the dispersion exponent vanishes and quasi-periodicity to some extent forms the 'order in between the chaos.' We deal with circle dynamics near rigid rotations and similar settings for area-preserving and holomorphic maps. Persistence of quasi-periodicity is part of Kolmogorov–Arnold–Moser (or KAM) theory. The major motivation to this subject has been the fact that the motion of the planets, to a very good approximation, is multiperiodic. We also discuss KAM theory in the context of coupled and periodically driven oscillators of various types. For instance we encounter families of quasi-periodic attractors in periodically driven and in coupled Van der Pol oscillators. In the main text we focus on the linear small divisor problem that can be solved directly by Fourier series. In an appendix we show how this linear problem is being used to prove a KAM theorem by Newton iterations.

In another appendix we discuss transitions (or bifurcations) from orderly to more complicated dynamics upon variation of system parameters, in particular Hopf, Hopf–Neĭmark–Sacker, and quasi-periodic bifurcations, indicating how this yields a unified theory for the onset of turbulence in fluid dynamics according to Hopf–Landau–Lifschitz–Ruelle–Takens. Also we indicate a few open problems regarding both conservative and dissipative chaos. In fact we are dealing with a living theory with many open ends, where it should be noted that the mathematics of nonlinear science is notoriously tough. Throughout the text we often refer to these ongoing developments.

This book aims at a wide audience. On the one hand, the first four chapters for a great many years have been used for an undergraduate course in dynamical systems. Material from the last two chapters and from the appendices has been used quite a lot for Masters and PhD courses. All chapters are concluded by an exercise section. Apart from a good knowledge of ordinary differential equations, some maturity in general mathematical concepts, such as metric spaces, topology, and measure theory will also help and an appendix on these subjects has been included. The book also is directed towards researchers, where one of the challenges is to help applied researchers acquire background for a better understanding of the data provided them by computer simulation or experiment.

A Brief Historical Note

Occasionally in the text a brief historical note is presented and here we give a bird's-eye perspective. One historical line starts a bit before 1900 with Poincaré, originating from celestial mechanics, in particular from his studies of the unwieldy 3-body problem [64, 158, 191], which later turned out to have chaotic evolutions. Poincaré, among other things, introduced geometry in the theory of ordinary differential equations; instead of studying only one evolution, trying to compute it in one way or another, he was the one who proposed considering the geometrical structure and the organisation of all possible evolutions of a dynamical system. This qualitative line of research was picked up later by the topologists and Fields medalists Thom and Smale and by many others in the 1960s and 1970s; this development leaves many traces in the present book.

Around this same time a really significant input to 'chaos' theory came from outside mathematics. We mention Lorenz's meteorological contribution with the celebrated Lorenz attractor [160, 161], the work of May on population dynamics [166], and the input of Hénon–Heiles from astronomy, to which in 1976 the famous Hénon attractor was added [137]. Also the remarks by Feynman et al. [124] Vol. 1, pp. 38–39 or Vol. 3, pp. 2–9 concerning 'uncertainty' in classical mechanics as a consequence of molecular chaos, are noteworthy. It should be said that during this time the computer became an important tool for lengthy computations, simulations, and graphical representations, which had a tremendous effect on the field. Many of these developments are also dealt with below.

From the early 1970s on these two lines merged, leading to the discipline of *nonlinear dynamical systems* as it is known now, which is an exciting area of research and which has many applications in the classical sciences and in the life sciences, meteorology, and so on.

Guide to the Literature

We maintain a wide scope, but given the wealth of material on this subject, we obviously cannot aim for completeness. However, the book does aim to be a guide to the dynamical systems literature. Below we present a few general references to textbooks, handbooks, and encyclopædia. We have ordered the bibliography at the end, subdividing as follows.

- The references [2–5, 7–13, 15–17] form general contributions to the theory of nonlinear dynamical systems at a textbook level.
- For textbooks on the ergodic theory of nonlinear dynamics see [18, 19].
- An important special class of dynamical system is formed by Hamiltonian systems, used for modelling the dynamics of frictionless mechanics. Although we pay some attention to this broad field, for a more systematic study we refer to [20–24].
- More in the direction of bifurcation and normal form theory we like to point to [25–32].

- A few handbooks on dynamical systems are [33–36], where in the latter reference we like to point to the chapters [37–41].
- We also mention the *Russian Encyclopædia* [42], in particular [43–45] as well as the *Encyclopædia on Complexity* [47], in particular to the contributions [48–54].

For a more detailed discussion of the bibliography we refer to Appendix E.

Acknowledgements

We thank M. Cristina Ciocci, Bob Devaney, Konstantinos Efstathiou, Aernout van Enter, Heinz Hanßmann, Sijbo-Jan Holtman, Jun Hoo, Igor Hoveijn, George Huitema, Tasso Kaper, Ferry Kwakkel, Anna Litvak-Hinenzon, Olga Lukina, Jan van Maanen, Jan van Mill, Vincent Naudot, Khairul Saleh, Mikhail Sevryuk, Alef Sterk, Erik Thomas, Gert Vegter, Ferdinand Verhulst, Renato Vitolo, Holger Waalkens, and Florian Wagener for their comments and helpful discussions. Konstantinos is especially thanked for his help with the graphics. The first author thanks Boston University, Pennsylvania State University, and the Universitat de Barcelona for hospitality during the preparation of this book.

Groningen   *Henk Broer*
July 2010   *Floris Takens*[†]

[†] Deceased.

Contents

Chapter 1
Examples and definitions of dynamical phenomena

A dynamical system can be any mechanism that evolves deterministically in time.
Simple examples can be found in mechanics, one may think of the pendulum or the
solar system. Similar systems occur in chemistry and meteorology. We should note
that in biological and economic systems it is less clear when we are dealing with
determinism.[1] We are interested in *evolutions*, that is, functions that describe the
state of a system as a function of *time* and that satisfy the *equations of motion* of the
system. Mathematical definitions of these concepts follow later.

The simplest type of evolution is *stationary*, where the state is constant in time.
Next we also know *periodic evolutions*. Here, after a fixed period, the system always
returns to the same state and we get a precise repetition of what happened in the
previous period. Stationary and periodic evolutions are very regular and predictable.
Here prediction simply is based on matching with past observations. Apart from
these types of evolutions, in quite simple systems one meets evolutions that are not
so regular and predictable. In cases where the unpredictability can be established in
a definite way, we speak of *chaotic* behaviour.

In the first part of this chapter, using systems such as the pendulum with or without damping or external forcing, we give an impression of the types of evolution
that may occur. We show that usually a given system has various types of evolution. Which types are typical or prevalent strongly depends on the kind of system
at hand; for example, in mechanical systems it is important whether we consider
dissipation of energy, for instance by friction or damping. During this exposition
the mathematical framework in which to describe the various phenomena becomes
clearer. This determines the language and the way of thinking of the discipline of
Dynamical Systems.

In the second part of this chapter we are more explicit, giving a formal definition of dynamical system. Then concepts such as *state*, *time*, and the like are also
discussed. After this, returning to the previous examples, we illustrate these concepts. In the final part of the chapter we treat a number of miscellaneous examples

---

[1] The problem of whether a system is deterministic is addressed later, as well as the question of how
far our considerations (partially) keep their value in situations that are not (totally) deterministic.
Here one may think of sensitive dependence on initial conditions in combination with fluctuations
due to thermal molecular dynamics or of quantum fluctuations; compare [203].

H.W. Broer and F. Takens, *Dynamical Systems and Chaos*,
Applied Mathematical Sciences 172, DOI 10.1007/978-1-4419-6870-8_1,
© Springer Science+Business Media, LLC 2011

that regularly return in later parts of the book. Here we particularly pay attention to examples that, historically speaking, have given direction to the development of the discipline. These examples therefore act as running gags throughout the book.

1.1   The pendulum as a dynamical system

As a preparation to the formal definitions we go into later in this chapter, we first treat the pendulum in a number of variations: with or without damping and external forcing. We present graphs of numerically computed evolutions. From these graphs it is clear that qualitative differences exist between the various evolutions and that the type of evolution that is typical or prevalent strongly depends on the context at hand. This already holds for the restricted class of mechanical systems considered here.

1.1.1   The free pendulum

The planar mathematical pendulum consists of a rod, suspended at a fixed point in a vertical plane in which the pendulum can move. All mass is thought of as being concentrated in a point mass at the end of the rod (see Figure 1.1), and the rod itself is massless. Also the rod is assumed stiff. The pendulum has mass m and length ℓ. We moreover assume the suspension to be such that the pendulum not only can oscillate, but also can go 'over the top'. In the case without external forcing, we speak of the *free* pendulum.

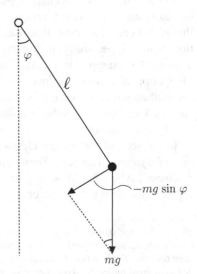

Fig. 1.1 Sketch of the planar mathematical pendulum. Note that the suspension is such that the pendulum can go 'over the top'.

1.1.1.1 The free undamped pendulum

In the case without damping and forcing the pendulum is only subject to gravity, with acceleration g. The gravitational force is pointing vertically downward with strength mg and has a component $-mg \sin \varphi$ along the circle described by the point mass; see Figure 1.1. Here φ is the angle between the rod and the downward vertical, often called 'deflection', expressed in radians. The distance of the point mass from the 'rest position' ($\varphi = 0$), measured along the circle of all its possible positions, then is $\ell\varphi$. The relationship between force and motion is determined by Newton's[2] famous law $F = m\, a$, where F denotes the force, m the mass, and a the acceleration.

By the stiffness of the rod, no motion takes place in the radial direction and we therefore just apply Newton's law in the φ-direction. The component of the force in the φ-direction is given by $-mg \sin \varphi$, where the minus sign accounts for the fact that the force is driving the pendulum back to $\varphi = 0$. For the acceleration a we have

$$a = \frac{d^2(\ell\varphi)}{dt^2} = \ell\frac{d^2\varphi}{dt^2},$$

where $(d^2\varphi/dt^2)(t) = \varphi''(t)$ is the second derivative of the function $t \mapsto \varphi(t)$. Substituted into Newton's law this gives

$$m\ell\varphi'' = -mg \sin \varphi$$

or, equivalently,

$$\varphi'' = -\frac{g}{\ell} \sin \varphi, \tag{1.1}$$

where obviously $m, \ell > 0$. So we derived the *equation of motion* of the pendulum, which in this case is an *ordinary differential equation*. This means that the evolutions are given by the functions $t \mapsto \varphi(t)$ that satisfy the equation of motion (1.1). In the sequel we abbreviate $\omega = \sqrt{g/\ell}$.

Observe that in the equation of motion (1.1) the mass m no longer occurs. This means that the mass has no influence on the possible evolutions of this system. According to tradition, as an experimental fact this was already known to Galileo,[3] a predecessor of Newton concerning the development of classical mechanics.

Remark (Digression on Ordinary Differential Equations I). In the above example, the equation of motion is an ordinary differential equation. Concerning the solutions of such equations we make the following digression.

1. From the theory of ordinary differential equations (e.g., see [16, 69, 118, 144]) it is known that such an ordinary differential equation, given initial conditions or an initial state, has a uniquely determined solution.[4] Because the differential

---

[2] Sir Isaac Newton 1642–1727.

[3] Galileo Galilei 1564–1642.

[4] For a more elaborate discussion we refer to §1.2.

equation has order two, the initial state at $t = 0$ is determined by the two data $\varphi(0)$ and $\varphi'(0)$, thus both the position and the velocity at the time $t = 0$. The theory says that, given such an initial state, there exists exactly one solution $t \mapsto \varphi(t)$, mathematically speaking a function, also called 'motion'. This means that position and velocity at the instant $t = 0$ determine the motion for all future time.[5] In the discussion later in this chapter, on the state space of a dynamical system, we show that in the present pendulum case the states are given by pairs (φ, φ'). This implies that the evolution, strictly speaking, is a map $t \mapsto (\varphi(t), \varphi'(t))$, where $t \mapsto \varphi(t)$ satisfies the equation of motion (1.1). The plane with coordinates (φ, φ') often is called the *phase plane*. The fact that the initial state of the pendulum is determined for all future time, means that the pendulum system is *deterministic*.[6]

2. Between existence and explicit construction of solutions of a differential equation there is a wide gap. Indeed, only for quite simple systems,[7] such as the harmonic oscillator with equation of motion

$$\varphi'' = -\omega^2 \varphi,$$

can we explicitly compute the solutions. In this example all solutions are of the form $\varphi(t) = A\cos(\omega t + B)$, where A and B can be expressed in terms of the initial state $(\varphi(0), \varphi'(0))$. So the solution is periodic with period $2\pi/\omega$ and with amplitude A, whereas the number B, which is only determined up to an integer multiple of 2π, is the phase at the time $t = 0$. Because near $\varphi = 0$ we have the linear approximation

$$\sin \varphi \approx \varphi,$$

in the sense that $\sin \varphi = \varphi + o(|\varphi|)$, we may consider the harmonic oscillator as a linear approximation of the pendulum. Indeed, the pendulum itself also has oscillating motions of the form $\varphi(t) = F(t)$, where $F(t) = F(t + P)$ for a real number P, the period of the motion. The period P varies for different solutions, increasing with the amplitude. It should be noted that the function F occurring here, is not explicitly expressible in elementary functions. For more information see [153].

3. General methods exist for obtaining numerical approximations of solutions of ordinary differential equations given initial conditions. Such solutions can only be computed for a restricted time interval. Moreover, due to accumulation of errors, such numerical approximations generally will not meet with reasonable criteria of reliability when the time interval is growing (too) long.

---

[5] Here certain restrictions have to be made, discussed in a digression later this chapter.

[6] In Chapter 6 we show that the concept of determinism for more general systems is not so easy to define.

[7] For instance, for systems that are both linear and autonomous.

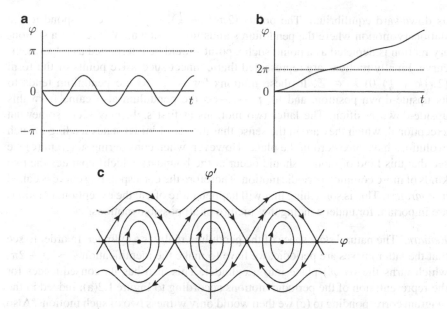

Fig. 1.2 Evolutions of the free undamped pendulum: (**a**) the pendulum oscillates; (**b**) the pendulum goes 'over the top'; (**c**) a few integral curves in the phase plane; note that φ should be identified with $\varphi + 2\pi$. Also note that the integral curves are contained in level curves of the energy function H (see (1.2)).

We now return to the description of motions of the pendulum without forcing or damping, that is, the free undamped pendulum. In Figure 1.2 we present a number of characteristic motions, or evolutions, as these occur for the pendulum.

In diagrams (a) and (b) we plotted φ as a function of t. Diagram (a) shows an oscillatory motion, that is, where there is no going 'over the top' (i.e., over the point of suspension). Diagram (b) shows a motion where this does occur. In the latter case the graph does not directly look periodic, but when realising that φ is an angle variable, which means that only its values modulo integer multiples of 2π are of interest for the position, we conclude that this motion is also periodic. In diagram (c) we represent motions in a totally different way. Indeed, for a number of possible motions we depict the integral curve in the (φ, φ')-plane, that is, the phase plane. The curves are of the parametrised form $t \mapsto (\varphi(t), \varphi'(t))$, where t varies over $\mathbb{R}$. We show six periodic motions, just as in diagram (a), noting that in this representation each motion corresponds to a closed curve. Also periodic motions are shown as in diagram (b), for the case where φ continuously increases or decreases. Here the periodic character of the motion does not show so clearly. This becomes more clear if we identify points on the φ-axis which differ by an integer multiple of 2π.

Apart from these periodic motions, a few other motions have also been represented. The points $(2\pi k, 0)$, with integer k, correspond to the stationary motion (which rather amounts to rest and not to motion) where the pendulum hangs in its downward equilibrium. The points $(2\pi(k + \frac{1}{2}), 0)$ $k \in \mathbb{Z}$ correspond to the stationary motion where the pendulum stands upside down. We see that a stationary motion is depicted as a point; such a point should be interpreted as a 'constant' curve. Finally motions are represented that connect successive points of the form $(2\pi(k + \frac{1}{2}), 0)$ $k \in \mathbb{Z}$. In these motions for $t \to \infty$ the pendulum tends to its upside-down position, and for $t \to -\infty$ the pendulum also came from this upside-down position. The latter two motions at first sight may seem somewhat exceptional, which they are in the sense that the set of initial states leading to such evolutions have area zero in the plane. However, when considering diagram (c) we see that this kind of motion should occur at the boundary which separates the two kinds of more common periodic motion. Therefore the corresponding curve is called a *separatrix*. This is something that will happen more often: the exceptional motions are important for understanding the 'picture of all possible motions.'

Remark. The natural state space of the pendulum really is a cylinder. In order to see that the latter curves are periodic, we have to make the identification $\varphi \sim \varphi + 2\pi$, which turns the (φ, φ')-plane into a cylinder. This also has its consequences for the representation of the periodic motions according to Figure 1.2(a); indeed in the diagram corresponding to (c) we then would only witness two of such motions. Also we see two periodic motions where the pendulum goes 'over the top'.

For the mathematical understanding of these motions, in particular of the representations of Figure 1.2(c), it is important to realise that our system is undamped. Mechanics then teaches us that *conservation of energy* holds. The energy in this case is given by

$$H(\varphi, \varphi') = m\ell^2 \left(\frac{1}{2}(\varphi')^2 - \omega^2 \cos \varphi \right). \tag{1.2}$$

Notice that H has the format 'kinetic plus potential energy.' Conservation of energy means that for any solution $\varphi(t)$ of the equation of motion (1.1), the function $H(\varphi(t), \varphi'(t))$ is constant in t. This fact also follows from more direct considerations; compare Exercise 1.3. This means that the solutions as indicated in diagram (c) are completely determined by level curves of the function H. Indeed, the level curves with H-values between $-m\ell^2\omega^2$ and $+m\ell^2\omega^2$ are closed curves corresponding to oscillations of the pendulum. And each level curve with $H > m\ell^2\omega^2$ consists of two components where the pendulum goes 'over the top': in one component the rotation is clockwise and in the other one counterclockwise. The level $H = m\ell^2\omega^2$ is a curve with double points corresponding to the exceptional motions just described: the upside-down position and the motions that for $t \to \pm\infty$ tend to this position.

We briefly return to a remark made in the above digression, saying that the explicit expression of solutions of the equation of motion is not possible in terms of elementary functions. Indeed, considering the level curve with equation $H(\varphi, \varphi') = E$, we solve to

$$\varphi' = \pm\sqrt{2\left(\frac{E}{m\ell^2} + \omega^2\cos\varphi\right)}.$$

The time it takes for the solution to travel between $\varphi = \varphi_1$ and $\varphi = \varphi_2$ is given by the integral

$$T(E, \varphi_1, \varphi_2) = \left| \int_{\varphi_1}^{\varphi_2} \frac{d\varphi}{\sqrt{2\left(\frac{E}{m\ell^2} + \omega^2 \cos\varphi\right)}} \right|,$$

which cannot be 'solved' in terms of known elementary functions. Indeed, it is a so-called *elliptic* integral, an important subject of study for complex analysis during the nineteenth century. For more information see [153]. For an oscillatory motion of the pendulum we find values of φ, where $\varphi' = 0$. Here the oscillation reaches its maximal values $\pm\varphi_E$, where the positive value usually is called the amplitude of oscillation. We get

$$\pm\varphi_E = \arccos\left(-\frac{E}{m\ell^2\omega^2}\right).$$

The period $P(E)$ of this oscillation then is given by $P(E) = 2T(E, -\varphi_E, \varphi_E)$.

1.1.1.2 The free damped pendulum

In the case that the pendulum has damping, dissipation of energy takes place and a possible motion is bound to converge to rest: we speak of the damped pendulum. For simplicity it is here assumed that the damping or friction force is proportional to the velocity and of opposite sign.[8] Therefore we now consider a pendulum, the equation of motion of which is given by

$$\varphi'' = -\omega^2 \sin\varphi - c\varphi', \tag{1.3}$$

where $c > 0$ is determined by the strength of the damping force and on the mass m. For nonstationary solutions of this equation of motion the energy H, given by (1.2), decreases. Indeed, if $\varphi(t)$ is a solution of (1.3), then

$$\frac{dH(\varphi(t), \varphi'(t))}{dt} = -cm\ell^2(\varphi'(t))^2. \tag{1.4}$$

Equation (1.4) confirms that friction makes the energy decrease as long as the pendulum moves (i.e., as long as $\varphi' \neq 0$). This implies that we cannot expect periodic motions to occur (other than stationary). We expect that any motion will tend to a

---

[8] To some extent this is a simplifying assumption, but we are almost forced to this. In situations with friction there are no simple first principles giving unique equations of motion. For an elementary treatment on the complications that can occur when attempting an exact modelling of friction phenomena, see [124]: Vol. 1, Chapter 12, §§12.2 and 12.3. Therefore we are fortunate that almost all qualitative statements are independent of the precise formula used for the damping force. The main point is that, as long as the pendulum moves, the energy decreases.

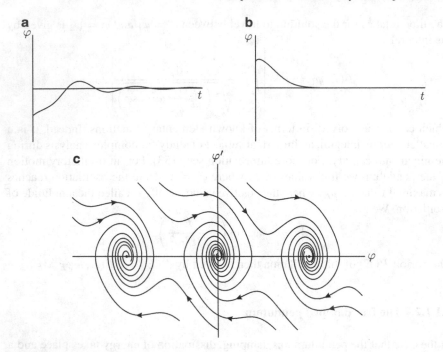

Fig. 1.3 Evolutions of the free damped pendulum: (**a**) the motion of the pendulum damps to the lower equilibrium; (**b**) As under (a), but faster: this case often is called 'overdamped'; (**c**) a few evolutions in the phase plane.

state where the pendulum is at rest. This indeed turns out to be the case. Moreover, we expect that the motion generally will tend to the stationary state where the pendulum hangs downward. This is exactly what is shown in the numerical simulations of Figure 1.3, represented in the same way as in Figure 1.2.

Also in this case there are a few exceptional motions, which, however, are not depicted in diagram (c). To begin with we have the stationary solution, where the pendulum stands upside down. Next there are motions $\varphi(t)$, in which the pendulum tends to the upside-down position as $t \to \infty$ or as $t \to -\infty$.

We noted before that things do not change too much when the damping law is changed. In any case this holds for the remarks made up to now. Yet there are possible differences in the way the pendulum tends to its downward equilibrium. For instance, compare the cases (a) and (b) in Figure 1.3. In case (a) we have that $\varphi(t)$ passes through zero infinitely often when tending to its rest position, whereas in diagram (b) this is not the case: now $\varphi(t)$ creeps to equilibrium; here one speaks of 'overdamping'. In the present equation of motion (1.3) this difference between (a) and (b) occurs when the damping strength c increases and passes a certain threshold value. In the linearised equation

$$\varphi'' = -\omega^2\varphi - c\varphi',$$

that can be solved explicitly, this effect can be directly computed. See Exercise 1.5.

We summarise what we have seen so far regarding the dynamics of the free pendulum. The undamped case displays an abundance of periodic motions, which is completely destroyed by the tiniest amount of damping. Although systems without damping in practice do not occur, still we often witness periodic behaviour. Among other things this is related to the occurrence of *negative* damping, that is directed in the same direction as the motion. Here the damping or friction force has to deliver work and therefore external energy is needed. Examples of this can be found among electrical oscillators, as described by Van der Pol [192, 193].[9] A more modern reference is [144] Chapter 10; also see below. Another example of a kind of negative damping occurs in a mechanical clockwork that very slowly consumes the potential energy of a spring or a weight. In the next section we deal with periodic motions that occur as a consequence of external forces.

1.1.2   The forced pendulum

Our equations of motion are based on Newton's law $F = m a$, where the acceleration a, up to a multiplicative constant, is given by the second derivative φ''. Therefore we have to add the external force to the expression for φ''. We assume that this force is periodic in the time t, even that it has a simple cosine shape. In this way, also adding damping, we arrive at the following equation of motion

$$\varphi'' = -\omega^2 \sin \varphi - c\varphi' + A \cos \Omega t. \tag{1.5}$$

To get an impression of the types of motion that can occur now, in Figure 1.4 we show a number of motions that have been computed numerically. For $\omega = 2.5$, $c = 0.5$, and $A = 3.8$, we let Ω vary between 1.4 and 2.1 with steps of magnitude 0.1. In all these diagrams time runs from 0 to 100, and we take $\varphi(0) = 0$ and $\varphi'(0) = 0$ as initial conditions.

We observe that in the diagrams with $\Omega = 1.4$, 1.5, 1.7, 1.9, 2.0, and 2.1 the motion is quite orderly: disregarding transient phenomena the pendulum describes a periodic oscillatory motion. For the values of Ω in between, the oscillatory motion of the pendulum is alternated by motions where it goes 'over the top' several times. It is not even clear from the observation with $t \in [0, 100]$, whether the pendulum will ever tend to a periodic motion. It indeed can be shown that the system, with equation of motion (1.5) with well-chosen values of the parameters ω, c, A, and Ω, has motions that never tend to a periodic motion, but that keep going on in a weird fashion. In that case we speak of *chaotic* motions. Below, in Chapter 2, we discuss chaoticity in connection with unpredictability.

In Figure 1.5 we show an example of a motion of the damped forced pendulum (1.5), that only after a long time (say about 100 time units) tends to a periodic motion. So, here we are dealing with a transient phenomenon that takes a long time.

---

[9] Balthasar van der Pol 1889–1959.

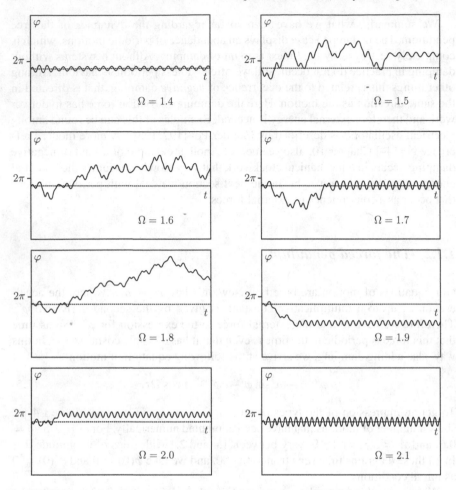

Fig. 1.4 Motions of the forced damped pendulum for various values of Ω for t running from 0 to 100 and with $\varphi(0) = 0$ and $\varphi'(0) = 0$ as initial conditions. $\omega = 2.5$, $c = 0.5$, and $A = 3.8$.

We depict the solution as a t-parametrised curve in the (φ, φ')-plane. The data used are $\omega = 1.43$, $\Omega = 1$, $c = 0.1$ and $A = 0.2$. In the left diagram we took initial values $\varphi(0) = 0.3$, $\varphi'(0) = 0$, and t traversed the interval $[0, 100]$. In the right diagram we took as initial state the end state of the left diagram.

We finish with a few remarks on the forced pendulum without damping; that is, with equation of motion (1.5) where $c = 0$:

$$\varphi'' = -\omega^2 \sin \varphi + A \cos \Omega t.$$

We show two motions in the diagrams of Figure 1.6. In both cases $\Omega = 1$, $A = 0.01$, $\varphi(0) = 0.017$, $\varphi'(0) = 0$, and the time interval is $[0, 200]$. Left, $\omega = \frac{1}{2}(1 + \sqrt{5}) \approx 1.61803$ (golden ratio). Right, $\omega = 1.602$, which is very close to

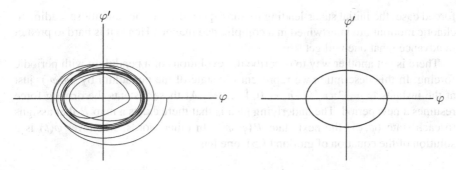

Fig. 1.5 Evolution of the damped and forced pendulum (1.5) as a t-parametrised curve in the (φ, φ')-plane. Left: $t \in [0, 100]$, the evolution tends to a periodic motion. Right: $t > 100$, continuation of the segment in the left diagram. We took $\omega = 1.43$, $\Omega = 1$, $A = 0.2$, and $c = 0.1$. Initial conditions in the left diagram are $\varphi(0) = 0.3$, $\varphi'(0) = 0$.

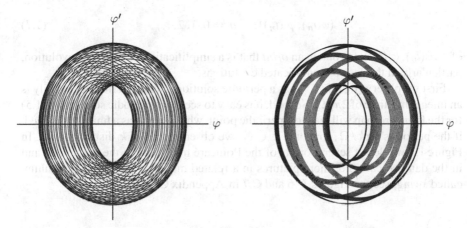

Fig. 1.6 Two multiperiodic evolutions of the periodically forced undamped pendulum. In both cases $\Omega = 1$, $A = 0.01$, $\varphi(0) = 0.017$, $\varphi'(0) = 0$ and the time interval is $[0, 200]$. Left, $\omega = \frac{1}{2}(1 + \sqrt{5}) \approx 1.61803$ (golden ratio). Right, $\omega = 1.602$, which is very close to the rational $8/5$. Notice how in the left figure the evolution 'fills up' an annulus in the (φ, φ')-plane almost homogeneously even after a relatively short time. In the right figure we have to wait much longer till the evolution homogeneously 'fills up' an annulus. This means that the former is easier to predict than the latter when using the principle *l'histoire se répète*; compare Chapter 2, §2.2.4.

the rational $8/5$. In both diagrams we witness a new phenomenon: the motion does not become periodic, but it looks rather regular. Later we return to this subject, but now we globally indicate what is the case here. The free pendulum oscillates with a well-defined frequency and also the forcing has a well-defined frequency. In the motions depicted in Figure 1.6 both frequencies remain visible. This kind of motion is quite predictable and therefore is not called chaotic. Rather one speaks of a *multi-* or *quasi-periodic* motion. However, it should be mentioned that in this undamped forced case, the initial states leading to quasi-periodic motions and those leading to chaotic motions are intertwined in a complicated manner. Hence it is hard to predict in advance what one will get.

There is yet another way to describe the evolutions of a pendulum with periodic forcing. In this description we represent the state of the pendulum by (φ, φ'), just at the instants $t_n = 2\pi n / \Omega$, $n = 0, 1, 2, \ldots$. At these instants the driving force resumes a new period. The underlying idea is that there exists a map P that assigns to each state (φ, φ') the next state $P(\varphi, \varphi')$. In other words, whenever $\varphi(t)$ is a solution of the equation of motion (1.5), one has

$$P(\varphi(t_n), \varphi'(t_n)) = (\varphi(t_{n+1}), \varphi'(t_{n+1})). \tag{1.6}$$

This map often is called the *Poincaré map*, or alternatively *period map* or *strobo-scopic map*. In §1.2 we return to this subject in a more general setting. By depicting the successive iterates

$$(\varphi(t_n), \varphi'(t_n)), \quad n = 0, 1, 2, \ldots, \tag{1.7}$$

of $P(\varphi(t_0), \varphi'(t_0))$, we obtain an *orbit* that is a simplification of the entire evolution, particularly in the case of complicated evolutions.

First we observe that if (1.5) has a periodic solution, then its period necessarily is an integer multiple of $2\pi / \Omega$. Second, it is easy to see that a periodic solution of (1.5) for the Poincaré map will lead to a periodic point, which shows as a finite set: indeed, if the period is $2\pi k / \Omega$, for some $k \in \mathbb{N}$, we observe a set of k distinct points. In Figure 1.7 we depict several orbits of the Poincaré map, both in the undamped and in the damped case. For more pictures in a related model of the forced pendulum, called *swing*, see the Figures. C.6 and C.7 in Appendix C, §C.3.

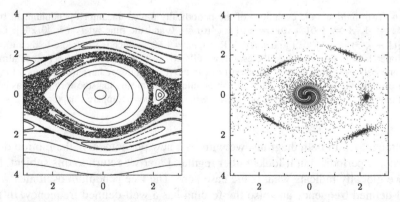

Fig. 1.7 Phase portraits of Poincaré maps for the undamped (left) and damped (right) pendulum with periodic forcing. In each case about 20 different initial states were chosen. The large cloud in the left picture consists of 10 000 iterations of one initial state.

Remarks.

- Iteration of the map (1.6) is the first example we meet of a dynamical system with *discrete* time: the state of the system only is considered for $t = 2\pi n / \Omega$, $n = 0, 1, 2, \ldots$. The orbits (1.7) are the evolutions of this system. In §1.2 and further we deal with this subject more generally.
- Oscillatory dynamical behaviour occurs abundantly in the world around us and many of the ensuing phenomena can be modelled by pendula or related oscillators, including cases with (periodic) driving or coupling. A number of these phenomena such as resonance are touched upon below. Further examples of oscillating behaviour can be found in textbooks on mechanics; for instance, see Arnold [21]. For interesting descriptions of natural phenomena that use driven or coupled oscillators, we recommend Minnaert [170], Vol. III, §§54 and 60. These include the swinging of tree branches in the wind and stability of ships.

1.1.3 Summary and outlook

We summarise as follows. For various models of the pendulum we encountered stationary, periodic, multi- or quasi-periodic, and chaotic motions. We also met with motions that are stationary or periodic after a transient phenomenon. It turned out that certain kinds of dynamics are more or less characteristic or typical for the model under consideration. We recall that the free damped pendulum, apart from a transient, can only have stationary behaviour. Stationary behaviour, however, is atypical for the undamped pendulum with or without forcing. Indeed, in the case without forcing the typical behaviour is periodic, whereas in the case with forcing we see both (multi- or quasi-) periodic and chaotic motions. Furthermore, for both the damped and the undamped free pendulum, the upside-down equilibrium is atypical. We here use the term 'atypical' to indicate that the initial states invoking this motion form a very small set (i.e., of area[10] zero). It has been shown that the quasi-periodic behaviour of the undamped forced pendulum also is typical; below we come back to this. The chaotic behaviour probably also is typical, but this fact has not been established mathematically. On the other hand, for the damped forced pendulum it has been proven that certain forms of chaotic motion are typical, however, multi- or quasi-periodic motion does not occur at all. For further discussion, see below.

In the following chapters we make similar qualitative statements, only formulated more accurately, and valid for more general classes of dynamical systems. To this purpose we first need to describe precisely what is meant by the concepts: dynamical system, (multi- or quasi-) periodic, chaotic, typical and atypical behaviour, independence of initial conditions, and so on. While doing this, we keep demonstrating their meaning on concrete examples such as the pendulum in all its variations,

---

[10] That is, 2-dimensional Lebesgue measure; see Appendix A.

and many other examples. Apart from the question of how the motion may depend on the initial state, we also investigate of how the total dynamics may depend on 'parameters' like ω, c, A, and Ω, in the equation of motion (1.5), or on more general perturbations. Again we speak in terms of more or less typical behaviour when the set of parameter values where this occurs is larger or smaller. In this respect the concept of *persistence* of dynamical properties is of importance. We use this term for a property that remains valid after small perturbations, where one may think both of perturbing the initial state and the equation of motion.

In these considerations the occurrence of chaotic dynamics in deterministic systems receives a lot of attention. Here we pay attention to a few possible scenarios according to which orderly dynamics can turn into chaos; see Appendix C. This discussion includes the significance that chaos has for physical systems, such as the meteorological system and its bad predictability. Also we touch upon the difference between chaotic deterministic systems and stochastic systems, that is, systems that are influenced by randomness.

1.2   General definition of dynamical systems

Essential for the definition of a dynamical system is *determinism*, a property we already met before. In the case of the pendulum, the present state, (i.e., both the position and the velocity) determines all future states. Moreover, the whole past can be reconstructed from the present state. The central concept here is that of *state*, in the above examples coined as position and velocity. All possible states together form the *state space*, in the free pendulum case the (φ, φ')-plane or the phase plane. For a function $\varphi = \varphi(t)$ that satisfies the equation of motion, we call the points $(\varphi(t), \varphi'(t))$ of the state space, seen as a function of the time t, an *evolution;* the curve $t \mapsto (\varphi(t), \varphi'(t))$ is called evolution as well. We now aim to express the determinism, that is, the unique determination of each future state by the present one, in terms of a map as follows. If (φ, φ') is the present state (at $t = 0$) and $t > 0$ is an instant in the future, then we denote the state at time t by $\Phi((\varphi, \varphi'), t)$. This expresses the fact that the present state determines all future states, that is, that the system is *deterministic*. So, if $\varphi(t)$ satisfies the equation of motion, then we have

$$\Phi((\varphi(0), \varphi'(0)), t) = (\varphi(t), \varphi'(t)).$$

The map

$$\Phi : \mathbb{R}^2 \times \mathbb{R} \to \mathbb{R}^2$$

constructed in this way, is called the *evolution operator* of the system. Note that such an evolution operator also can 'reconstruct the past.'

We now generally define a dynamical system as a structure, consisting of a *state space*, also called *phase space*, indicated by M, and an *evolution operator*

$$\Phi : M \times \mathbb{R} \to M. \tag{1.8}$$

Later on we also have to deal with situations where the evolution operator has a smaller domain of definition, but first, using the pendulum case as a leading example, we develop a few general properties for evolution operators of dynamical systems.

The first property discussed is that for any $x \in M$ necessarily

$$\Phi(x, 0) = x. \tag{1.9}$$

This just means that when a state evolves during a time interval of length 0, the state remains unchanged. The second property we mention is

$$\Phi(\Phi(x, t_1), t_2) = \Phi(x, t_1 + t_2). \tag{1.10}$$

This has to do with the following. If we consider the free pendulum, then, if $t \mapsto \varphi(t)$ satisfies the equation of motion, this also holds for $t \mapsto \varphi(t + t_1)$, for any constant t_1.

These two properties (1.9) and (1.10) together often are called the *group property*. To explain this terminology we rewrite $\Phi^t(x) = \Phi(x, t)$. Then the group property is equivalent to saying that the map $t \mapsto \Phi^t$ is a group morphism of $\mathbb{R}$, as an additive group, to the group of bijections of M, where the group operation is composition of maps. The latter also can be expressed in a more common form:

1. $\Phi^0 = \text{Id}_M$ (here Id_M is the identity map of M).
2. $\Phi^{t_1} \circ \Phi^{t_2} = \Phi^{t_1 + t_2}$ (where $\circ$ indicates composition of maps).

In the above terms we define an *evolution* of a general dynamical system as a map (or curve) of the form $t \mapsto \Phi(x, t) = \Phi^t(x)$, for a fixed $x \in M$, called the *initial state*. The image of this map also is called evolution. We can think of an *evolution operator* as a map that defines a flow in the state space M, where the point $x \in M$ flows along the orbit $t \mapsto \Phi^t(x)$. This is why the map Φ^t sometimes also is called *flow* over time t. Notice that in the case where the evolution operator Φ is differentiable, the maps Φ^t are *diffeomorphisms*. This means that each Φ^t is differentiable with a differentiable inverse, in this case given by Φ^{-t}. For more information on concepts such as diffeomorphisms, the reader is referred to Appendix A or to Hirsch [142].

1.2.1 Differential equations

Dynamical systems as just defined are almost identical with systems of first-order ordinary differential equations. In fact, if for the state space M we take a vector space and Φ of class at least C^2 (i.e., twice continuously differentiable) then we define the C^1-map $f : M \to M$ by

$$f(x) = \frac{\partial \Phi}{\partial t}(x, 0).$$

Then it is not hard to check that a curve $t \mapsto x(t)$ is an evolution of the dynamical system defined by Φ, if and only if it is a solution of the differential equation

$$x'(t) = f(x(t)). \tag{1.11}$$

For the theory of (ordinary) differential equations we refer to [16, 69, 118, 144]. The evolution operator $\Phi : M \times \mathbb{R} \to M$ in this case often is called the 'flow' of the ordinary differential equation (1.11). A few remarks are in order on the relationship between dynamical systems and ordinary differential equations.

Example 1.1 (On the pendulum cases). In the case of the free pendulum with damping we were given a second-order differential equation

$$\varphi'' = -\omega^2 \sin\varphi - c\varphi'$$

and not a first-order equation. However, there is a standard way to turn this equation into a system of first-order equations. Writing $x = \varphi$ and $y = \varphi'$, we obtain

$$x' = y$$
$$y' = -\omega^2 \sin x - cy.$$

In the forced pendulum case we not only had to deal with a second-order equation, but moreover, the time t also occurs explicitly in the 'right-hand side':

$$\varphi'' = -\omega^2 \sin\varphi - c\varphi' + A\cos\Omega t$$

We notice that here it is not the case that for a solution $t \mapsto \varphi(t)$ of the equation of motion, $t \mapsto \varphi(t + t_1)$ is also a solution, at least not if t_1 is not an integer multiple of $2\pi/\Omega$. To obtain a dynamical system we just add a state variable z that 'copies' the role of time. Thus, for the forced pendulum we get:

$$x' = y$$
$$y' = -\omega^2 \sin x - cy + A\cos z$$
$$z' = \Omega, \tag{1.12}$$

It should be clear that $(x(t), y(t), z(t))$ is a solution of the system (1.12) of first-order differential equations, if and only if $\varphi(t) = x(t - z(0)/\Omega)$ satisfies the equation of motion of the forced pendulum. So we have eliminated the explicit time-dependence by raising the dimension of the state space by one.

Remark (Digression on Ordinary Differential Equations II). In continuation of a remark in §1.1.1.1 we now review a few more issues on ordinary differential equations, for more information again referring to [16, 69, 118, 144].

1. In the pendulum with periodic forcing, the time t, in the original equation of motion, occurs in a periodic way. This means that a shift in t over an integer multiple of the period $2\pi/\Omega$ does not change the equation of motion. In turn this means

Fig. 1.8 A circle that originates by identifying the points of $\mathbb{R}$, the distance of which is an integer multiple of 2π.

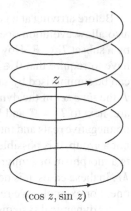

$(\cos z, \sin z)$

that also in the system (1.12) for (x, y, z) a shift in z over an integer multiple of 2π leaves everything unchanged. Mathematically this means that we may consider z as a variable in $\mathbb{R}/(2\pi\mathbb{Z})$, that is, in the set of real numbers in which two elements are identified whenever their difference is an integer multiple of 2π. As indicated in Figure 1.8, the real numbers by such an identification turn into a circle.

In this interpretation the state space is no longer $\mathbb{R}^3$, but rather

$$\mathbb{R}^2 \times \mathbb{R}/(2\pi\mathbb{Z}).$$

Incidentally note that in this example the variable x also is an angle and can be considered as a variable in $\mathbb{R}/(2\pi\mathbb{Z})$. In that case the state space would be

$$M = (\mathbb{R}/(2\pi\mathbb{Z})) \times \mathbb{R} \times (\mathbb{R}/(2\pi\mathbb{Z})).$$

This gives examples of dynamical systems, the state space of which is a *manifold*, instead of just a vector space. For general reference see [74, 142, 215].

2. Furthermore it is not entirely true that any first-order ordinary differential equation $x' = f(x)$ always defines a dynamical system. In fact, examples exist, even with continuous f, where an initial condition $x(0)$ does not determine the solution uniquely. However, such anomalies do not occur whenever f is (at least) of class C^1.

3. Another problem is that differential equations exist with solutions $x(t)$ that are not defined for all $t \in \mathbb{R}$, but that for a certain t_0 'tend to infinity'; that is,

$$\lim_{t \to t_0} \|x(t)\| = \infty.$$

In our definition of dynamical systems, such evolutions are not allowed. We may include such systems as *local dynamical systems*; for more examples and further details see below.

Before arriving at a general definition of dynamical systems, we mention that we also allow evolution operators where in (1.8) the set of real numbers $\mathbb{R}$ is replaced by a subset $T \subset \mathbb{R}$. However, we do require the group property. This means that we have to require that $0 \in T$, and for $t_1, t_2 \in T$ we also want that $t_1 + t_2 \in T$. This can be summarised by saying that $T \subseteq \mathbb{R}$ should be an *additive semigroup*. We call T the *time set* of the dynamical system. The most important examples that we meet are, next to $T = \mathbb{R}$ and $T = \mathbb{Z}$, the cases $T = \mathbb{R}_+$ and $T = \mathbb{Z}_+$, being the sets of nonnegative reals and integers, respectively. In the cases where T is a semigroup and not a group, it is possible that the maps Φ^t are non-invertible: the map $t \mapsto \Phi^t$ then is a morphism of semigroups, running from T to the endomorphisms semigroup of M. In these cases, it is not always possible to reconstruct past states from the present one. For examples we refer to the next section.

As dynamical systems with time set $\mathbb{R}$ are usually given by differential equations, so dynamical systems with time set $\mathbb{Z}$ are given by an invertible map (or automorphism) and systems with time set $\mathbb{Z}_+$ by an endomorphism. In the latter two cases this map is given by Φ^1.

Definition 1.1 (Dynamical System). A *dynamical system* consists of a state space M, a time set $T \subseteq \mathbb{R}$, being an additive semigroup, and an evolution operator $\Phi : M \times T \to M$ satisfying the group property; that is, $\Phi(x, 0) = x$ and

$$\Phi(\Phi(x, t_1), t_2) = \Phi(x, t_1 + t_2)$$

for all $x \in M$ and $t_1, t_2 \in T$.

Often the dynamical system is denoted by the triple (M, T, Φ). If not explicitly said otherwise, the space M is assumed at least to be a topological space and the operator Φ is assumed to be continuous.

Apart from systems strictly satisfying Definition 1.1, we also know *local dynamical systems*. Here the evolution operator Φ is not defined on the entire product $M \times T$. We keep the property that $M \times \{0\}$ belongs to the domain of definition of Φ, that $\Phi(x, 0) = x$ for all $x \in M$ and that $\Phi(\Phi(x, t_1), t_2) = \Phi(x, t_1 + t_2)$, as far as both (x, t_1) and $(\Phi(x, t_1), t_2)$ are in the domain of definition of Φ.[11]

Remarks.

- We already met the system of ordinary differential equations (1.11) as the generator of a dynamical system (M, T, Φ) with time set $T = \mathbb{R}$, namely of the (solution) flow $\Phi : M \times \mathbb{R} \to M$, which acts as the evolution operator.
- We encountered a first example of a dynamical system with time set $T = \mathbb{Z}$ in §1.1.2, when discussing the Poincaré map $P : \mathbb{R}^2 \to \mathbb{R}^2$ of the pendulum

---

[11] In the cases where M is a topological space and Φ continuous we want the domain of Φ in $M \times T$ to be open.

with periodic forcing. In that case for the state space we have $M = \mathbb{R}^2$ and the evolution operator can be defined as $\Phi : \mathbb{R}^2 \times \mathbb{Z} \to \mathbb{R}^2$ by

$$\Phi((\varphi, \varphi'), n) = P^n(\varphi, \varphi').$$

Also see Exercise 1.6.

- One of the aims of the general Definition 1.1 of a dynamical system is to unify these two examples: ordinary differential equations (with their solution flows) and maps. Below we also introduce noninvertible maps, which similarly generate a dynamical system with time set $T = \mathbb{Z}_+$. Moreover, when discussing partial differential equations, we encounter an example with time set $T = \mathbb{R}_+$.

1.2.2 Constructions of dynamical systems

In this part we deal with a few elementary, but important, constructions that turn a given dynamical system into another one. In all these constructions we start with a dynamical system as in Definition 1.1, thus with state space M, time set $T \subset \mathbb{R}$, and evolution operator Φ.

1.2.2.1 Restriction

A subset $A \subset M$ is called *invariant* under the given dynamics if $\Phi(A \times T) = A$. In that case we define the restriction of Φ to A as a dynamical system with state space A, time set T, and evolution operator $\Phi_A = \Phi|_{A \times T}$. Two important special cases for application of this idea occur when A is an *invariant submanifold* or an *attractor*. For details see below.

A variation of this construction occurs where one only gets a local dynamical system. This happens when we restrict to a subset A that is not invariant. In general we then get an evolution operator that is not defined on the entire product $A \times T$. Restriction to A only makes sense when the domain of definition is 'not too small.' As an example think of the following. Suppose that $a \in M$ is a stationary point, which means that $\Phi(a, t) = a$ for all $t \in T$. Assuming that Φ is continuous, then the domain of definition of its restriction to a neighbourhood A of a consists of pairs (x, t) such that both $x \in A$ and $\Phi(x, t) \in A$, thus forming a neighbourhood of $\{a\} \times T$ in $A \times T$.

A restriction as constructed in the previous example is used to study the dynamics in a small neighbourhood of a stationary point, in this case the point a. Often we restrict even more, inasmuch as we are not interested in evolutions that start in A and, after a long transient through M, return to A. In that case we restrict the domain of definition of Φ even further, namely to the set of pairs (x, t) where $x \in A, t \in T$ and where for all $t' \in T$ with $0 \le t' \le t$ (or if $t < 0$, with $0 > t' \ge t$), one has $\Phi(x, t') \in A$.

Both forms of restriction are used and often it is left implicit which is the case. In the latter case we sometimes speak of *restriction in the narrower sense* and in the former of *restriction in the wider sense*.

1.2.2.2 Discretisation

In the case where $T = \mathbb{R}$ (or $T = \mathbb{R}_+$) we can introduce a new system with time set $\tilde{T} = \mathbb{Z}$ (or $\tilde{T} = \mathbb{Z}_+$) by defining a new evolution operator $\tilde{\Phi} : M \times \tilde{T} \to M$ by

$$\tilde{\Phi}(x, n) = \Phi(x, hn),$$

where $h \in \mathbb{R}$ is a positive constant. This construction has some interest in the theory of numerical simulations, which often are based on (an approximation of) such a discretisation. Indeed, one computes subsequent states for instants (times) that differ by a fixed value h. Such an algorithm therefore is based on (an approximation of) Φ^h and its iterations. In the case where Φ is determined by a differential equation

$$x' = f(x), \quad x \in \mathbb{R}^n, \tag{1.13}$$

the simplest approximation of Φ^h is given by

$$\Phi^h(x) = x + hf(x).$$

The corresponding method to approximate solutions of the differential equation (1.13) is called the Euler [12] method, where h is the step size.

Remark. Not all dynamical systems with $T = \mathbb{Z}$ can be obtained in this way, namely by discretising an appropriate system with time set $\mathbb{R}$ and continuous evolution operator. A simple example of this occurs when the map $\varphi : \mathbb{R} \to \mathbb{R}$, used to generate the system by iteration, has a negative derivative φ' everywhere. (For example, just take $\varphi(x) = -x$.) Indeed, for any evolution operator Φ of a continuous dynamical system with state space $M = \mathbb{R}$ and time set $T = \mathbb{R}$, we have that for $x_1 < x_2 \in \mathbb{R}$ necessarily $\Phi^t(x_1) < \Phi^t(x_2)$ for all t. This is true, because Φ^t is invertible, which implies that $\Phi^t(x_1) \neq \Phi^t(x_2)$. Moreover, the difference $\Phi^t(x_2) - \Phi^t(x_1)$ depends continuously on t, and has a positive sign for $t = 0$. Therefore, this sign necessarily is positive for all $t \in \mathbb{R}$. Now observe that this is contrary to what happens when applying the map φ. Indeed, inasmuch as $\varphi'(x) < 0$ everywhere, for $x_1 < x_2$ it follows that $\varphi(x_1) > \varphi(x_2)$. Therefore, for any constant h it follows that $\varphi \neq \Phi^h$ and the dynamics generated by φ cannnot be obtained by discretisation.

---

[12] Leonhard Euler 1707–1783.

1.2.2.3 Suspension and poincaré map

We now discuss two constructions, the suspension and the Poincaré map, which relate systems with discrete and continuous time (i.e., maps with flows). We already met an example of a Poincaré map in §1.1.2

Suspension. We start with a dynamical system (M, T, Φ) with time set $T = \mathbb{Z}$, the suspension of which is a dynamical system with time set $\mathbb{R}$. The state space of the suspended system is defined as the quotient space

$$\tilde{M} = M \times \mathbb{R}/\sim,$$

where $\sim$ is the equivalence relation given by

$$(x_1, s_1) \sim (x_2, s_2) \Leftrightarrow s_2 - s_1 \in \mathbb{Z} \quad \text{and} \quad \Phi^{s_2 - s_1}(x_2) = x_1.$$

Another way to construct $\tilde{M}$ is to take $M \times [0, 1]$, and to identify the 'boundaries' $M \times \{0\}$ and $M \times \{1\}$ such that the points $(x, 1)$ and $(\Phi^1(x), 0)$ are 'glued together.' The fact that the two constructions give the same result can be seen as follows. The equivalence class in $M \times \mathbb{R}$ containing the point (x, s) contains exactly one element (x', s') with $s' \in [0, 1)$. The elements in $M \times \{1\}$ and $M \times \{0\}$ are pairwise equivalent, such that $(x, 1) \sim (\Phi^1(x), 0)$, which explains the 'gluing prescription.' The evolution operator $\tilde{\Phi} : \tilde{M} \times \mathbb{R} \to \tilde{M}$ now is defined by

$$\tilde{\Phi}([x, s], t) = [x, s + t],$$

where $[-]$ indicates the $\sim$-equivalence class.

Remark (Topological complexity of suspension). Let M be connected, then after suspension, the state space is not simply connected, which means that there exists a continuous curve $\alpha : \mathbb{S}^1 \to \tilde{M}$ that, within $\tilde{M}$, cannot be deformed continuously to a point. For details see the exercises; for general reference see [142].

Example 1.2 (Suspension to cylinder or to Möbius strip). As an example we consider the suspension of two dynamical systems both with $M = (-1, +1)$. We recall that $T = \mathbb{Z}$, so in both cases the system is generated by an invertible map on M. In the first case we take $\Phi_1^1 = \text{Id}$, the identity map, and in the second case $\Phi_2^1 = -\text{Id}$. To construct the suspension we consider $M \times [0, 1] = (-1, 1) \times [0, 1]$. In the former case we identify the boundary points $(x, 0)$ and $(x, 1)$: in this way $\tilde{M}$ is the cylinder. In the latter case we identify the boundary points $(x, 0)$ and $(-x, 1)$, which leads to $\tilde{M}$ being the Möbius strip. Compare Figure 1.9. In both cases the suspended evolutions are periodic. For the suspension in the case where $\Phi_1^1 = \text{Id}$, we follow the evolution that starts at $[x, 0]$. After a time t this point has evolved to $[x, t]$. In this case $\Phi^n(x) = x$, thus the pairs (x, t) and (x', t') belong to the same equivalence class if and only if $x = x'$ and $t - t'$ is an integer. This implies that the evolution is periodic with period 1. In the second case where $\Phi_1^1 = -\text{Id}$, we have that

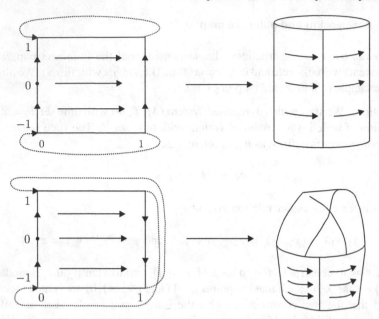

Fig. 1.9 By 'gluing' two opposite edges of a rectangle we can either obtain a cylinder or a Möbius strip. On both the cylinder and Möbius strip we indicated how the new evolutions run in the cases where $\Phi_1^1 = \text{Id}$ and $\Phi_2^1 = -\text{Id}$, in both cases with $M = (-1, +1)$.

$\Phi^n(x) = (-1)^n x$. This means that the evolution starting at $[0, 0]$, is periodic with period 1, because here also the pairs $(0, t)$ and $(0, t')$ belong to the same equivalence class if and only if $t - t'$ is an integer. However, any evolution starting at $[x, 0]$, with $x \neq 0$, is periodic with period 2. Indeed, the pairs $(x, 0)$ and $(x, 2)$ are equivalent, but for no $0 < t < 2$ are the pairs $(x, 0)$ and (x, t) equivalent. The fact that evolutions which do not start 'right in the middle' have a double period can also be seen in Figure 1.9.

We finally note the following. If we first suspend a dynamical system with time set $\mathbb{Z}$ to a system with time set $\mathbb{R}$, and we discretise subsequently (with constant $h = 1$ in the discretisation construction), after which we restrict to $\{[x, s] \mid s = 0\}$, we have returned to our original system, where we do not distinguish between x and $[x, 0]$.

Poincaré map. We now turn to the construction of a Poincaré map: a kind of inverse of the suspension construction described above. See the example at the end of §1.1.2. We start with a dynamical system with time set $\mathbb{R}$, given by a differential equation $x' = f(x)$, assuming that in the state space M there is a submanifold $S \subset M$ of codimension 1 (meaning that $\dim M = \dim S + 1$), with the following properties.

1. For all $x \in S$ the vector $f(x)$ is transversal to S (i.e., not tangent to S).
2. For any point $x \in S$ there exist real numbers $t_-(x) < 0 < t_+(x)$, such that $\Phi(x, t_-(x)) \in S$ and $\Phi(x, t_+(x)) \in S$.
   We call $t_-(x), t_+(x)$ the return times and in the sequel we assume that both have been chosen with minimal absolute value. This means that, due to transversality (and the implicit function theorem) both $t_\pm(x)$ are differentiable in $x \in S$; both are times of first return in S, either for negative or for positive time.

In the above setting we define a diffeomorphism

$$\tilde{\varphi} : S \to S \quad \text{by} \quad \tilde{\varphi}(x) = \Phi(x, t_+(x)).$$

Note that this indeed defines a diffeomorphism, because an inverse of this map simply is given by $x \mapsto \Phi(x, t_-(x))$. We call $\tilde{\varphi}$ the Poincaré map of S.[13] The corresponding dynamical system on S has an evolution operator given by $\tilde{\Phi}(x, n) = \tilde{\varphi}^n(x)$.

Remarks.

– Observe that the suspension construction, for any dynamical system with time set $\mathbb{Z}$, provides a dynamical system with time set $\mathbb{R}$. However, we like to point out that not every dynamical system with time set $\mathbb{R}$ can be obtained in this way. In fact, a system obtained by suspension cannot have stationary evolutions. This means that dynamical systems with time set $\mathbb{R}$ having stationary evolutions, such as the examples of the free pendulum with or without damping, cannot be obtained as the outcome of a suspension construction.
– Observe that the variable $t_\pm(x)$ can be made constant by reparametrising the time t.
– One may well ask whether the requirement of the existence of a submanifold $S \subset M$ is not too restrictive. That the requirement is restrictive can be seen as follows. For $M = \mathbb{R}^n$ [14] and Φ continuous, a connected submanifold $S \subset \mathbb{R}^n$ with the above requirements does not exist; this can be seen using some algebraic topology (e.g., see [165], but we do not discuss this subject here). Related to this, a flow with state space $\mathbb{R}^n$ cannot be obtained by suspension. See Exercise 1.23.

We now discuss for which differential equations a Poincaré map construction is applicable. For this we need to generalise the concept of differential equation to the case where the state space is a general *manifold*; again see Appendix A and [74, 142, 215]. In the differential equation $\xi' = f(\xi)$ we then interpret the map f as a vector field on the state space. We do not enter into all the complications here, but refer to the above example of the forced pendulum (see (1.12)) where the state space has the form $\mathbb{R}^2 \times (\mathbb{R}/(2\pi\mathbb{Z}))$ and its points are denoted by (x, y, z). This is an example of a 3-dimensional manifold, because it locally cannot be distinguished from $\mathbb{R}^3$. In this

---

[13] Sometimes also called (first) return map.

[14] Vector spaces are simply connected.

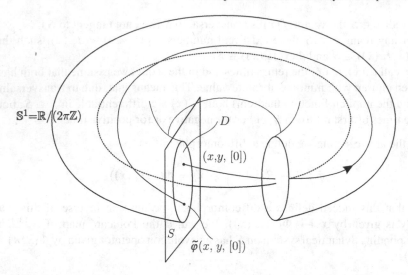

Fig. 1.10 State space $\mathbb{R}^2 \times \mathbb{S}^1$, with $\mathbb{S}^1 = \mathbb{R}/(2\pi\mathbb{Z})$, where D is a disc in $\mathbb{R}^2$. Also we indicate the submanifold S as well as an evolution of the 3-dimensional system. Here two subsequent passages through S are visible, as well as the image of one of the points of S under the Poincaré map $\tilde{\varphi}$.

example it is easy to construct a submanifold S with the desired properties: we can just take $S = \{(x, y, [0])\}$. In this case the Poincaré map also is called a period map or stroboscopic map; see §1.1.2, where we already met this particular situation. Compare Figure 1.10. For a given point $(x, y, [0]) \in S$ we depict the evolution of the 3-dimensional system starting at $(x, y, [0])$ and also the image of $(x, y, [0])$ under the Poincaré map.

Often the term Poincaré map is used in a wider sense. To explain this take a submanifold S of codimension 1, for which we still require that for each $\xi \in S$ the vector $f(\xi)$ is transversal to S. However, we do not require the return times to be defined for all points of S. The Poincaré map then has the same domain of definition as t_+. In this way we obtain a *local dynamical system*. The evolution operator then can be defined by $\tilde{\Phi}^{t_+} = \tilde{\varphi}$, where $\tilde{\varphi}^{-1}$ has the same domain of definition as t_-.

One of the most important cases where such a locally defined Poincaré map occurs, is 'near' periodic evolutions of (autonomous) differential equations $\xi' = f(\xi)$. In the state space the periodic evolution determines a closed curve γ. For S we now take a codimension 1 submanifold that intersects γ transversally. This means that γ and S have one point ξ_0 in common, that is, $\{\xi_0\} = \gamma \cap S$, where γ is a transversal to S. In that case $f(\xi_0)$ is transverse to S. By restricting S to a (possibly) smaller neighbourhood of ξ_0 can achieve that for any point $\xi \in S$ the vector $f(\xi)$ is transverse to S. In this case the Poincaré map is defined on a neighbourhood of ξ_0 in S. Later on we return to this subject.

1.3 Further examples of dynamical systems

In this section we discuss a few examples of dynamical systems from a widely diverging background. These examples frequently occur in the literature and the concepts of the previous section can be well illustrated by them. We also meet new phenomena.

1.3.1 A Hopf bifurcation in the Van der Pol equation

The equation of Van der Pol was designed as a model for an electronic oscillator; compare [192, 193]. We do not go into the derivation of this model. In any case it is a mathematical description of obsolete technology based on radio tubes, the predecessors of our present transistors. We first describe the equation, which turns out to have a periodic attractor and next introduce a parameter dependence in the system.

1.3.1.1   The Van der Pol equation

The Van der Pol equation serves as an example of a (nonlinear) oscillator with a partially 'negative damping'. The equation can be written in the form

$$x'' = -x - x'(x^2 - 1), \tag{1.14}$$

as usual leading to the system

$$x' = y$$
$$y' = -x - y(x^2 - 1).$$

We recognise the equation of a damped harmonic oscillator, where the damping coefficient c has been replaced by $(x^2 - 1)$. This means that for $|x| < 1$ the damping is negative, increasing the energy instead of decreasing. For $|x| > 1$ we have the usual positive damping. To make these remarks more precise we define the energy of a solution $x(t)$ by

$$E(t) = \frac{1}{2}\left(x(t)^2 + x'(t)^2\right),$$

in analogy with the undamped harmonic oscillator. By differentiation we then find that

$$E' = xx' + x'x'' = xx' + x'(-x - x'(x^2 - 1)) = -(x')^2(x^2 - 1),$$

from which our assertions easily follow.

Remark. Let us briefly discuss the present terminology. In the Van der Pol equation (1.14) we consider a position-dependent damping $c(x) = x^2 - 1$. Notice that without any damping we would just be dealing with the harmonic oscillator $x'' = -x$, which conserves the energy $E(x, x') = \frac{1}{2} \left(x^2 + (x')^2 \right)$. In this way we observe the effect of an unconventional damping on the conventional energy.

From the properties of this damping it follows that very small oscillations gain in amplitude because they occur in an area with negative damping. On the other hand very large oscillations will decrease in amplitude because these mainly occur in an area with positive damping. One can show that each solution of this equation, with the exception of the zero solution, converges to one and the same periodic solution [16, 130]. The Van der Pol equation (1.14) therefore describes a system that, independent of the initial state different from $(x, x') = (0, 0)$, eventually will end up in the same periodic behaviour. It can even be shown that this property is robust or persistent, in the sense that small changes of the right-hand side of the differential equation will not change this property. Of course, amplitude and frequency of the oscillation can shift by such a small change. Extensive research has been done on the behaviour of the solutions of (1.14) when a periodic forcing is applied; for more information see [5, 27] and §2.2.3.2, below.

1.3.1.2   Hopf bifurcation

We now introduce a *parameter* in the Van der Pol equation (1.14), assuming that this parameter can be tuned in an arbitrary way. The question is how the dynamics depends on such a parameter More in general *bifurcation theory* studies the dynamics as a function of one or more parameters; for a discussion and many references see Appendices C and E. In this example we take variable damping, so that we can make a continuous transition from the Van der Pol equation to an equation with only positive damping. Indeed we consider

$$x'' = -x - x'(x^2 - \mu). \tag{1.15}$$

The value of the parameter μ determines the region where the friction is negative: for given $\mu > 0$ this is the interval given by $|x| < \sqrt{\mu}$. We now expect the following.

1. For $\mu \le 0$ the equation (1.15) has no periodic solutions, but all solutions converge to the zero solution.
2. For $\mu > 0$ all solutions of (1.15) (with the exception of the zero solution) converge to a fixed periodic solution, the amplitude of which decreases as μ tends to 0. See the Figures 1.11 and 1.12.

These expectations indeed are confirmed by the numerical phase portraits of Figure 1.12, and it is possible to prove these facts [145, 207] in an even more general context. The transition at $\mu = 0$, where the behaviour of the system changes from nonoscillating to oscillating behaviour is an example of a *bifurcation*, in particular a

Fig. 1.11 Phase portrait of
the Van der Pol oscillator
(1.14).

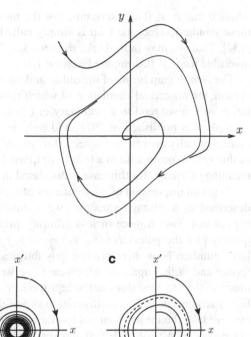

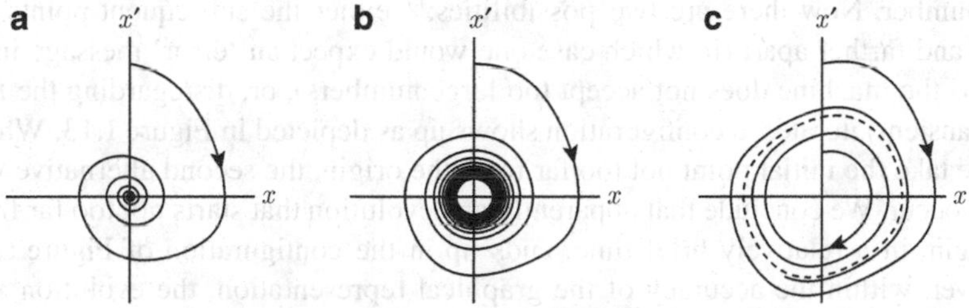

Fig. 1.12 A few evolutions of the parametrised Van der Pol equation (1.15). a: $\mu < 0$, b: $\mu = 0$,
and c: $\mu > 0$. Compare Figure 1.11.

Hopf bifurcation.[15] For a brief description see Appendix C. In general a bifurcation
occurs at parameter values where the behaviour of the system changes *qualitatively*,
this as opposed to *quantitative* changes, where for instance, the amplitude or the
frequency of the oscillation shifts in a continuous way.

1.3.2   The Hénon map: Saddle points and separatrices

The Hénon[16] map defines a dynamical system with state space $\mathbb{R}^2$ and with time set
$T = \mathbb{Z}$. The evolution operator $\Phi : \mathbb{R}^2 \times \mathbb{Z} \to \mathbb{R}^2$ is determined by the map

$$\Phi^1 = H_{a,b} : (x, y) \mapsto (1 - ax^2 + y, bx),$$

---

[15] Eberhard Hopf 1902–1983.
[16] Michel Hénon 1931–.

where a and $b \neq 0$ are constants. For the moment we fix $a = 1.4, b = 0.3$. For these standard values the map is simply called H; this is the original Hénon map [137]. That we may indeed take the time set $\mathbb{Z}$, follows from the fact that $H_{a,b}$ is invertible for $b \neq 0$, compare Exercise 1.10.

The Hénon map is one of the oldest and simplest examples of an explicitly given system, the numerical simulation of which raises the conjecture that a general evolution will never tend to a stationary or (multi- or quasi-) periodic evolution. The example was published in 1976, and only in 1991, that is, 15 years later, was it mathematically proven that indeed, for 'many' values of a and b, most evolutions of this system never tend to a (multi- or quasi-) periodic evolution [67]. The precise meaning of 'many' for this case is discussed in a later chapter.

To get an impression of the dynamics of this system and of what Hénon in [137] described as a 'strange attractor,' we should undertake the following computer experiment. Take a more or less arbitrary initial point $(x_0, y_0) \in \mathbb{R}^2$ and subsequently plot the points $H^j(x_0, y_0) = (x_j, y_j)$ for $j = 1, \ldots, N$, where N is a large number. Now there are two possibilities:[17] either the subsequent points get farther and farther apart (in which case one would expect an 'error' message inasmuch as the machine does not accept too large numbers), or, disregarding the first few (transient) iterates, a configuration shows up as depicted in Figure 1.13. Whenever we take the initial point not too far from the origin, the second alternative will always occur. We conclude that apparently any evolution that starts not too far from the origin, in a relatively brief time, ends up in the configuration of Figure 1.13. Moreover, within the accuracy of the graphical representation, the evolution also visits each point of this configuration. Because the configuration attracts all these evolutions, it is called an *attractor*. In Chapter 4 we give a mathematical definition of attractors. In the magnification we see that a certain fine structure seems present, which keeps the same complexity. This property is characteristic for what is nowa-

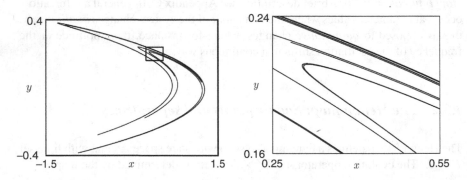

Fig. 1.13 The attractor of the Hénon map and a magnification by a factor of 10.

---

[17] Theoretically there are more possibilities, but these are probably just as exceptional as the motions related to the upside-down free pendulum.

days called a *fractal*; for general background see [12, 163]. The existence of such fractals has led to new *dimension concepts*, which also can attain noninteger values. For general reference see Falconer [122, 123]; also see Chapter 6.

We surely cannot give a complete exposition of all properties of the Hénon map which can be mathematically proven, but we have to mention here a single aspect. For more information and references to the literature see Chapter 4, §§4.2 and 4.3 and the Appendices C and D. To start with, we look for stationary evolutions, that is, points $(x, y) \in \mathbb{R}^2$ for which $H(x, y) = (x, y)$. The solutions are

$$x = \frac{-0.7 \pm \sqrt{0.49 + 5.6}}{2.8}, \quad y = 0.3x,$$

or, in approximation, $(0.63, 0.19)$ and $(-1.13, -0.34)$. The former of these two points, called p, turns out to be situated inside the attractor. We investigate the nearby dynamics around p by expansion in a Taylor series

$$H(p + (\tilde{x}, \tilde{y})) = p + \mathrm{d}H_p \begin{pmatrix} \tilde{x} \\ \tilde{y} \end{pmatrix} + o(|\tilde{x}|, |\tilde{y}|).$$

To first approximation we have to deal with the linear map $\mathrm{d}H_p$, the derivative of H at p. A direct computation yields that this matrix has real eigenvalues, one of which is smaller than -1 and the other inside the interval $(0, 1)$. First we consider the dynamics when neglecting the higher-order terms $o(|\tilde{x}|, |\tilde{y}|)$ (i.e., of the linearised Hénon map). We can choose new coordinates ξ and η in such a way that p in these coordinates becomes the origin, and the coordinate axes are pointing in the direction of the eigenvectors of $\mathrm{d}H_p$; see Figure 1.14.

In the (ξ, η)-coordinates the linearised Hénon map has the form

$$H_{\mathrm{appr}}(\xi, \eta) = (\lambda_1 \xi, \lambda_2 \eta),$$

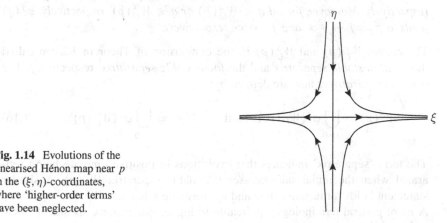

Fig. 1.14 Evolutions of the linearised Hénon map near p in the (ξ, η)-coordinates, where 'higher-order terms' have been neglected.

with $\lambda_1 < -1$ and $\lambda_2 \in (0,1)$. This means first of all that both the ξ- and the η-axis are invariant by iteration of H_{appr}. Moreover, points on the ξ-axis, by repeated iteration of H_{appr} get further and further elongated (by repeated application of the inverse H_{appr}^{-1} points on the ξ-axis converge to p). For points on the η-axis the opposite holds: by iteration of H these converge to p (and under iteration by H_{appr}^{-1} get elongated). Points that belong to neither axis move away both by iteration of H_{appr} and of H_{appr}^{-1}. Compare this with the behaviour of linear ordinary differential equations [16,69,118,144]. All of this holds when neglecting the higher order terms. We now indicate, without proofs, what remains of this when the higher order terms are no longer neglected. These statements hold for general 2-dimensional diffeomorphisms, and not only for the Hénon map, therefore we give a general formulation.

Remarks (Digression on saddle points and separatrices).

– Let $\varphi : \mathbb{R}^2 \to \mathbb{R}^2$ be a diffeomorphism. We call $p \in \mathbb{R}^2$ a *fixed point* whenever $\varphi(p) = p$, in other words if the constant map $n \in \mathbb{Z} \mapsto p \in \mathbb{R}^2$ is a stationary evolution of the dynamical system generated by φ. Now suppose that the eigenvalues λ_1, λ_2 of the derivative map $\mathrm{d}\varphi_p$ are real, where moreover $|\lambda_1| > 1 > |\lambda_2|$; in this case p is called a *saddle point* of φ. Compare the situation for the Hénon map as sketched above.

  In the sequel $v_1, v_2 \in T_p(\mathbb{R}^2)$ are eigenvectors of $\mathrm{d}(\varphi_p)$, belonging to the eigenvalues λ_1 and λ_2, respectively. For a proof of the following theorem we refer to, for example, [9].

Theorem 1.2 (Local Separatrices). *For a diffeomorphism φ, with saddle point p, eigenvalues λ_1, λ_2, and eigenvectors v_1, v_2 as above, there exists a neighbourhood U of p such that*

$$W_U^u(p) = \{q \in U \mid \varphi^j(q) \in U \text{ for all } j \leq 0\} \quad and$$
$$W_U^s(p) = \{q \in U \mid \varphi^j(q) \in U \text{ for all } j \geq 0\}$$

are smooth curves that contain the point p and that in p are tangent to v_1 and v_2, respectively. Moreover, for all $q \in W_U^u(p)$ or $q \in W_U^s(p)$, respectively, $\varphi^j(q)$ tends to p as $j \to -\infty$ and $j \to \infty$, respectively.

The curves $W_U^u(p)$ and $W_U^s(p)$ in the conclusion of Theorem 1.2 are called the *local unstable separatrix* and the *local stable separatrix*, respectively. The *(global) separatrices* now are defined by:

$$W^u(p) = \bigcup_{j>0} \varphi^j(W_U^u(p)) \quad \text{and} \quad W^s(p) = \bigcup_{j<0} \varphi^j(W_U^s(p)). \qquad (1.16)$$

– The term 'separatrix' indicates that evolutions in positive time are being separated when the initial state crosses the stable separatrix $W^s(p)$. A similar statement holds for negative time and the unstable separatrix $W^u(p)$.
– A more general terminology, applicable to higher dimensions, speaks of 'local stable' and 'local unstable' manifolds.

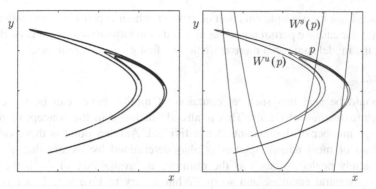

Fig. 1.15 Left: Hénon attractor; compare Figure 1.13. Right: Pieces of the stable and unstable separatrices $W^s(p)$ and $W^u(p)$ of the saddle point $p = (.63, .19)$ of the Hénon map. Note the resemblance between the attractor and $W^u(p)$; later we come back to this.

Numerical computation of separatrices. We now return to our discussion on the Hénon family. To get a better heuristic idea, we now describe how such separatrices, or a finite segment of these, can be approximated numerically. Again we consider the saddle point p of the diffeomorphism φ, with eigenvalues $|\lambda_1| > 1 > |\lambda_2|$ and corresponding eigenvectors v_1 and v_2. For the approximation of the unstable separatrix we initially take a brief segment $[p - \varepsilon v_1, p + \varepsilon v_1]$. The smaller we choose ε, the better this segment approximates $W^u(p)$ (the error made here is order $O(\varepsilon^2)$). In this segment we choose N equidistant points. In order to find a longer segment of the unstable separatrix $W^u(p)$, we apply the m-fold iterate φ^m to all subsequent points and connect the resulting N points by segments. The result evidently depends on the choice of the parameters ε, N, and m. A segment of $W^u(p)$ of any prescribed length in this way can be obtained to arbitrary precision by choosing these parameters appropriately. A proof of this statement is not too complicated, but is does use ideas in the proof of Theorem 1.2. Therefore, we do not go into details here.

By numerical approximation of the unstable separatrix $W^u(p)$ of the saddle point p of the Hénon map, we obtain the same picture as the Hénon attractor; compare Figure 1.15 and also see Exercise 1.24. We come back to this in Chapter 4, in particular in §4.2.

1.3.3 The logistic system: Bifurcation diagrams

The example to be treated now originally comes from mathematical biology, in particular from population dynamics. The intention was to give a model for the succesive generations of a certain animal species. The simplest model is the so-called 'model with exponential growth', where it is assumed that each individual on average gives rise to μ individuals in the next generation. If we denote the size of the nth generation by x_n, we would expect the *linear* relationship

$$x_{n+1} = \mu x_n, \quad n = 0, 1, 2, \ldots. \tag{1.17}$$

This indeed is a very simple equation of motion, which is provably nonrealistic. In this particular case we permit ourselves an excursion into modelling aspects. Before making the model somewhat more realistic, we first give a few remarks.

Remarks.

– It should be said that such an equation of motion never can be more than an *approximation* of reality. This is already included in the concept of μ, the average number of descendants per individual. Another point is that evidently the fate of most animal species is also determined by factors that we here completely neglected, such as the climate, the availability of food, the presence of natural enemies, and so on. What we try to give here is a model in which only influence of the size of the previous generation on the next is being expressed.
– As already said the model given by (1.17) is quite unrealistic. To see this we compute its evolutions. For a given initial population x_0 it directly follows that the nth generation should have size

$$x_n = \mu^n x_0.$$

Here one speaks of *exponential growth*, inasmuch as the time n is in the exponent. We now distinguish three cases, namely $\mu < 1$, $\mu = 1$ and $\mu > 1$. By the way, by its definition, necessarily $\mu \geq 0$, where the case $\mu = 0$ is too uninteresting for further consideration. In the former case $\mu < 1$, the population decreases stepwise and will become extinct. Indeed, the 'real' size of the population is an integer and therefore a decreasing population after some time disappears completely. In the second case $\mu = 1$, the population always keeps the same size. In the latter case $\mu > 1$, the population increases ad infinitum. So it would seem that only the second case $\mu = 1$ is somewhat realistic. However, this situation turns out to be very unstable: already very small changes in the reproduction factor μ lead us into one of the other two cases.

One main reason why the model (1.17) is unrealistic is that the effects of overpopulation are not taken into account. For this we can include a correction in the model by making the reproduction factor dependent on the size of the population, for example, by replacing

$$\mu \text{ by } \mu\left(1 - \frac{x_n}{K}\right).$$

Here K is the size of the population which implies such a serious overpopulation that it leads to immediate extinction. Probably it will be clear that overpopulation should have a negative effect, but it may be less clear why the correction proposed here is so relevant. This claim of relevance also is not made, but the most important reason for this choice is that it is just about the most simple one. Moreover, it turns out that the results discussed below do not strongly depend on the precise form of the correction for overpopulation. Therefore, from now on we deal with the dynamics generated by the map

$$x_{n+1} = \mu x_n \left(1 - \frac{x_n}{K}\right). \tag{1.18}$$

We deal here with a system in which the parameters μ and K still have to be specified. Both have a clear biological interpretation: μ is the reproduction factor for very small populations, and K indicates for which population size overpopulation becomes disastrous. In principle we would like to know the dynamics for all different values of these parameters. It now turns out that we can simplify this investigation. First of all the parameter K can be 'transformed away' as follows. If we define $y_n = 1/K x_n$, the equation (1.18) turns into

$$y_{n+1} = \frac{x_{n+1}}{K} = \frac{\mu x_n}{K}\left(1 - \frac{x_n}{K}\right) = \mu y_n(1 - y_n),$$

which has the same form as (1.18), but now with $K = 1$. So we can get rid of K by not dealing with the population size, but by expressing this as the fraction of the critical size K. It turns out that the parameter μ cannot be removed. So we now deal with the dynamics given by

$$y_{n+1} = \Phi^1(y_n) = \mu y_n(1 - y_n), \tag{1.19}$$

called the *logistic system*. The state space is $M = [0, 1]$, indeed, any value $y > 1$ under iteration would immediately be succeeded by a negative value. Moreover, we see that it is not enough to require $\mu > 0$, because for $\mu > 4$ we would have that $\Phi^1(\frac{1}{2}) \notin M$. So we take $0 < \mu \leq 4$. Finally it should be clear that the time set should be $T = \mathbb{Z}_+$, because the map Φ^1 is not invertible.

Remarks.

- Although population sizes necessarily are integers, this was already ignored in (1.18) and even more in (1.19), where we turn to fractions of the maximal population.
- Note that this description makes sense only when the generations are clearly distinguishable. For the human population, for example, it makes no sense to express that two individuals who do not belong to the same family have a generation difference of, say, 5. For many species of insects, however, each year or day a new generation occurs and so the generations can be numbered. In experimental biology, in this respect the *drosophila melanogaster*, a special kind of fruit fly, is quite popular, because here the dynamics evolves more rapidly and produces one (new) generation every day.
- A related quadratic demographic model with continuous time was introduced by Verhulst[18] in 1838. The above counterpart with discrete time was extensively

---

[18] Pierre François Verhulst 1804–1849.

studied by May [166].[19] For further historical remarks compare Mandelbrot [163][20] and Peitgen, jurgens, and Saupe [12].[21]

An obvious question now is how we may investigate dynamics of the logistic system (1.19). In particular it is not so clear what kind of questions should be posed here. The burst of interest in this model was partly due to the biological implications, but also to the numerical experiments that generally preceded the mathematical statements. Indeed, the numerical experiments gave rise to enigmatic figures that badly needed an explanation. We indicate some results of these numerical experiments. A complete mathematical analysis of this system would exceed the present scope by far. For an introductory study we refer to [4], whereas an in-depth analysis can be found in [167].

In Figure 1.16 we illustrate how iteration $y_{n+1} = F(y_n), n \in \mathbb{Z}_+$ generally can be visualised using the graph of F and the diagonal. Note that the intersections of the graph of F and the diagonal exactly give the fixed points (i.e., the stationary evolutions) of the corresponding dynamics. The right-hand picture of Figure 1.16 illustrates this for the case $F(y_n) = \mu y_n$, where $y = 0$ is a repelling fixed point.

As an experiment with the logistic system (1.19) we apply this graphical method for $F = \Phi^1$ computing evolutions for a number of μ-values. In Figure 1.17 we depicted y_{n+1} as a function of y_n, for $n = 0, 1, \ldots, 15$, and where the initial value was always $y_0 = 0.7$. For lower values of μ nothing unexpected happens: for $\mu \le 1$ the population gets extinct and for $1 < \mu < 3$ the population tends to a stable value that is monotonically increasing with μ. For $\mu > 3$, however, various things can happen. So it is possible that an evolution, after an irregular transient, gets periodic.

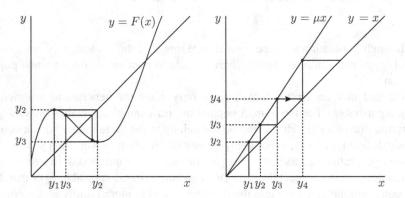

Fig. 1.16 Left: Iteration of $y_{n+1} = F(y_n)$ using the graph of the function F and the diagonal. Right: The case of $y_{n+1} = \mu y_n$, with $\mu > 1$, giving rise to the exponential growth $y_n = \mu^n y_0$, $n \in \mathbb{Z}_+$.

---

[19] Robert May 1936–.

[20] Benoît B. Mandelbrot 1924–.

[21] Heinz-Otto Peitgen 1945–.

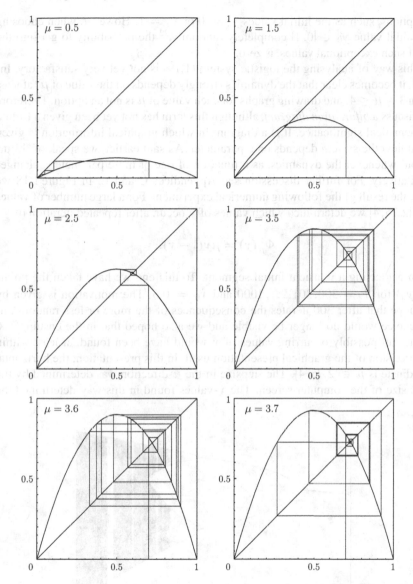

Fig. 1.17 Six evolutions of the logistic system (1.19), where each time the result of 15 successive iterations is shown, beginning with $y_0 = 0.7$. The subsequent μ-values are $\mu = 0.5, 1.5, 2.5, 3.5, 3.6$, and 3.7.

However, it is also possible that the motion never tends to anything periodic, although such a statement evidently cannot be made on the basis of finitely many numerical iterations. Moreover, it can be shown that the observed dynamical behaviour almost solely depends on the value of μ and is only influenced a little by the choice of the initial state y_0. The word 'little' implies that there are exceptions, such as the initial values $y_0 = 0$ or $y_0 = 1$. However, when choosing the initial value $y_0 \in [0, 1]$ completely randomly,[22] the probability to get into the set of such exceptional values, is zero.

This way of analysing the logistic system (1.19) is not yet very satisfactory. Indeed, it becomes clear that the dynamics strongly depends on the value of μ, at least when $3 \leq \mu \leq 4$, and drawing graphs for each value of μ is not an option. Therefore we discuss a *bifurcation diagram*, although this term has not yet been given a formal mathematical significance. It is a diagram in which graphical information is given about how the system depends on a parameter. As said earlier, we speak of 'bifurcation' whenever the dynamics, as a function of one or more parameters, changes qualitatively. For further discussion see Appendices C and E. In Figure 1.18 we show the result of the following numerical experiment. For a large number of values $\mu \in [2.75, 4]$ we determined which values of y occur, after repeated iteration of

$$\Phi_\mu^1(y) = \mu y(1 - y),$$

when neglecting a transient initial segment. To this end we have taken the points $\Phi_\mu^j(y_0)$ for $j = 300, 301, \ldots, 1000$ and $y_0 = 0.5$. The motivation is given by the hope that after 300 iterates the consequences of the more or less random initial choice would no longer be visible and we also hoped that in the ensuing 700 iterates all possibly occurring values of y would have been found, at least within the precision of the graphical presentation used. In this presentation, the horizontal coordinate is $\mu \in [2.75, 4]$. The stepsize in the μ-direction was determined by the pixel size of the computer screen. The y-values found in this way determined the

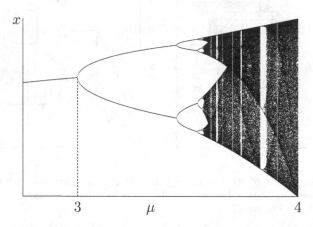

Fig. 1.18 Bifurcation diagram of the logistic system (1.19) for $2.75 < \mu < 4$. For each value of μ, on the corresponding vertical line, we indicated the y-values in $[0, 1]$ that belong to the associated 'attractor'. The various 'gaps' or 'windows' correspond to periodic attractors.

---

[22] One may think of the uniform distribution associated to the Lebesgue measure on $[0, 1]$.

vertical coordinates of the points represented. The sets detected in this way, for the various values of μ, can be viewed as numerical approximations of the attractors as these occur in dependence on μ. Compare this with the way we found a numerical approximation for the Hénon attractor.

In the 'middle' part of Figure 1.18, for each value of μ we witness two values of y. We conclude that the corresponding dynamics is periodic with period 2. For a higher value of μ we observe a splitting, after which for each value of μ four values of y occur: here the dynamics has period 4. This continues over all powers of 2, which cannot be totally derived from Figure 1.18, but at present this is known from theoretical investigations. The *period doubling bifurcation* is discussed in Appendix C. After the infinite period doubling sequence has been passed, approximately at $\mu = 3.57$, the diagram gets even more involved. For many values of μ it seems as if we have one or more intervals in the y-direction, whereas for other values of μ only finitely many y-values are obtained. The latter μ-values occur where one observes 'gaps' or 'windows' in the diagram, which correspond to periodic attractors. This is most clearly witnessed in the μ-interval with period 3, but also the period 5 area is clearly visible. We refer to a discussion in Chapter 4, in particular §4.2.

1.3.4 The Newton algorithm

We next turn to the Newton algorithm as used for determining a zero of a function. We show how this algorithm gives rise to a dynamical system, investigate how this algorithm can fail (even persistently), and show one of the miraculous pictures provoked by such a dynamical system.

We start by giving a short description of the Newton algorithm, which aims at finding the zero of a real function f, that is, at solving the equation $f(x) = 0$. Although we might consider more general cases, we assume that $f : \mathbb{R} \to \mathbb{R}$ is a polynomial map. The finding of the zero then runs as follows. First we look for a rough approximation x_0 of the zero, for instance by a graphical analysis. Next we apply a method explained below and that we hope provides us with a better approximation of the zero. This improvement can be repeated ad libitum, by which an arbitrarily sharp precision can be obtained.

The method to obtain a better approximation of the zero consists of replacing the intial guess x_0 by

$$x_1 = N_f(x_0) = x_0 - \frac{f(x_0)}{f'(x_0)}. \tag{1.20}$$

We call N_f the Newton operator associated with f. First we explain why we expect that this replacement gives a better approximation. Indeed, we construct an approximation of f by the linear polynomial $L_{x_0,f} : \mathbb{R} \to \mathbb{R}$, determined by requiring that both the value of the function and its first derivative of f and of $L_{x_0,f}$ coincide in x_0. This means that

$$L_{x_0,f}(x) = f(x_0) + (x - x_0)f'(x_0).$$

Fig. 1.19 Newton operator
for a polynomial f applied to
a point x_0.

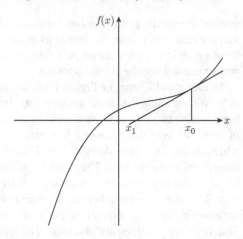

In a small neighbourhood of x_0 this is a good approximation, because by Taylor's formula

$$f(x) - L_{x_0, f}(x) = O(|x - x_0|^2)$$

as $x \to x_0$. Although in general the detection of a zero of f may be an impossible assignment, finding a zero of $L_{x_0, f}$ is simple; it is just given by

$$0 = f(x_0) + (x - x_0)f'(x_0) \quad \text{or} \quad x = x_0 - f(x_0)/f'(x_0),$$

which is exactly the value $x_1 = N_f(x_0)$ proposed in (1.20) as the better approximation of the zero of f; also see Figure 1.19. So, instead of a zero of f we determine a zero of an approximation of f, which gets ever better when we repeat the approximation. It can indeed be proven, that when we choose x_0 sufficiently close to the zero at hand, the points x_j, defined by

$$x_j = N_f(x_{j-1}) = (N_f)^j(x_0),$$

converge to that zero. See Exercise 1.17.

It now seems feasible to define an evolution operator by

$$\Phi(x, n) = (N_f)^n(x).$$

However, there is a problem: in a point $x \in \mathbb{R}$ where $f'(x) = 0$, the expression $(N_f)(x)$ is not defined. We can solve this problem by choosing as a state space $M = \mathbb{R} \cup \{\infty\}$ and extending the definition of N_f as follows.

1. $N_f(\infty) = \infty$.
2. If $0 = f'(x) = f(x)$, then $N_f(x) = x$.
3. If $0 = f'(x) \neq f(x)$, then $N_f(x) = \infty$.

In this way N_f is defined on the whole state space M and we have that $x \in M$ is a fixed point of N_f, that is, that $N_f(x) = x$, if and only if either $f(x) = 0$

or $x = \infty$. In the case where f is a polynomial of degree at least 2, the map N_f is even *continuous*. To make this statement precise, we have to endow M with a topology. To this end we show that M in a one-to-one way can be mapped onto the circle and how this circle can be endowed with a metric, and hence a topology.

1.3.4.1 $\mathbb{R} \cup \{\infty\}$ as a circle: Stereographic projection

We view the line of real numbers $\mathbb{R}$ as the x-axis of the plane $\mathbb{R}^2 = \{x, y\}$. In the same plane we also have the unit circle $\mathbb{S}^1 \subset \mathbb{R}^2$ given by $x^2 + y^2 = 1$. Naming $NP = (0, 1) \in \mathbb{S}^1$, the North Pole of $\mathbb{S}^1$, we now project $\mathbb{S}^1 \backslash \{NP\} \to \mathbb{R}$ as follows. For any point $P = (x, y) \in \mathbb{S}^1 \backslash \{NP\}$ consider the straight line that connects P with the North Pole NP. The intersection P' of this line with the x-axis is the image of $P = (x, y)$ in $\mathbb{R}$; compare Figure 1.20.

In this way we get a one-to-one correspondence between the points of $\mathbb{R}$ and $\mathbb{S}^1 \backslash \{NP\}$. We call this map the *stereographic projection*.

Remark. Stereographic projection exists in higher dimensions as well. Indeed, completely analogously we can consider the unit sphere $\mathbb{S}^{n-1} \subset \mathbb{R}^n$, define the North Pole $NP = (0, 0, \dots, 0, 1)$, and define a map $\mathbb{S}^{n-1} \backslash \{NP\} \to \mathbb{R}^{n-1}$ in a completely similar way.

Returning to the case $n = 2$, we now get the one-to-one correspondence between M and $\mathbb{S}^1$ by associating ∞ with NP. On $\mathbb{S}^1$ we can define the distance between two points in different ways. One way is the Euclidean distance inherited from the plane: for points (x_1, y_1) and $(x_2, y_2) \in \mathbb{S}^1$ we define

$$d_E((x_1, y_1), (x_2, y_2)) = \sqrt{(x_1 - x_2)^2 + (y_1 - y_2)^2}.$$

Another way is to take for two points $(\cos \varphi_1, \sin \varphi_1)$ and $(\cos \varphi_2, \sin \varphi_2) \in \mathbb{S}^1$ the distance along the unit circle by

$$d_{\mathbb{S}^1}((\cos \varphi_1, \sin \varphi_1), (\cos \varphi_2, \sin \varphi_2)) = \min_{n \in \mathbb{N}} |\varphi_1 - \varphi_2 + 2n\pi|.$$

It is not hard to show that these two metrics are equivalent in the sense that

$$1 \le \frac{d_{\mathbb{S}^1}}{d_E} \le \frac{1}{2}\pi.$$

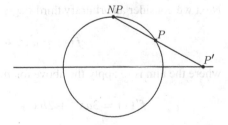

Fig. 1.20 Stereographic projection $\mathbb{S}^1 \backslash \{NP\} \to \mathbb{R}$ which maps P to P'.

We discuss these metrics in some detail, inasmuch as they are also useful later.

In summary, the stereographic projection is a natural way to pass from the line $\mathbb{R}$ to $\mathbb{R} \cup \{\infty\}$, where the latter is identified with the circle $\mathbb{S}^1$. In this way, $\mathbb{R} \cup \{\infty\}$ inherits a metric from $\mathbb{S}^1$. It turns out that for any real polynomial f of degree at least 2; the Newton operator N_f, seen as a map on $\mathbb{R} \cup \{\infty\}$, is continuous. See Exercise 1.15.

1.3.4.2 Applicability of the Newton algorithm

In the case of an algorithm such as the Newton algorithm N_f (see (1.20)) it is important to know whether for an arbitrary choice of the initial point there is a reasonable probability that repeated iteration leads to convergence to a zero of f. We already stated that, when the initial point is close enough to a given zero of f, we are sure to converge to this zero. It is easily seen that for any polynomial f of degree at least 1, the fixed point ∞ of N_f is repelling. In fact one can show that if $|x|$ is sufficiently large, then $|N_f(x)| < |x|$. This means that the only way for a sequence $x_j = (N_f)^j(x_0)$, $j \in \mathbb{Z}_+$ to escape to ∞, is by landing on a point x where $f'(x) = 0$. (For a polynomial of positive degree, this only holds for a finite number of points.) Another question now is whether it is possible for such a sequence x_j, $j \in \mathbb{Z}_+$ not to escape to ∞, but neither ever to converge to a zero of f. Moreover one may wonder how exceptional such behaviour would be. Again compare Exercise 1.15.

1.3.4.3 Nonconvergent Newton algorithm

We here construct examples where the Newton algorithm (1.20) does not converge. The simplest case would be to take a polynomial without any (real) zeroes. We do not discuss this case further. Another obvious way to invoke failure of convergence is to ensure that there are two points p_1 and p_2, such that

$$N_f(p_1) = p_2 \quad \text{and} \quad N_f(p_2) = p_1,$$

where $\{p_1, p_2\}$ is an attracting orbit of period 2. To achieve this, we first observe that in general

$$N_f'(x) = \frac{f(x) f''(x)}{(f'(x))^2}.$$

Next we consider an arbitrary third-degree polynomial

$$f(x) = ax^3 + bx^2 + cx + d,$$

where the aim is to apply the above for $p_1 = -1$ and $p_2 = +1$. We have

$$f'(x) = 3ax^2 + 2bx + c \quad \text{and} \quad f''(x) = 6ax + 2b.$$

Direct computation now yields that

$$N_f(1) = -1 \Leftrightarrow 5a + 3b + c - d = 0 \quad \text{and}$$
$$N_f(-1) = 1 \Leftrightarrow 5a - 3b + c + d = 0,$$

where attractivity of this period 2 orbit $\{-1, 1\}$ can be simply achieved by requiring that $N'_f(1) = 0$, which amounts to

$$f''(1) = 0 \Leftrightarrow 6a + 2b = 0.$$

All these conditions are fulfilled for

$$a = 1, \quad b = -3, \quad c = -5 \quad \text{and} \quad d = -9,$$

thereby yielding the polynomial

$$f(x) = x^3 - 3x^2 - 5x - 9; \tag{1.21}$$

also see Figure 1.21.

A subsequent question is whether this phenomenon, of persistent nonconvergence of the Newton algorithm to a zero, is exceptional in the set of all polynomials. This is not the case. Indeed, it turns out that in the space of all polynomials of degree 3 there exists an open set of polynomials for which the above phenomenon occurs.

We now show that for any third-degree polynomial, the coefficients of which are sufficiently close to those of (1.21), there is an open set of initial points such that the corresponding Newton algorithm does not converge to any zero of f. Note that thus we have established another form of persistence, in the sense that the property at hand remains valid under sufficiently small perturbations of f within the space of third-degree polynomials.

To construct our open set of third-degree polynomials, take an interval $[a, b]$ such that:

  i. $a < 1 < b$.
  ii. $(N_f)^2([a, b]) \subset (a, b)$.
  iii. $|((N_f)^2)'(x)| < 1$ for all $x \in [a, b]$.
  iv. f has no zeroes in $[a, b]$.

This implies that $(N_f)^2$ maps the interval $[a, b]$ into itself and is a contraction here. Again, by the contraction principle (see Appendix A), it follows that the points of

Fig. 1.21 N_f-orbit of the polynomial $f(x) = x^3 - 3x^2 - 5x - 9$, asymptotic to the period 2 orbit $\{-1, +1\}$.

$[a, b]$, as an initial point for the Newton algorithm, do not give convergence to a zero of f. Indeed, as before, iterations of $(N_f)^2$ always converge to a fixed point inside $[a, b]$.

We now define the neighbourhood $\mathcal{U}$ of f in the space of third-degree polynomials by requiring that $\tilde{f} \in \mathcal{U}$ if and only if the following properties hold.

1. $(N_{\tilde{f}})^2[a, b] \subset (a, b)$.
2. For all $x \in [a, b]$ one has $|((N_{\tilde{f}})^2)'(x)| < 1$.
3. $\tilde{f}$ has no zeroes in $[a, b]$.

It is not hard to show that $\mathcal{U}$, as defined thus, is open in the space of polynomials of degree 3 and that $f \in \mathcal{U}$. The contraction principle then implies that for any $\tilde{f} \in \mathcal{U}$ and for all $x \in [a, b]$ the limit

$$\lim_{j \to \infty} (N_{\tilde{f}})^{2j}(x)$$

exists. Finally, because $\tilde{f}$ has no zeroes inside $[a, b]$, it follows that the Newton algorithm does not converge to any zero of $\tilde{f}$.

Remarks.

- This same argument also holds for polynomials of any fixed degree greater than or equal to 3.
- An example such as (1.21) does not exist in the space of second-degree polynomials. This is a direct consequence of Exercise 1.12.

1.3.4.4 Newton algorithm in higher dimensions

We conclude this part with a few remarks on the Newton algorithm in higher dimensions. First we think of the search of zeroes of maps $f : \mathbb{R}^n \to \mathbb{R}^n$. On the same grounds as in the 1-dimensional case we define the Newton operator as

$$x_1 = N_f(x_0) = x_0 - (D_{x_0} f)^{-1}(f(x_0)).$$

The most important difference with (1.20) is that here $(D_{x_0} f)^{-1}$ is a matrix inverse. Such a simple extension for maps between vector spaces of different dimensions does not exist, because then the derivative would never be invertible.

Special interest should be given to the case of holomorphic maps $f : \mathbb{C} \to \mathbb{C}$. Now the Newton operator is given by (1.20)

$$x_1 = N_f(x_0) = x_0 - \frac{f(x_0)}{f'(x_0)},$$

with complex quantities and complex differentiation. In the complex case it also holds true that, if f is a complex polynomial of degree at least 2, the Newton operator can be extended continuously to $\mathbb{C} \cup \{\infty\}$. The latter set, by stereographic projection, can be identified with the unit sphere $\mathbb{S}^2 \subset \mathbb{R}^3 = \{x, y, z\}$, given by the equation $x^2 + y^2 + z^2 = 1$. This complex Newton algorithm gives rise to very beautiful pictures; compare [12]. The most well known of these is related to the Newton operator for the polynomial $f(z) = z^3 - 1$. Of course it is easy to identify the zeroes of f as

$$p_1 = 1 \quad \text{and} \quad p_{2,3} = \exp\left(\pm\frac{2}{3}\pi\right)i.$$

The problem, however, is how to divide the complex plane into four subsets B_1, B_2, B_3, and R, where B_j is the (open) subset of points that under the Newton algorithm converge to p_j, and where R is the remaining set. In Figure 1.22 the three regions $B_j, j = 1, 2$, and 3 are indicated in black, grey, and white. Note that the areas can be obtained from one another by rotating over $(2/3)\pi$ radians to the left or to the right.

From the figure it already shows that the boundary of the white region B_1 is utterly complicated. It can be shown that each point of this boundary is a boundary point of all three regions B_1, B_2, and B_3. So we have a subdivision of the plane in three territories, the boundaries of which only contain triple points. For further reference, in particular regarding the construction of pictures such as Figure 1.22, see [12, 187].

Remarks.

- This problem of dividing the complex plane in domains of attraction of the three zeroes of $z^3 - 1$ under the Newton operator is remarkably old: in [100] Cayley[23]

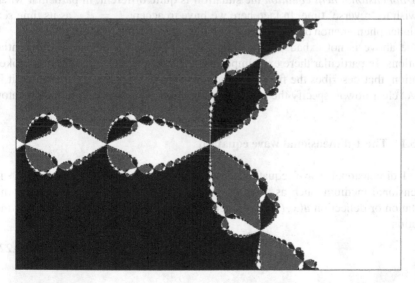

Fig. 1.22 Sets of points in $\mathbb{C}$ that under the Newton algorithm of $f(z) = z^3 - 1$ converge to the zeroes 1 (white), $e^{(2/3)\pi i}$ (grey), and $e^{-(2/3)\pi i}$ (black).

---

[23] Arthur Cayley 1821–1895.

proves that the corresponding problem for $z^2 - 1$ has a simple solution: the zeroes are ± 1 and the corresponding domains of attraction are the left and the right half-plane (the remaining set being the imaginary axis). He already notes that the 'next', cubic case $z^3 - 1$ turns out to be more difficult. In 1910 Brouwer[24] in [97], without having any dynamical interpretation in mind, shows that it is possible to divide the plane into three domains, such that the boundaries consist solely of triple points. For a long period of time, this result was viewed as a pathology, born in the curious mind of a pure mathematician. Here we see that such examples have a natural existence. The question may well be asked whether, without computer graphics, we would ever have gotten the idea that this phenomenon occurs here.

- In Appendix B a Newtonian iteration process is being used on a space of functions, therefore in an ∞-dimensional space.

1.3.5 Dynamical systems defined by partial differential equations

We announced that mostly we deal with dynamical systems, the state space of which is finite-dimensional. Still, here we treat two examples of dynamical systems described by partial differential equations, the state space of which therefore is infinite-dimensional, because it is a function space. We show that the 1-*dimensional wave equation* determines a dynamical system, the evolution operator of which can be well defined in all respects, admitting the time set $T = \mathbb{R}$. Next we show that for the 1-*dimensional heat equation* the situation is quite different, in particular when we wish to 'reverse' time. In fact, here we have to accept $T = \mathbb{R}_+$ as its time set. The latter phenomenon occurs generally in the case of diffusion equations.

The above is not exhaustive regarding the peculiarities of partial differential equations. In particular there exist important equations (such as the Navier–Stokes equation, that describes the flow of incompressible viscous fluids), for which it is not yet clear how to specify the state space to get a well-defined evolution operator.

1.3.5.1 The 1-dimensional wave equation

The 1-dimensional wave equation describes the wave propagation in a 1-dimensional medium, such as a string or an organ pipe. For such a medium the excitation or deflection $u(x, t)$, as a function of position x and time t, satisfies the equation

$$\frac{\partial^2 u}{\partial t^2} = V^2 \frac{\partial^2 u}{\partial x^2},$$
(1.22)

---

[24] Luitzen Egbertus Jan Brouwer 1881–1966.

where V is the propagation velocity of the medium at hand. The choice of units can be made such that $V = 1$. Also we use an abbreviated notation for partial derivatives and so rewrite equation (1.22) to

$$\partial_{tt} u = \partial_{xx} u. \qquad (1.23)$$

When considering such equations for a string, then it is feasible to impose *boundary conditions*. For example, if we deal with a string of length L, then for u we take as the domain of definition $[0, L] \times \mathbb{R} = \{x, t\}$ and we require that for all $t \in \mathbb{R}$ one has $u(0, t) = u(L, t) = 0$. First, however, we discuss the case where the 'string' is unbounded and where u therefore is defined on all of $\mathbb{R}^2$. To determine the state at the instant $t = 0$, it is not sufficient to only specify the function $x \mapsto u(x, 0)$. Indeed, just as for mechanical systems like the pendulum, where we also have an equation of motion for the second-order derivative(s) with respect to time, here too we have to specify the initial velocity $x \mapsto \partial_t u(x, 0) = v(x, 0)$. If the initial state is thus specified, we can determine the corresponding solution of the wave equation (1.23).

1.3.5.2 Solution of the 1-dimensional wave equation

First note that whenever f and g are real functions of one variable, it follows that

$$u_{f+}(x, t) = f(x + t) \quad \text{and} \quad u_{g-}(x, t) = g(x - t)$$

are solutions of the wave equation: they represent a traveling wave running to the left and to the right, respectively. Because the wave equation is linear, the sum of such solutions is also a solution. The initial state of $u = u_{f+} + u_{g-}$ now is given by

$$u(x, 0) = f(x) + g(x) \quad \text{and} \quad \partial_t u(x, 0) = f'(x) - g'(x).$$

On the other hand, if the initial state is given by the functions $u(x, 0)$ and $v(x, 0)$, then we can find corresponding functions f and g, and hence the associated solution, by taking

$$f'(x) = \tfrac{1}{2}(\partial_x u(x, 0) + v(x, 0)) \quad \text{and} \quad g'(x) = \tfrac{1}{2}(\partial_x u(x, 0) - v(x, 0)),$$

after which f and g are obtained by taking primitives. We have to include appropriate additive constants to make the solution fit with $u(x, 0)$.

The method described above needs some completion. The wave equation is of second order, thus it is feasible to require that the solutions should be at least of class C^2. This can be obtained by only considering initial states for which $u(x, 0)$ and $v(x, 0)$ are C^2 and C^1, respectively. The solution as constructed then is automatically C^2. Moreover, it can be shown that the given solution is unique.[25]

---

[25] The proof is not very complicated, but uses a transformation $\xi = x + t, \eta = x - t$, which turns the wave equation into $\partial_{\xi\eta} u = 0$.

Finally, to find solutions that are only defined on the strip $[0, L] \times \mathbb{R}$, satisfying the boundary conditions $u(0, t) = u(L, t) = 0$ for all $t \in \mathbb{R}$, we have to take initial states $u(x, 0)$ and $v(x, 0)$ that satisfy $u(0, 0) = u(L, 0) = v(0, 0) = v(L, 0) = 0$. However, this is not sufficient. To ensure that also $\partial_{tt} u(0, 0) = 0$ and $\partial_{tt} u(0, L) = 0$, the initial state also has to satisfy $\partial_{xx} u(0, 0) = \partial_{xx} u(L, 0) = 0$. Now to construct the solution, we first extend $u(x, 0)$ and $v(x, 0)$ to all values of $x \in \mathbb{R}$, explained now. Indeed, we require that both functions are periodic, with period $2L$. Next we take for $x \in [0, L]$: $u(L + x, 0) = -u(L - x, 0)$ and $v(L + x, 0) = -v(L - x, 0)$. It is easily checked that the thus extended functions $u(x, 0)$ and $v(x, 0)$ are of class C^2 and C^1, respectively. The reader is invited to fill in further details; also see [105].

1.3.5.3   The 1-dimensional heat equation

We now consider a somewhat different equation, namely

$$\partial_t u(x, t) = \partial_{xx} u(x, t). \tag{1.24}$$

This equation describes the evolution of the temperature in a 1-dimensional medium (think of a rod), as a consequence of heat transport. In general, the right-hand side has to be multiplied by a constant, that depends on the heat conductivity in the medium and on the so-called heat capacity of the medium. We, however, assume that the units have been chosen in such a way that this constant equals 1. Also, from the start we now assume that the medium has a finite length and that the temperature at the endpoints is kept constantly at 0. Mathematically it turns out to be convenient to take as the x-interval $[0, \pi]$, although this choice is not essential. With an argument completely analogous to the previous case of the wave equation, we now see that an initial state is given by a C^2 temperature distribution $u(x, 0)$ at the instant $t = 0$, that should satisfy

$$u(0, 0) = u(\pi, 0) = \partial_{xx} u(0, 0) = \partial_{xx} u(\pi, 0) = 0.$$

That the initial state does not have to contain information on $\partial_t u(x, 0) = v(x, 0)$ comes from the fact that the equation of motion just contains the first derivative with respect to the time t. As in the wave equation, the initial distribution $u(x, 0)$ can be extended all over $x \in \mathbb{R}$, such that the result is periodic with period 2π and such that $u(\pi + x, 0) = -u(\pi - x, 0)$; and as in the case of the wave equation, this extension is of class C^2.

For the heat equation it turns out to be handy to use Fourier series. The simplest properties of such series can be found in any good textbook on calculus (e.g., [105]). We now give a brief summary as far as needed. Any C^2-function f that is 2π-periodic and that satisfies $f(\pi + x) = -f(\pi - x)$, and hence also $f(0) = f(\pi) = 0$, can be expressed uniquely by a series

$$f(x) = \sum_{k=1}^{\infty} c_k \sin(kx). \tag{1.25}$$

The advantage of such a representation becomes clear when we realise that

$$\partial_{xx} f(x) = -\sum_{k=1}^{\infty} c_k k^2 \sin(kx). \tag{1.26}$$

For the moment we do not worry about convergence of infinite sums. Continuing in the same spirit, we now write down the solution of the heat equation. To this end we first expand as a Fourier series the initial state $u(x, 0)$, properly extended over $\mathbb{R}$ as a periodic function:

$$u(x, 0) = \sum_{k=1}^{\infty} c_k \sin(kx).$$

We express the solution, for each value of t, again as a Fourier series:

$$u(x, t) = \sum_{k=1}^{\infty} c_k(t) \sin(kx).$$

We next use formula (1.26) to conclude that necessarily

$$c_k(t) = c_k(0) \exp(-k^2 t)$$

and as the general (formal) solution we thus have found

$$u(x, t) = \sum_{k=1}^{\infty} c_k(0) \exp(-k^2 t) \sin(kx). \tag{1.27}$$

We now just have to investigate the convergence of these series. In this way it turns out that the solutions in general are not defined for negative t.

For the investigation of the convergence of functions defined by infinite sums as in (1.25), we have to go a little deeper into the theory of Fourier series; compare [105, 200]. From this theory we need the following. For further details also see §5.5.3.

Theorem 1.3 (Fourier series). *For any C^2-function $f : \mathbb{R} \to \mathbb{R}$ the Fourier series in (1.25) is absolutely convergent and even such that, for a positive constant C and for all k, one has*

$$c_k \leq C k^{-2}.$$

On the other hand, if the series in (1.25) is absolutely convergent, the limit function is well defined and continuous and for the coefficients c_k one has

$$c_k = \frac{2}{\pi} \int_0^{\pi} \sin(kx) f(x) dx.$$

By ℓ-fold differentiation of the equation (1.25) with respect to x, we conclude that if for a constant $C > 0$ and for all k one has

$$c_k \leq Ck^{-(\ell+2)},$$

then the corresponding function f has to be of class C^ℓ. From this, and from the formula (1.24), it follows that if the initial function $u(x, 0)$ is of class C^2, then the corresponding solution $u(x, t)$, for each $t > 0$, is a C^∞-function of x. It also should be said here that there exist initial functions $u(x, 0)$ of class C^2, and even of class C^∞, for which the corresponding expression (1.27) does not converge for any $t < 0$. An example of this is

$$u(x, 0) = \sum_{k=1}^{\infty} \exp(-k)\sin(kx).$$

Indeed, for such an initial condition of the heat equation, no solution is defined for $t < 0$, at least not if we require that such a solution should be of class C^2.

1.3.6   The Lorenz attractor

In 1963 Lorenz[26] [160] published a strongly simplified set of equations for the circulation in a horizontal fluid layer that is being heated from below. (Although we keep talking of a fluid, one may also think of a layer of gas, e.g., of the atmosphere.) If the lower side of such a layer is heated sufficiently, an instability arises and a convection comes into existence, the so-called Rayleigh–Bénard convection. The equations for this phenomenon are far too complicated to be solved analytically, and, certainly during the time that Lorenz wrote his paper, they were also too complicated for numerical solution. By strong simplification, Lorenz was able to get some insight in possible properties of the dynamics which may be expected in fluid flows at the onset of turbulence.

1.3.6.1   The Lorenz system; the Lorenz attractor

This simplification, now known as the Lorenz system, is given by the following differential equations on $\mathbb{R}^3$.

$$\begin{aligned}
x' &= \sigma y - \sigma x \\
y' &= rx - y - xz \\
z' &= -bz + xy.
\end{aligned} \tag{1.28}$$

---

[26] Edward N. Lorenz 1917–2008.

The standard choice for the coefficients is $\sigma = 10$, $b = 8/3$ and $r = 28$. In Appendix D we discuss how the simplification works and how the choice of constants was made.

Here two phenomena showed up. First, as far as can be checked, the solutions in general do not converge to anything periodic. Compare Figure 1.23, which is obtained numerically and shows a typical evolution of the Lorenz system (1.28). Second, however close together we take two initial states, after a certain amount of time there is no longer any connection between the behaviour of the ensuing evolutions. This phenomenon is illustrated in Figure 1.24. The latter is an indication that the evolutions of this system will be badly predictable. In the atmospheric dynamics,

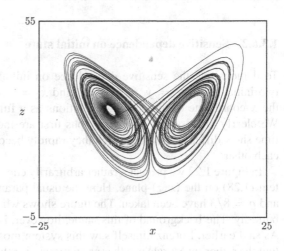

Fig. 1.23 Typical evolution of the Lorenz system (1.28), projected on the (x, z)-plane.

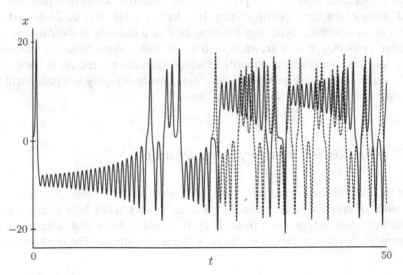

Fig. 1.24 The x-coordinate of two evolutions of the Lorenz system (1.28) with nearby initial states, as a function of time: an illustration of sensitive dependence on initial state.

convection plays an important role. So the unpredictability of the evolution of the Lorenz system may be related to the fact that prediction of the future state of the atmospheric system (such as weather or climate forecasts) over long time intervals is so unreliable. The theoretical foundation of the bad predictability of the weather has been greatly expanded over the last years [60].

We have witnessed comparable behaviour already for the logistic system, but the systematic study of that system really started in the 1970s. In Lorenz's earlier work, for the first time it was shown by numerical simulations that deterministic systems by this *sensitive dependence on initial state* can be highly unpredictable. Next we show some results of numerical simulations. In Chapter 4, in particular in §4.2, we continue the mathematical discussion on the Lorenz system.

1.3.6.2 Sensitive dependence on initial state

To demonstrate the sensitive dependence on initial state, in Figure 1.24 we take two initial states $x = y = z = 0.2$ and $x = y = 0.2$, $z = 0.20001$, plotting the x-coordinate of the ensuing evolutions as a function of the time $t \in [0, 50]$. We clearly witness that the two graphs first are indistinguishable, but after some time show differences, after which they rapidly become completely independent of each other.

In Figure 1.23 we project a rather arbitrarily chosen evolution of the Lorenz system (1.28) on the (x, z)-plane. Here the usual parameter values $\sigma = 10$, $r = 28$, and $b = 8/3$ have been taken. The figure shows why one is referring to the Lorenz butterfly. The background of this name, however, is somewhat more complicated. As said earlier, Lorenz himself saw this system mostly as supportive for the thesis that the rather inaccurate weather predictions are related to sensitive dependence on initial states of the atmospheric system. In a lecture [160, 161] he illustrated this by way of an example, saying that the wing beat of a butterfly in the Amazon jungle, after a month, can have grown to such a size, that it might 'cause' a tornado in Texas. It is this catastrophical butterfly to which the name relates. As in the case of the Hénon attractor (see Figure 1.13), this configuration is highly independent of the initial state, at least if we disregard a transient.

1.3.7   The Rössler attractor; Poincaré map

As in the case of the Lorenz attractor (1.23), also for the Rössler [27] attractor we deal with differential equations in $\mathbb{R}^3$. Rössler's work dates later than Lorenz's and we borrowed the equations from [201]. It should be noted that at that time the occurrence of chaotic attractors no longer was a great surprise. The most important

---

[27] Otto E. Rössler 1940–.

contribution of Rössler has been that he developed a systematic method to find such a system. We include the example here merely as an illustration of the construction of a Poincaré map, which differs a little from our original definition.

1.3.7.1 The Rössler system

To be more precise, we now present the equations of Rössler as

$$x' = -y - z$$
$$y' = x + ay$$
$$z' = bx - cz + xz, \qquad (1.29)$$

where we use the standard values $a = 0.36$, $b = 0.4$, and $c = 4.5$.

In Figure 1.25 we see the projection of an evolution curve on the (x, y)-plane. Again it is true that such an evolution curve of (1.29), apart from an initial transient that is strongly dependent on the chosen initial state, always yields the same configuration. As reported earlier, such a subset of the state space which attracts solutions, is called an attractor. In this case we call the attractor $A \subset \mathbb{R}^3$. From Figure 1.25 we derive that this set A forms a kind of band, wrapped around a 'hole' and the evolution curves in A wind around this 'hole'. We consider a half-plane of the form $L = \{x < 0, y = 0\}$, such that the boundary $\partial L = \{x = 0, y = 0\}$ sticks through the hole in the attractor. It is clearly suggested by the numerics that in each point of the intersection $A \cap L$, the evolution curve of (1.29) is not tangent to L and also that the evolution curve, both in forward and backward time, returns to $A \cap L$. This means that we now have a Poincaré map

$$P : A \cap L \to A \cap L,$$

that assigns to each point $p \in A \cap L$ the point where the evolution curve of (1.29) starting at p hits $A \cap L$ at the first future occasion. So this is a Poincaré map of (1.29), restricted to A.

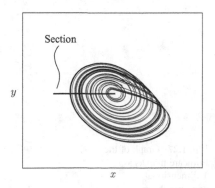

Fig. 1.25 Projection of an evolution of the Rössler system (1.29) on the (x, y)-plane. The domain of both x and y is the interval $[-16, 16]$.

1.3.7.2   The attractor of the Poincaré map

We now can get an impression of $A \cap L$, which is the attractor of the Poincaré map corresponding to A, by plotting the successive passings of the evolution of Figure 1.25 through the plane L. The result is depicted in Figure 1.26. We 'see' now that the attractor of the Poincaré map is part of the curve which is the graph of a function $z = z(x)$. Unfortunately we can 'prove' that the structure of this attractor cannot be that simple. Also see Exercise 1.27.

For the moment we disregard the theoretical complications concerning the structure of the Poincaré map attractor, and pretend that it is indeed part of the graph of a smooth function $z = z(x)$. We can then get an impression of the dynamics inside this attractor, by computing, from the evolution in Figure 1.25, the approximate graph of the Poincaré map. This can be done as follows. As we already saw, the x-coordinate yields a good parametrisation of the attractor. If now $x_1, x_2, \ldots$ are the values of the x-coordinate at the successive passings of the evolution through $L = \{x < 0, y = 0\}$, then the points $(x_i, x_{i+1}) \in \mathbb{R}^2$ should lie on the graph of the Poincaré map. In this way we get Figure 1.27. When replacing (x_n, x_{n+1}) by

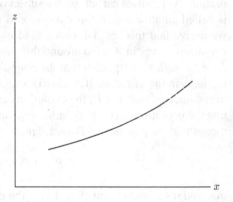

Fig. 1.26 Attractor of the Poincaré map; horizontally we have the x-axis with domain $[-6, 0]$ and vertically the z-axis with domain $[-0.3, 0.1]$.

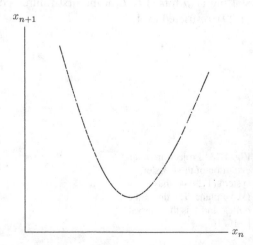

Fig. 1.27 Graph of the Poincaré map as a 1-dimensional endomorphism.

$(-x_n, -x_{n+1})$, this graph reminds us of the graph of the endomorphism that determines the logistic system; see §1.3.3: it has one maximum in the interior of its domain of definition and also two (local) minima at the boundaries. The mathematical theory developed for the logistic map largely applies also to such a (so-called unimodal) map. We thus have a strong 'experimental' indication that the behaviour of the Poincaré map of the Rössler system (1.29) may strongly resemble that of the logistic system.

Remark. It should be said that this kind of thinking is an important object of current research that is developed elsewhere in the literature and is beyond the scope of the present book. Also compare the remarks and references in §§4.2 and 4.3 and in Appendix C.

1.3.8 The doubling map and chaos

This chapter is concluded by an example which is not directly connected to any kind of application or to any interpretation outside mathematics. Its interest, however, is that it can be handled mathematically rigourously. The example occurs in three variations, that mutually differ only little. We now describe all three of them.

1.3.8.1 The doubling map on the interval

We first deal with the doubling map on the interval. In this case for the state space we have $M = [0, 1)$, and the dynamics is defined by the map

$$\varphi(x) = \begin{cases} 2x & \text{if } 0 \le x < \dfrac{1}{2} \\ 2x - 1 & \text{if } \dfrac{1}{2} \le x < 1; \end{cases} \tag{1.30}$$

see Figure 1.28. We also like to use the notation $\varphi(x) = 2x \bmod 1$ for this, or even

$$\varphi(x) = 2x \bmod \mathbb{Z}.$$

It should be clear that this map is not invertible, whence for the time set we have to choose $T = \mathbb{Z}_+$. The map φ generating the dynamics, clearly is not continuous: there is a discontinuity at the point $x = \frac{1}{2}$. We can 'remove' this discontinuity by replacing the state space by the circle. In this way we get the second version of our system.

Fig. 1.28 Graph of the
doubling map on the interval
$[0, 1)$.

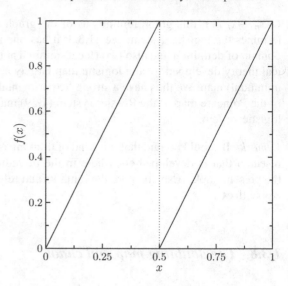

1.3.8.2 The doubling map on the circle

We met the circle as a state space before, when considering the Newton algorithm;
see §1.3.4. There we thought of the unit circle in the plane. By parametrising this
circle as $\{(\cos\sigma, \sin\sigma) \mid \sigma \in \mathbb{R}\}$, it should be clear that the circle can also be
viewed as $\mathbb{R}/(2\pi\mathbb{Z})$. The value of 2π is quite arbitrary and just determined by our
convention to measure angles in radians. Therefore we may now replace 2π by 1
and consider the circle as the quotient $\mathbb{R}/\mathbb{Z}$ of additive groups. In summary, we
chose $M' = \mathbb{R}/\mathbb{Z}$. In this version, the transformation that generates the dynamics
is almost the same, namely

$$\varphi'([x]) = [2x], \tag{1.31}$$

where, as before , we use the notation $[-]$ to indicate the equivalence class. It is not
difficult to show that this transformation is continuous. The map

$$x \mapsto [x]$$

determines a one-to-one correspondence between the elements of M and M'.

1.3.8.3 The doubling map in symbolic dynamics

The third version of the doubling map has a somewhat different appearance. As the
state space we consider all infinite sequences

$$s = (s_0, s_1, s_2, \ldots),$$

where each s_j takes on either the value 0 or 1. The set of all such sequences is denoted by Σ_2^+. Here the subscript 2 refers to the choice of each s_j from the set $\{0, 1\}$, whereas the superscript $+$ indicates that the set for the indices j is $\mathbb{Z}_+$. The dynamics now is defined by the *shift map*

$$\sigma : \Sigma_2^+ \to \Sigma_2^+, s = (s_0, s_1, s_2, \ldots) \mapsto \sigma(s) = (s_1, s_2, s_3, \ldots). \qquad (1.32)$$

The relationship between this system and the previous ones can be seen by expressing the elements of $M = [0, 1)$ in the binary number system. For each $x \in M$ this yields an expression

$$x = 0.x_0 x_1 x_2 \ldots \Leftrightarrow x = \sum_{j=1}^{\infty} x_{j-1} 2^{-j},$$

where $x_j \in \{0, 1\}$, for all $j \in \mathbb{Z}_+$. In this way for each point $x \in M$ we have a corresponding point in Σ_2^- Moreover, for $x = 0.x_0 x_1 x_2 \ldots$ we see that $\varphi(x) = 0.x_1 x_2 x_3 \ldots$, which indicates the relationship between the dynamical systems

$$(M, T, \varphi) \quad \text{and} \quad (\Sigma_2^+, T, \sigma).$$

A few remarks are in order.

Remarks.

- It is clear now that the choice of the binary number system is determined by the fact that we deal with a *doubling* map. For the *tenfold* map $x \mapsto 10 x$ mod $\mathbb{Z}$ we would have to use the decimal number system. See Exercise 1.28.
- The relation between the first and the last version can also be expressed without using the binary number system. To this end we take $M = [0, 1)$, dividing it in two parts $M = M_0 \cup M_1$, where

$$M_0 = \left[0, \frac{1}{2}\right) \quad \text{and} \quad M_1 = \left[\frac{1}{2}, 1\right).$$

Now if a point $x \in M$ happens to be in M_0, we say that its address is 0, and similarly the address is 1 for M_1. Thus for $x \in M$ we can define a sequence $s(x) \in \Sigma_2^+$, by taking $s_j(x)$ as the address of $\varphi^j(x)$, $j \in \mathbb{Z}_+$, in other words by defining

$$s_j(x) = \begin{cases} 0 & \text{if} \quad \varphi^j(x) \in M_0, \\ 1 & \text{if} \quad \varphi^j(x) \in M_1, \end{cases}$$

for $j \in \mathbb{Z}_+$. In this way we get, apart from a small problem that we discuss in a moment, exactly the same sequences as with the binary expansions. These addresses are also called *symbols* and this is how we arrive at the term *symbolic dynamics*.

– The small problem just mentioned is that there is an ambiguity in assigning elements of Σ_2^+ to elements of M. This problem is illustrated by the fact that in the decimal system one has the ambiguity

$$0.3 = \frac{3}{10} = 0.299999\ldots,$$

related to an 'infinite tail of nines'. This problem in general can be overcome by excluding such tails of nines in all decimal expansions.

In the binary system the analogous problem is constituted by 'tails of ones'. This makes the assignment

$$x \mapsto s(x) = (s_0(x), s_1(x), s_2(x), \ldots)$$

slightly problematic. However, if we exclude tails of ones, then the map

$$s : M \to \Sigma_2^+ \tag{1.33}$$

is well-defined and injective, but not surjective (because we avoided all tails of ones). This can be seen when we use the method of 'addresses' of the previous item. Similarly the map

$$s = (s_0, s_1, s_2, \ldots) \mapsto \sum_j s_j 2^{-(j+1)}$$

is well-defined but not injective, for the same kind of reason. For further discussion also see Exercise 1.18.

It is customary to define on Σ_2^+ a metric as follows: if $s = (s_0, s_1, s_2, \ldots)$ and $t = (t_0, t_1, t_2, \ldots) \in \Sigma_2^+$ are two sequences, then

$$d(s, t) = 2^{-j(s,t)},$$

where $j(s, t)$ is the smallest index for which $s_{j(s,t)} \neq t_{j(s,t)}$. It is easy to show that this indeed defines a metric.

We conclude that the three versions of the doubling map, as constructed here, are basically identical, but that there are still small troubles (i.e., tails of ones) which means that we have to be careful when translating results from one version to another.

1.3.8.4 Analysis of the doubling map in symbolic form

Now we show that the symbolic version of the doubling map

$$\sigma : \Sigma_2^+ \to \Sigma_2^+$$

is very appropriate for proving a number of results on the dynamics generated by it. Here we recall that the time set is $\mathbb{Z}_+$. In each case we indicate what the corresponding results are for the other versions.

Periodic points. We recall that a point $s \in \Sigma_2^+$ is periodic with period k if $\sigma^k(s) = s$. In the next chapter we treat periodic evolutions in greater generality, but we now indicate the general situation in the present case of time set $\mathbb{Z}_+$.

Observe that if a periodic point has period k, then automatically it also has periods nk, for all $n \in \mathbb{Z}_+$. For any periodic point there exists a smallest period $k_0 > 0$, called the *prime period*. A point with prime period k_0 then has as all its periods $\{nk_0\}_{n \in \mathbb{Z}_+}$. and no others. Moreover, a point of period 1 is a fixed point.

For our shift map $\sigma : \Sigma_2^+ \to \Sigma_2^+$ it is clear that the sequence s is a periodic point with period k if and only if the sequence s is periodic of period k. This means that

$$s_j = s_{j+k}$$

for all $j \in \mathbb{Z}_+$. Such a periodic point therefore is determined completely by the initial part $(s_0, \ldots, s_{k-1})$ of the sequence. From this comes the following.

Proposition 1.4 (Periodic points of shift map). *Let $\sigma : \Sigma_2^+ \to \Sigma_2^2$ be the shift map* (1.32). *Then the number of points in Σ_2^+ which are periodic of period k under σ is equal to 2^k.*

When transferring this result to the other versions of the doubling map, we have to make a correction: for any period we include here the sequence $s = (1, 1, 1, \ldots)$, that completely consists of digits 1. This point should not be counted in the other versions, where it is identified with $(0, 0, 0, \ldots)$. Note that this is the only case with a tail of ones for a periodic point. Thus we find

Proposition 1.5 (Periodic points of doubling map). *Let $\varphi : [0, 1) \to [0, 1)$ be the doubling map of the interval* (1.30). *Then the number of points in $[0, 1)$ that is periodic of period k under φ is equal to $2^k - 1$.*

A similar statement holds for the doubling map on the circle (1.31).

Next we show that the set of all periodic points, therefore of any period, is dense in the set Σ_2^+. In general we say that a subset $A \subset M$ is dense if its closure is equal to the total set (i.e., when $\overline{A} = M$). This means that for any point $s \in M$ and for any neighbourhood U of s in M, the neighbourhood U contains points of A. For an explanation of topological concepts, see Appendix A.

Therefore consider an arbitrary point $s \in \Sigma_2^+$. For an arbitrary neighbourhood U of s we can say that for a certain index $j \in \mathbb{Z}_+$, the subset U contains all sequences t that are at most at a distance 2^{-j} from s, which means all sequences t the first j elements of which coincide with the first j elements of s. We now show that U contains a periodic point of σ. Such a periodic point t can be obtained as follows. We take

$$t_0 = s_0, t_1 = s_1, \ldots, t_{j-1} = s_{j-1} \quad \text{and further} \quad t_{k+j} = t_k \quad \text{for all } k \in \mathbb{Z}_+.$$

We leave it to the reader to transfer this argument to the other two versions of the doubling map. We thus showed the following.

Proposition 1.6 (The set of periodic points is dense).

1. *Let* $\sigma : \Sigma_2^+ \to \Sigma_2^2$ *be the shift map* (1.32). *Then the set of all points in* Σ_2^+ *which are periodic under* σ *is dense in* Σ_2^+.
2. *Let* $\varphi : [0, 1) \to [0, 1)$ *be the doubling map of the interval* (1.30). *Then the set of all points in* $[0, 1)$ *which are periodic under* φ *is dense in* $[0, 1)$.

A similar statement holds for the doubling map on the circle (1.31).

Construction of dense evolutions. If $s \in \Sigma_2^+$ is a periodic point, then its evolutions as a subset $\{\sigma^j(s) | j \in \mathbb{Z}_+\} \subset \Sigma_2^+$ is finite, which therefore will not be dense in Σ_2^+. We now show that a sequence s also exists, the evolution of which as a subset of Σ_2^+ is dense. We give two constructions, the first of which is very explicit. In the second construction we show that not only do such sequences exist, but moreover, to some extent almost every sequence has this property. In the latter case, however, we do not construct any of these sequences.

For the first construction, we first give a list of all finite sequences consisting of the symbols 0 and 1. We start with sequences of length 1, next of length 2, and so on. We also can determine the order of these sequences, by using a kind of lexicographic ordering. Thus we get

$$0 \ 1$$

$$00 \ 01 \ 10 \ 11$$

$$000 \ 001 \ 010 \ 011 \ 100 \ 101 \ 110 \ 111$$

et cetera. The announced sequence s we get by joining all these blocks, one behind the other, into one sequence. To prove that the evolution of s, as a subset $\{\sigma^j(s) | j \in \mathbb{Z}_+\} \subset \Sigma_2^+$ is dense, we show the following. For any sequence $t \in \Sigma_2^+$ and any $j \in \mathbb{Z}_+$, we aim to find an element of the form $\sigma^N(s)$ with

$$d(t, \sigma^N(s)) \leq 2^{-j}.$$

As we saw before, this would mean that both sequences are identical as far as the first j elements are concerned. We now look up in the sequence s the block of length j that contains the first elements of t. This block exists by the construction of s. To be precise, we look at the first occurrence of such a block and call the index of the first element of this N. It may now be clear that the sequence $\sigma^N(s)$, obtained from s by deleting the first N elements starts as

$$(s_N, s_{N+1}, \ldots, s_{N+j-1}, \ldots)$$

and coincides with t at the first j digits. This means that its distance to t is at most 2^{-j}, as required. This proves that the evolution of s under σ is dense in Σ_+^2.

The second 'construction' resorts to probability theory. We determine a sequence s by chance: each of the s_j is determined by tossing a fair coin such that 0 and

1 occur with equal probability $\frac{1}{2}$. Also we assume mutual independence of these events for all $j \in \mathbb{Z}_+$. Our aim is to show that with probability 1 we will find a sequence with a dense evolution.

Consider an arbitrary finite sequence $H = (h_0, \ldots, h_{k-1})$ of length k, where all h_j are either 0 or 1. The question is what is the probability that such a sequence will occur somewhere as a finite subsequence in the infinite sequence s. In fact, for any j_0 the probability that

$$(s_{j_0}, \ldots, s_{j_0+k-1}) \neq H$$

is equal to $1 - 2^{-k}$. Moreover, for $j_0 = 0, k, 2k, 3k, \ldots$ these events are mutually independent. This means that the probability that such a finite sequence does not occur starting at the indices $j_0 = 0, k, 2k, \ldots, Mk$, is equal to $(1 - 2^{-k})^M$. Because

$$\lim_{M \to \infty} (1 - 2^{-k})^M = 0,$$

we conclude that, with probability 1, the finite sequence H occurs somewhere as a finite subsequence in s.

Because the total number of finite sequences of the form H is *countable* we thus have shown that with probability 1 any finite sequence occurs somewhere as a finite subsequence in s. Returning to the original discussion, the occurrence of any finite sequence as a subsequence of s implies that the evolution

$$\{\sigma^j(s) \mid j \in \mathbb{Z}_+\} \subset \Sigma_2^+$$

is dense.

What do the above considerations mean for the other versions of the doubling map? It is not hard to see that the probability distribution we introduced on Σ_2^+ by the assignment map (1.33) is pulled back to the uniform distribution (or Lebesgue measure) on $[0, 1)$. This means that a point of $[0, 1)$, which is chosen at random according to the uniform distribution, will have a dense evolution with probability 1. Also see Appendix A. Summarising, we state the following proposition.

Proposition 1.7 (A dense evolution). *For the doubling map (1.30), (1.31) initial states exist, the evolution of which, as a subset of the state space $[0, 1)$ or the circle, is dense.*

In an appropriate sense, dense evolutions even occur with probability 1, also see Appendix A.

1.3.9 General shifts

The method of analysing certain kinds of dynamics in terms of symbolic dynamics and shift operators has been considerably extended, for instance, to all so-called hyperbolic attractors, see Chapter 4, §4.2.1.

To discuss this we need a few generalisations. One considers sequences

$$(s_0, s_1, s_2, \ldots), \tag{1.34}$$

with all s_j in some finite set, which we may assume to be $\{1, 2, \ldots, n\}$.[28] Here it is important that not all possible sequences are allowed, but only the ones satisfying *transition conditions*, which are determined by an $n \times n$-matrix A, the entries of which can be only 0 or 1. The sequence (1.34) then only is allowed if for all $j \in \mathbb{Z}_+$ one has

$$A_{s_j, s_{j+1}} = 1 \tag{1.35}$$

for the corresponding entry of the matrix A. The set of all sequences (1.34) satisfying condition (1.35) often is denoted by Σ_A. The shift operator

$$\sigma : \Sigma_A \to \Sigma_A$$

now is defined as before.

We can illustrate this more general method on the doubling map $\varphi : [0, 1) \to [0, 1)$, but with a different partition

$$Y_1 = \left[0, \frac{1}{4}\right), \quad Y_2 = \left[\frac{1}{4}, \frac{1}{2}\right), \quad Y_3 = \left[\frac{1}{2}, \frac{3}{4}\right), \quad Y_4 = \left[\frac{3}{4}, 1\right).$$

The Doubling map maps Y_1 onto $Y_1 \cup Y_2$, Y_2 onto $Y_3 \cup Y_4$, and so on, see Figure 1.28. This means that the transition conditions are given by the matrix

$$A = \begin{pmatrix} 1 & 1 & 0 & 0 \\ 0 & 0 & 1 & 1 \\ 1 & 1 & 0 & 0 \\ 0 & 0 & 1 & 1 \end{pmatrix}.$$

Then for each $x \in [0, 1)$ such that

$$\varphi^j(x) \in Y_{s_j},$$

the corresponding sequence $(s_0, s_1, s_2, \ldots) \in \Sigma_A$, that is, satisfies the transition conditons. Apart from exceptional cases (of zero measure), there is a one-one correspondence between the sequences in Σ_A and the points of $[0, 1)$. For other elementary examples see [8], Part 1, Section 2.5.

---

[28] These considerations apply to the case where the dynamics is generated by an *endomorphism*; in the case of a *diffeomorphism* one uses two-sided sequences $(\ldots, s_{-2}, s_{-1}, s_0, s_1, s_2, \ldots)$ (compare Chapter 4).

1.4 Exercises

Exercises that we found more difficult are indicated by an asterisk ($*$).

Exercise 1.1 (Integration constants and initial state). For the harmonic oscillator, with equation of motion

$$x'' = -\omega^2 x,$$

we saw that the general solution has the form $x(t) = A\cos(\omega t + B)$. Determine the values of A and B in terms of the initial conditions $x(0)$ and $x'(0)$.

Exercise 1.2 (Pendulum time to infinity). Consider the free undamped pendulum in the area of the phase plane close to the region where the pendulum starts to go 'over the top.' We already mentioned that the period of the oscillation grows. Now show that, as the amplitude tends to π, the period tends to ∞. (Hint: Use the fact that solutions of ordinary differential equations depend continuously on their initial conditions.)

Exercise 1.3 (Conservative mechanical systems). For a differentiable function $V : \mathbb{R}^n \to \mathbb{R}$ consider the second-order differential equation

$$x''(t) = -\mathrm{grad}\,V(x(t)).$$

Show that for solutions of such an equation the law of conservation of energy holds. Here the energy is defined as

$$H(x, x') = \frac{1}{2}\|x'\|^2 + V(x),$$

where $\| - \|$ denotes the Euclidean norm.

Exercise 1.4 (Mechanical systems with damping). In the setting of Exercise 1.3 we introduce a damping term in the differential equation as follows,

$$x'' = -\mathrm{grad}\,V(x(t)) - cx'(t)$$

with $c > 0$. Show that the energy H decreases in the sense that along any solution $x(t)$ of the latter equation

$$\frac{\mathrm{d}H(x(t), x'(t))}{\mathrm{d}t} = -c\|x'(t)\|^2,$$

where H is as above.

Exercise 1.5 (Linear damped oscillator). Consider the equation of motion of the damped harmonic oscillator

$$x'' = -\omega^2 x - cx', \quad x \in \mathbb{R},$$

where $c > 0$ is a constant. We already noted that there can be several ways in which solutions $x(t)$ tend to zero: the solution either can or cannot pass infinitely often through the value zero; see the diagrams (a) and (b) of Figure 1.3. Show that the type of damping which occurs here depends on the eigenvalues of the matrix

$$\begin{pmatrix} 0 & 1 \\ -\omega^2 & -c \end{pmatrix}$$

of the corresponding linear system

$$x' = y$$
$$y' = -\omega^2 x - cy.$$

Indicate for which eigenvalues these kinds of damping occur. Moreover show that the function $x(t)$ either passes infinitely often, only once, or never through zero.

Remark. Analogous conclusions can be drawn when nonlinear terms are included.

Exercise 1.6 (A stroboscopic map). Consider the second-order differential equation $x'' = f(x, x', \Omega t)$, where f is 2π-periodic in the third argument. The corresponding vector field on $\mathbb{R}^3 = \{x, y, z\}$ can be given by

$$x' = y$$
$$y' = f(x, y, z)$$
$$z' = \Omega.$$

Let $P : \mathbb{R}^2 \to \mathbb{R}^2$ be the corresponding stroboscopic or Poincaré map. If $\Psi : \mathbb{R}^3 \times \mathbb{R} \to \mathbb{R}^3$ denotes the evolution operator (or solution flow) of this vector field, then express P in terms in terms of Ψ.

Exercise 1.7 (Shift property). Consider a map $\Phi : M \times \mathbb{R} \to M$ with $\Phi(x, 0) = x$ for all $x \in M$. We call a map of the form $t \mapsto \Phi(x, t)$, where $x \in M$ is fixed, an evolution (corresponding to Φ). Show that the following two properties are equivalent.

  i. For any evolution $t \mapsto \varphi(t)$ and $\bar{t} \in \mathbb{R}$ also $t \mapsto \varphi(t + \bar{t})$ is an evolution.
 ii. For any $x \in M$ and $t_1, t_2 \in \mathbb{R}$ we have $\Phi(\Phi(x, t_1), t_2) = \Phi(x, t_1 + t_2)$.

Exercise 1.8 (Shift property for time-dependent systems). As in Exercise 1.7 we consider a map $\Phi : M \times \mathbb{R} \to M$, where now M is a vector space. We moreover assume that Φ is a C^2-map, where for any t the map Φ^t, defined by $\Phi^t(x) = \Phi(x, t)$, is a diffeomorphism.[29] We also assume that $\Phi^0 = \mathrm{Id}_M$. Again we call

---

[29] Which means that a smooth inverse map exists.

maps of the form $t \mapsto \Phi(x,t)$, for fixed $x \in M$, evolutions (corresponding to Φ). Show that there exists a differential equation of the form

$$x'(t) = f(x(t),t),$$

with a C^1-map $f : M \times \mathbb{R} \to M$, such that every solution of this differential equation is an evolution (corresponding to Φ). Give the connection between Φ and f. Furthermore show that in this setting the following two properties are equivalent.

i. The function f is independent of t; in other words that for any $x \in M$ and $t_1, t_2 \in \mathbb{R}$ one has $f(x,t_1) = f(x,t_2)$.
ii. For any $x \in M$ and $t_1, t_2 \in \mathbb{R}$ one has $\Phi(\Phi(x,t_1),t_2) = \Phi(x, t_1 + t_2)$.

Exercise 1.9 (To ∞ in finite time). Consider the differential equation

$$x' = 1 + x^2, \quad x \in \mathbb{R}.$$

Show that the corresponding dynamical system is only locally defined. Specify its evolution operator and its domain of definition.

Exercise 1.10 (Inverse of the Hénon map). Show that the Hénon map, given by

$$H_{a,b} : (x,y) \mapsto (1 - ax^2 + y, bx),$$

for $b \neq 0$ is an invertible map. Compute the inverse.

Exercise 1.11 (Inverse of a diffeomorphism near a saddle point). Let p be a saddle fixed point of a diffeomorphism

$$\varphi : \mathbb{R}^2 \to \mathbb{R}^2.$$

Show that p is also a saddle point of the inverse map φ^{-1} and that the unstable separatrix at p for φ is the stable separatrix at p for φ^{-1} and vice versa. Express the eigenvalues of $d\varphi^{-1}$ at p in the corresponding eigenvalues for $d\varphi$ at p.

Exercise 1.12 (A normal form for the logistic system). When dealing with the logistic system in §1.3.3 we saw how the 'maximal population' parameter K could be transformed away. Now show more generally that for any quadratic map

$$x \mapsto ax^2 + bx + c,$$

thus with $a \neq 0$, an affine substitution always exists of the form $y = \alpha x + \beta$, with $\alpha \neq 0$, such that the given transformation gets the form $y \mapsto y^2 + \gamma$.

Exercise 1.13 (The Hénon family for $b = 0$). When dealing with the Hénon family

$$H_{a,b} : (x,y) \mapsto (1 - ax^2 + y, bx),$$

we assumed that $b \neq 0$, to ensure invertibility. However, if we take $b = 0$, we obtain a dynamical system with time set $\mathbb{Z}_+$. We aim to show here that this system essentially equals the logistic system

$$\Phi_\mu(x) = \mu x(1 - x),$$

for an appropriately chosen value of μ.

To start with, all of $\mathbb{R}^2$ by $H_{a,0}$ is mapped onto the x-axis so we don't lose any information by restricting to that x-axis. For $a \neq 0$, the map $H_{a,0}$ defines a quadratic map. Now determine for which μ-value, as a function of a, the maps

$$H_{a,0}|_{x-\text{axis}} \quad \text{and} \quad \Phi_\mu$$

are taken into each other by affine substitutions. Give these substitutions.

Exercise 1.14 (Fattening of noninvertible maps by diffeomorphisms). In the previous Exercise 1.13 we saw how the maps $H_{a,b}$, for $b \to 0$ give approximations by invertible maps on the plane, of the noninvertible map $H_{a,0}$ on the x-axis. Now give a general method to approximate noninvertible maps of $\mathbb{R}^n$ by invertible maps of $\mathbb{R}^{2n}$.

Exercise 1.15 (Continuity of Newton operator). Show that, whenever f is a real polynomial of degree at least 2, the Newton operator N_f on $\mathbb{R} \cup \{\infty\}$ is a continuous map. Moreover show that for sufficiently large $|x|$ one has $|N_f(x)| < |x|$. What exceptions show up for polynomials of degree 0 and 1?

Exercise 1.16 (The Newton operator of quadratic polynomials). Consider the polynomial map
$$f(x) = x^2 - 1.$$

Specify the corresponding Newton operator N_f and consider its dynamics on the interval $(0, \infty)$.

1. Show that for $0 < x < 1$ one has $N_f(x) > x$.
2. Show that for $1 < x$ one has $1 < N_f(x) < x$.

What is your conclusion for Newton iterations with initial points in $(0, \infty)$?

Exercise 1.17 (Convergence of Newton iteration to a zero). Let $f : \mathbb{R} \to \mathbb{R}$ be a polynomial map and consider the Newton iteration $x_0, x_1, x_2, \ldots$ where

$$x_j = N_f(x_{j-1}),$$

for $j \geq 1$. Let p be such that $f(p) = 0$ and $f'(p) \neq 0$.

1. Show that
$$\frac{dN_f}{dx}(p) = 0.$$

2. Show that for x_0 sufficiently close to p, the iteration converges to p. (Hint: Invoke the contraction principle; see Appendix A.)

3. What happens when $f(p) = f'(p) = 0$, but when for some $k \in \mathbb{N}$ one has $f^{(k)}(p) \neq 0$?

Exercise 1.18 ((Dis-) continuity of the doubling map). When treating the doubling map, we saw

$$M = [0, 1) \to \Sigma_2^+$$

(see formula (1.33)), where we use the definition by 'addresses' as introduced in a Remark in §1.3.8.3.

1. Show that in the image of this map there are no sequences with a 'tail of digits 1.'
2. Show that this map is not continuous.

Exercise 1.19 (Periodic points of the doubling map). Determine how many points exist on the circle $\mathbb{R}/\mathbb{Z}$ that have prime period 6 under the doubling map.

Exercise 1.20 (The doubling map in $\mathbb{C}$). Consider the circle $\mathbb{S}^1$ as the unit circle in the complex plane $\mathbb{C}$. Also consider the polynomial map $f : \mathbb{C} \to \mathbb{C}$ defined by $f(z) = z^2$. Show that $\mathbb{S}^1$ is invariant under f and that the restriction $f|_{\mathbb{S}^1}$ coincides with the doubling map in the appropriate version. Is $\mathbb{S}^1$ an attractor of f?

Exercise 1.21 (Linear parts). Given the time-dependent differential equation

$$x' = f(x, t), \quad x \in \mathbb{R}^n,$$

with $f(0, t) \equiv 0$ and $f(x, t) = Ax + O(|x|^2)$, for an $n \times n$-matrix A and where the $O(|x|^2)$-terms contain all t-dependence, which we assume to be periodic of period T. As the state space we take $M = \mathbb{R}^n \times (\mathbb{R}/T\mathbb{Z})$. Consider the Poincaré map $P : \mathbb{R}^n \to \mathbb{R}^n$, corresponding to the section $t = 0 \mod T\mathbb{Z}$. Show that $P(0) = 0$ and that

$$dP_0 = e^{TA}.$$

Exercise 1.22 (Measure zero in symbol space). Show that Σ_2^+ is compact. Also show that the set of $s \in \Sigma_2^+$, which ends in a 'tail of digits 1', in $[0, 1)$ corresponds with a set of Lebesgue measure zero. Also show that in Σ_2^+ it has zero measure for the measure corresponding to random choices. Also see Appendix A.

Exercise 1.23 ((∗) Suspension lives on noncontractible state space). Consider the state space $\tilde{M}$ of a dynamical system which has been generated by suspension of a dynamical system the state space M of which is a vector space. The aim of this exercise is to show that $\tilde{M}$ cannot be a vector space. To this end we make use of some new concepts.

Whenever Y is a topological space, we call a continuous map $f : \mathbb{S}^1 = \mathbb{R}/\mathbb{Z} \to Y$ a closed curve in Y. Such a closed curve is contractible if it can be continuously deformed into a constant map; that is, if there exists a continuous map

$$F : \mathbb{S}^1 \times [0, 1] \to Y,$$

such that $F(s, 0) = f(s)$, for each $s \in \mathbb{S}^1$ and such that $F(\mathbb{S}^1 \times \{1\})$ consists of one point.

1. Show that each closed curve in a vector space is contractible.
2. Show that the identity map $f : \mathbb{S}^1 \to \mathbb{S}^1$, as a closed curve in $\mathbb{S}^1$, is not contractible.
3. Next construct a closed curve in $\tilde{M}$ that is not contractible. (Hint: Use the fact that the state space of the suspension is $M \times \mathbb{R} / \sim$, with $\sim$ as introduced before. This state space has a canonical projection on $\mathbb{S}^1 = \mathbb{R}/\mathbb{Z}$; $M \times \mathbb{R} / \sim$ contains a closed curve projecting to $\mathbb{S}^1$ as the identity.)

Exercise 1.24 ((*) **Attractor** $\subseteq \overline{W^u(p)}$, [30]). Let $\varphi : \mathbb{R}^2 \to \mathbb{R}^2$ be a diffeomorphism with a saddle point p, such that

i. $|\det(\mathrm{d}\varphi)| \leq c < 1$ everywhere.
ii. $W^u(p) \cap W^s(p)$ contains a (homoclinic) point $q \neq p$.

Let U be an open bounded domain with boundary $\partial U \subseteq W^u(p) \cup W^s(p)$. Then show

1. Area $(\varphi^n(U)) \leq c^n$ Area U.
2. $\lim_{n \to \infty}$ length $(W^s(p) \cap \partial (\varphi^n(U))) = 0$.
3. For all $r \in U$ the distance from $\varphi^n(r)$ to $W^u(p)$ tends to 0 as $n \to \infty$.

Exercise 1.25 ((*) **Numerical approximation of** $W^u(p)$). In §1.3.2 on the Hénon map an algorithm was treated to find numerical approximations of a segment of the unstable separatrix. Here certain parameters ε, N, and m had to be chosen. How would you change these parameters to increase the precision? You may ignore rounding-off errors that occur in numerical floating point operations.

Exercise 1.26 ((*) **Attractor** $= \overline{W^u(p)}$?). In Figure 1.15 we witness numerical output regarding the Hénon attractor and a piece of the unstable separatrix $W^u(p)$. Check that these compare well. Carry out a similar comparison program for the attractor of the Poincaré map of the forced damped swing; see Figure C.7 in Appendix C.

Exercise 1.27 ((*) **The Poincaré map of the Rössler system**). When considering the Rössler attractor in §1.3.7 we saw that the attractor of the Poincaré map as constructed there, appears to lie inside a (smooth) curve. Now numerical simulations are always unreliable, in the sense that it may be hard to get an idea of what the possible errors can be. It therefore may be difficult to base rigourous mathematical proofs on numerical simulations. Here we ask to give *arguments*, based on the represented simulation of an evolution of the Rössler system, why the structure of the attractor of the Poincaré map in this case can't be as simple as the curve depicted there.

Exercise 1.28 ((*) **The tenfold map**). Adapt the theory of §1.3.8 for the tenfold map

$$\varphi : [0, 1) \to [0, 1), \quad \text{given by } x \mapsto 10\,x \bmod \mathbb{Z}.$$

Chapter 2
Qualitative properties and predictability of evolutions

In the previous chapter we already met various types of evolutions, including stationary and periodic ones. In the present chapter we investigate these evolutions more in detail and we also consider more general types. Here qualitative properties are our main interest. The precise meaning of 'qualitative' in general is hard to give; to some extent it is the contrary or opposite of 'quantitative'. Roughly one could say that *qualitative* properties can be described in words (or integers), whereas *quantitative* properties usually are described in terms of (real) numbers.

It turns out that for the various types of evolutions, it is not always equally easy to predict the future course of the evolution. At first sight this may look strange: once we know the initial state and the evolution operator, we can just predict the future of the evolution in a unique way. However, in general, we only know the initial state approximately. Moreover, often the evolution operator is not known at all or only approximately. We ask the question of predictability mainly in the following form. Suppose the evolution operator is unknown, but that we do know a (long) piece of the evolution; is it then possible to predict the future course of the evolution at least partially? It turns out that for the various kinds of evolution, the answer to this question strongly differs. For periodic, in particular for stationary, evolutions, the predictability is not a problem. Indeed, based on the observed regularity we predict that this will also continue in the future. As we show, the determination of the exact period still can be a problem. This situation occurs, to an even stronger degree, with quasi-periodic solutions. However, it turns out that for the so-called chaotic evolutions an essentially worse predictability holds. In Chapter 6 we return to predictability problems in a somewhat different context.

2.1 Stationary and periodic evolutions

Stationary and periodic evolutions have been already widely discussed in the previous chapter.

Definition 2.1 (Stationary and periodic evolutions). Let (M, T, Φ) be a dynamical system with state space M, time set T, and evolution operator $\Phi : M \times T \to M$.

H.W. Broer and F. Takens, *Dynamical Systems and Chaos*,
Applied Mathematical Sciences 172, DOI 10.1007/978-1-4419-6870-8_2,
© Springer Science+Business Media, LLC 2011

We say that $x \in M$ is *periodic* (or *stationary*) if there exists a positive $t \in T$ such that $\Phi(x, t) = x$.

1. Moreover, x is *stationary* whenever for all $t \in T$ one has $\Phi(x, t) = x$.
2. If $\Phi(x, \tau) = x$, we call τ a period of x.

The periodic or stationary evolution of x as a set is given by

$$\{\Phi(x, s) \mid s \in T\} \subset M.$$

We already noted in the example of the doubling map (see §1.3.8) that each periodic point has many periods: if τ is a period, then so are all integer multiples of τ. As before we introduce the notion of *prime period* for periodic, nonstationary, evolutions. This is the smallest positive period. This definition appears to be without problems, but this is not entirely true. In the case where $T = \mathbb{R}$, such a prime period does not necessarily exist. First of all, this is the case for stationary evolutions, which can be excluded by just considering proper, that is, nonstationary, periodic evolutions. However, even then there are pathological examples; see Exercise 2.4. In the sequel we assume that prime periods of nonstationary periodic evolutions will always be well defined. It can be shown that this assumption is valid in all cases where M is a metric (or metrisable) space and where the evolution operator, considered as a map $\Phi : M \times T \to M$, is continuous. If such a prime period equals τ, it is easy to show that the set of periods is given by $\tau \mathbb{Z} \cap T$.[1]

2.1.1 *Predictability of periodic and stationary motions*

The prediction of future behaviour in the presence of a stationary evolution is completely trivial: the state does not change and the only reasonable prediction is that this will remain so. In the case of periodic evolution, the prediction problem does not appear much more exciting. In the case where $T = \mathbb{Z}$ or $T = \mathbb{Z}_+$, this is indeed so. Once the period has been determined and the motion has been precisely observed over one period, we are done. We show, however, that in the case where $T = \mathbb{R}$ or $T = \mathbb{R}_+$, when the period is a real number, the situation is less clear. We illustrate this when dealing with the prediction of the yearly motion of the seasons, or equivalently, the yearly motion of the Sun with respect to the vernal equinox (the beginning of Spring). The precise prediction of this cycle over a longer period is done with a 'calendar'. In order to make such a calendar, it is necessary to know the length of the year very accurately and it is exactly here where the problem occurs. Indeed, this period, measured in days, or months, is not an integer and we cannot expect to do better than find an approximation for this. By the 'calendar' here we

---

[1] Often one speaks of 'period' when 'prime period' is meant.

should understand the set of rules according to which this calendar is made, where one should think of the leap year arrangements. Instead of 'calendar' one also speaks of *time tabulation*.

Remark (Historical digression). We give some history. In the year 45 BC the Roman dictator Julius Cæsar [2] installed the calendar named after him, the Julian Calendar. The year was thought to consist of 365.25 days. On the calendar this was implemented by adding a leap day every four years. However, the real year is a bit shorter and slowly the seasons went out of phase with the calendar. Finally Pope Gregorius XIII [3] decided to intervene. First, to 'reset' the calendar with nature, he decided that in the year 1582 (just at this one occasion) immediately after October 4, October 15 would follow. This means that the *predictions* based on the Julian system in this period of more than 1600 years already show an error of 10 days. Apart from this unique intervention and in order to prevent such shifts in the future, the leap year convention was changed as follows. In the future, year numbers divisible by 100, but not by 400, no longer will be leap years. This led to the present, Gregorian, calendar. Compare [158].

Of course we also understand that this intervention is still based on an approximation: real infallibility cannot be expected here! In fact, a correction of 3 days per 400 years amounts to 12 days per 1600 years, whereas only 10 days were necessary.

Our conclusion from this example is that when predicting the course of a periodic motion, we expect an error that in the beginning is proportional to the *interval of prediction*, that is, the time difference between the present moment of prediction and the future instant of state we predict. When the prediction errors increase, a certain *saturation* occurs: when predicting seasons the error cannot exceed one half year. Moreover, when the period is known to a greater precision (by longer and/or more precise observations), the growth of the errors can be made smaller.

As said earlier, the problem of determination of the exact period only occurs in the case of time sets $\mathbb{R}$ or $\mathbb{R}_+$. In the cases where the time sets are $\mathbb{Z}$ or $\mathbb{Z}_+$ the periods are integers, which by sufficiently accurate observation can be determined with absolute precision. These periods then can be described by a counting *word*; they can be viewed as qualitative properties. In cases where the time set is $\mathbb{R}$ or $\mathbb{R}_+$, however, the period is a quantitative property of the evolution.

On the other hand the difference is not that sharp. If the period is an integer, but very large, then it can be practically impossible to determine this by observation. As we show when discussing so-called *multiperiodic* evolutions, it may even be impossible to determine by observation whether we are dealing with a periodic evolution.

Also, regarding the more complicated evolutions dealt with below, we ask ourselves how predictable these are and which errors we can expect as a function of the prediction interval. We then merely consider predictions based on a (sufficiently) long observation of the evolution.

---

[2] Gaius Julius Cæsar 100–44 BC.

[3] Pope Gregorius XIII 1502–1585.

2.1.2   Asymptotically and eventually periodic evolutions

Returning to the general context of dynamical systems we now give the following definition.

Definition 2.2 (Asymptotic periodicity and stationarity). Let (M, T, Φ) be a dynamical system, where M is a metric space and $\Phi : M \times T \to M$ is continuous. We call the evolution $x(t) = \Phi(x, t)$ *asymptotically periodic (or stationary)* if for $t \to \infty$ the point $x(t)$ tends to a periodic (or stationary) evolution $\tilde{x}(t)$, in the sense that the distance between $x(t)$ and the set $\{\tilde{x}(s)|s \in T\}$ tends to zero as $t \to \infty$.

Remarks.

– This definition can be extended to the case where M only has a topology. In that case one may proceed as follows. We say that the evolution $x(t)$ is asymptotically periodic if there exists a periodic orbit $\gamma = \{\tilde{x}(s) \mid s \in T\}$ such that for any neighbourhood U of γ there exists $T_U \in \mathbb{R}$, such that for $t > T_U$ one has that $x(t) \in U$.

  We speak of asymptotic periodicity in the strict sense if there exists a periodic orbit $\gamma = \{\tilde{x}(s) \mid s \in T\}$ of period τ and a constant c, such that

$$\lim_{\mathbb{N} \ni n \to \infty} x(t + n\tau) = \tilde{x}(t + c).$$

In that case either the number c/τ modulo $\mathbb{Z}$ or the point $\tilde{x}(c)$ can be interpreted as the phase associated with $x(t)$.

  Note that in the case without topology, the concept of 'asymptotic' no longer makes sense, because the concept of limit fails to exist. The eventually periodic or stationary evolutions, discussed later in this section, are an exception to this.

– Assuming continuity of Φ, it follows that the set $\{\tilde{x}(s)|s \in T\}$ corresponding to the periodic evolution $\tilde{x}(s)$ is compact. Indeed, in the case where $T = \mathbb{Z}$ or $\mathbb{Z}_+$ the set is even finite, whereas in the cases where $T = \mathbb{R}$ or $\mathbb{R}_+$ the set $\{\tilde{x}(s)|s \in T\}$ is the continuous image of the circle. The reader can check this easily.

We have seen examples of asymptotically periodic or asymptotically stationary evolutions several times in the previous chapter: for the free pendulum with damping (see §1.1.2), every nonstationary evolution is asymptotically stationary. For the damped pendulum with forcing (see §1.1.2), and in the Van der Pol system (see §1.3.1) we saw asymptotically periodic evolutions. When discussing the Hénon family (in §1.3.2) we saw another type of stationary evolution, given by a saddle point. For such a saddle point p we defined stable and unstable separatrices (see (1.16)). From this it may be clear that the stable separatrix $W^s(p)$ exactly consists of points that are asymptotically stationary and of which the limit point is the saddle point p.

  The definition of asymptotically periodic or asymptotically stationary applies for any of the time sets we are using. In the case of time set $\mathbb{Z}_+$ another special type occurs.

Definition 2.3 (Eventual periodicity and stationarity). Let $(M, \mathbb{Z}_+, \Phi)$ be a dynamical system with an evolution $x(n) = \Phi(x, n)$. We say that $x(n)$ is *eventually periodic* (or *eventually stationary*) if for certain $N > 0$ and $k > 0$,

$$x(n) = x(n + k) \quad \text{for all } n > N.$$

So this is an evolution that after some time is exactly periodic (and not only in a small neighbourhood of a periodic evolution). Note that in this case neither distance function nor topology plays any role.

This type of evolution is abundantly present in the system generated by the doubling map $x \mapsto 2x \bmod \mathbb{Z}$, defined on $[0, 1)$; see §1.3.8. Here 0 is the only stationary point. However, any rational point p/q with q a power of 2 is eventually stationary. Compare Exercise 2.3. Similarly, any rational point is eventually periodic. Also for the logistic system (see §1.3.3), such eventually periodic evolutions occur often.

As said before, such evolutions do not occur in the case of the time sets $\mathbb{R}$ and $\mathbb{Z}$. Indeed, in these cases, the past can be uniquely reconstructed from the present state, which clearly contradicts the existence of an evolution that 'only later' becomes periodic. (Once the evolution has become periodic, we cannot detect how long ago this has happened.)

Regarding the predictability of these evolutions the same remarks hold as in the case of periodic and stationary evolutions. As said earlier, when discussing predictability, we always assume that the evolution to be predicted has been observed for quite some time. This means that it is no further restriction to assume that we already are in the stage that no observable differences exist between the asymptotically periodic (or stationary) evolutions and the corresponding periodic (or stationary) limit evolution.

2.2 Multi- and quasi-periodic evolutions

In the previous chapter we already met with evolutions where more than one period (or frequency) plays a role. In particular this was the case for the free pendulum with forcing; see §1.1.2. Also in daily life we observe such motions. For instance, the motion of the Sun as seen from the Earth knows two periodicities: the daily period, of sunrise and sunset, and the yearly one that determines the change of the seasons.

We describe these types of evolution as a generalised type of periodic motions, starting out from our dynamical system (M, T, Φ). We assume that M is a differentiable manifold and that $\Phi : M \times T \to M$ is a differentiable map. For periodic evolutions, the whole evolution $\{x(t) \mid t \in T\}$ as a subset of the state space M consists either of a finite set of points (in cases where $T = \mathbb{Z}_+$ or $\mathbb{Z}$) or of a closed curve (in the case where $T = \mathbb{R}$ or $\mathbb{R}_+$). Such a finite set of points we can view as $\mathbb{Z}/s\mathbb{Z}$, where s is the (prime-) period: the closed curve can be viewed as the differentiable image of the circle $\mathbb{S}^1 = \mathbb{R}/\mathbb{Z}$. Such parametrisations (i.e., with $\mathbb{Z}/s\mathbb{Z}$ or $\mathbb{R}/\mathbb{Z}$) can be chosen such that the dynamics when restricted to the periodic evolution consists of *translations*.

- In the case where the time set is $\mathbb{Z}$ or $\mathbb{Z}_+$:

$$\Phi([x], k) = [x + k],$$

where $[x] \in M$ and where $[-]$ denotes the equivalence class modulo $s\mathbb{Z}$.
- In the case where the time set is $\mathbb{R}$ or $\mathbb{R}_+$:

$$\Phi([x], t) = [x + t\omega],$$

where $[x] \in M$ and where again $[-]$ denotes the equivalence class modulo $\mathbb{Z}$ and where ω^{-1} is the (prime) period of the periodic motion.

In both cases we parametrised the periodic evolution by an Abelian group, namely $\mathbb{Z}/s\mathbb{Z}$ or $\mathbb{R}/\mathbb{Z}$, respectively. Translations then consist of 'adding everywhere the same constant,' just as for a translation on a vector space.

In the case of several frequencies, the motion takes place on the (differentiable) image of the product of a number of circles (and, possibly, of a finite set). The product of a number of circles is called a *torus*. A torus also can be seen as an Abelian group: the multiperiodic dynamics then consists of translations on such a torus. In the sections that follow we study these tori and translations on them in detail.

We already point out a complication, which is most simply explained with the help of a motion with two periods. As an example we take the motion of the Sun, as discussed before, with its daily and its yearly period. If the ratio of these two periods were rational, then the whole motion again would be periodic. For instance, if the yearly period would be exactly 365.25 daily periods, then the whole motion described here would be periodic with a period of 4 years. As we saw in §2.1.1, this ratio is only an imperfect fit, which is the reason for introducing the Gregorian calendar. If the Gregorian calendar were correct, the whole system would yet be periodic with a period of 400 years. And a further refinement would correspond to a still longer period. This phenomenon is characteristic for motions with more than one period: they can be well approximated by periodic motions, but as the approximation gets better, the period becomes longer. This might be an argument for saying: let us restrict ourselves to periodic motions. However, for periodic motions of very long period (say longer than a human life as would be the 400 years we just discussed), it can be less meaningful to speak of periodic motion: they simply are not experienced as such. This is the reason why we introduce the concept of *multiperiodic evolutions*. These are motions involving several periods (or frequencies), but where it is not excluded that, due to incidental rationality of frequency ratios, the system can also be described with fewer periods.

2.2.1 The n-dimensional torus

The n-dimensional torus is a product of n circles. This means that we can define
the n-dimensional torus as $\mathbb{R}^n/\mathbb{Z}^n$, namely as the set of n-dimensional vectors in
$\mathbb{R}^n$, where we identify vectors whenever their difference only has integer-valued
components. The n-dimensional torus is denoted by $\mathbb{T}^n$, so

$$\mathbb{T}^n = \mathbb{R}^n/\mathbb{Z}^n = (\mathbb{R}/\mathbb{Z})^n.$$

Remark (On terminology and notation). For $n = 1$ we get $\mathbb{T}^1 = \mathbb{R}/\mathbb{Z}$, identified
in §1.3.8 as the circle; also see §1.3.4. We already mentioned that often, instead of
$\mathbb{R}/\mathbb{Z}$ in the definition $\mathbb{R}/(2\pi\mathbb{Z})$, is being used. These two possibilities correspond
to the following two parametrisations

$$s \mapsto (\cos(2\pi s), \sin(2\pi s)) \quad \text{or} \quad s \mapsto (\cos s, \sin s)$$

of the unit circle in $\mathbb{R}^2$. The difference is just a matter of 'length' or 'angle' scales.
Observe that $\mathbb{S}^1 = \mathbb{T}^1$ is the special case for $n = 1$ of the manifolds $\mathbb{S}^n$ (the
n-dimensional unit sphere in $\mathbb{R}^{n+1}$) and $\mathbb{T}^n = (\mathbb{T}^1)^n, n = 1, 2, \ldots$. In the sequel
we use either $\mathbb{R}/\mathbb{Z}$ or $\mathbb{R}/(2\pi\mathbb{Z})$ in the definition of $\mathbb{S}^1 = \mathbb{T}^1$, depending on the
context, but always make sure that no confusion is possible.

The case where $n = 2$ can be described a bit more geometrically. Here we deal
with $\mathbb{T}^2 = \mathbb{R}^2/\mathbb{Z}^2$, so we start out with 2-dimensional vectors. Because in the
equivalence relation we identify two vectors whenever their difference is in $\mathbb{Z}^2$, we
can represent each point of $\mathbb{T}^2$ as a point in the square $[0, 1] \times [0, 1]$. Also in this
square there are still points that have to be identified. The left boundary $\{0\} \times [0, 1]$
is identified with the right boundary $\{1\} \times [0, 1]$ (by identifying points with the same
vertical coordinate), and similarly for the upper and the lower boundary. This means
that the four vertex points are all four identified. Compare Figure 2.1, where sides
with similar arrows have to be identified. We can realise the result of all this in the
3-dimensional space as the surface parametrised by 'angles' φ and ψ, both in $\mathbb{R}/\mathbb{Z}$,
as follows,

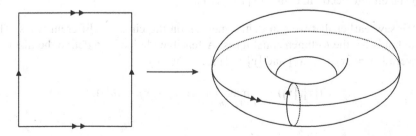

Fig. 2.1 The 2-torus as a square where the boundaries have to be identified two by two (a), and a
realisation of this in 3-dimensional space (b).

Fig. 2.2 Rosette orbit like
the orbit of Mercury. In
reality the major axis of the
Mercurian 'almost' ellipse
rotates much slower than
indicated here; it is only
1.4 in. per revolution (i.e., per
Mercurian year).

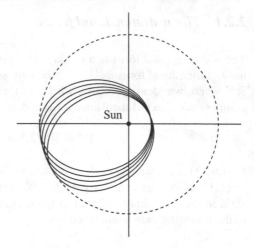

$$(\varphi, \psi) \mapsto ((1 - a\cos(2\pi\psi))\sin(2\pi\phi), (1 - a\cos(2\pi\psi))\cos(2\pi\phi), a\sin(2\pi\psi)),$$

where $0 < a < 1$ is a constant.

Example 2.1 (Rosette orbit of Mercury). The motion of the planet Mercury around
the Sun according to Kepler should take place along an ellipse. However, Newtonian
(and Einsteinian) universal gravitation has it that the Keplerian ellipse changes
slowly due to the influence of the other planets, in particular of Jupiter. The result-
ing motion is a so-called rosette orbit, as indicated in Figure 2.2 (in an exaggerated
way). Here two periods can be distinguished. The first is the period of the 'almost'
ellipse: one Mercurian year (88 days), given by successive instants when the distance between Mercury and the Sun is maximal.[4] As a second period we have the
period of the major axis, that is, for completing a full rotation which takes approximately 225 780 years. Compare Misner, Thorne, and Wheeler [171], p. 1113.

Looking at Figure 2.2 it is not hard to imagine that we are dealing with a curve on
a 2-dimensional torus, that is projected onto a plane. Interestingly, the latter period,
obtained from detailed observations, is somewhat smaller than can be explained by
the Newtonian theory alone. For a satisfying explanation relativistic effects also had
to be taken into account; compare [158, 171].

We saw earlier that we can define metrics on the circle in different ways. The
same holds for the n-dimensional torus. A feasible choice is to define the distance
between $[x] = [x_1, \ldots, x_n]$ and $[y] = [y_1, \ldots, y_n]$ by

$$d([x], [y]) = \max_{j=1,\ldots,n} (\min_{m \in \mathbb{Z}} (|x_j - y_j + m|)). \tag{2.1}$$

---

[4] Called the apohelium.

As usual, the square brackets indicate that we are dealing with equivalence classes modulo integers.

2.2.2 Translations on a torus

For a vector $\Omega = (\omega_1, \ldots, \omega_n)$, with $\omega_i \in \mathbb{R}$, we define the corresponding translation (or rotation)

$$R_\Omega : \mathbb{T}^n \to \mathbb{T}^n \quad \text{by} \quad R_\Omega[x_1, \ldots, x_n] = [x_1 + \omega_1, \ldots, x_n + \omega_n], \qquad (2.2)$$

where square brackets again refer to the equivalence class.

With the help of translations (2.2) we now can define dynamical systems with state space $M = \mathbb{T}^n$ by

$$\Phi^t = R_{t\Omega}. \qquad (2.3)$$

Here we can take as a time set both $T = \mathbb{Z}$ and $T = \mathbb{R}$. In the former case the dynamics is given by iterations of the map R_Ω and in the latter case by the flow of the constant vector field

$$x'_j = \omega_j, \qquad (2.4)$$

$j = 1, 2, \ldots, n$.[5] For a single translation R_Ω, only the values mod $\mathbb{Z}$ of the ω_j are of interest. However, when considering time set $\mathbb{R}$, where t takes arbitrary values, the components ω_j of Ω are of interest as real numbers. We call the dynamical system (2.3) the *translation system* defined by Ω. Before going to arbitrary dimension, we first consider a few simpler cases.

2.2.2.1 Translation systems on the 1-dimensional torus

In the case where $n = 1$ and with time set $T = \mathbb{R}$, the dynamics of the translation system (2.3) defined by Ω is just a periodic flow. This statement holds in the sense that any point lies on a periodic evolution, except in the case where $\Omega = 0$, in which case all points are stationary. So nothing new is to be found here.

Taking $n = 1$ with $T = \mathbb{Z}$ or $\mathbb{Z}_+$, Definition (2.3) gives something completely different. Now the dynamics is given by iterations of the map R_Ω, which strongly depends on $\Omega \in \mathbb{R}$, in the sense that it becomes important whether Ω is rational or not.[6] Indeed, if $\Omega = p/q$, for certain integers p and q, with $q > 0$, it then follows that

$$R_\Omega^q = R_{q\Omega} = \mathrm{Id}_{\mathbb{T}^1}$$

---

[5] Note that $x'_j = [x_j]'$.

[6] In this 1-dimensional case we do not make a difference between the vector $\Omega = (\omega_1)$ and the value of ω_1 as a real number.

is the identity map of $\mathbb{T}^1$. This implies that each point of $\mathbb{T}^1$ is periodic with period q. Conversely, if there exists a point $[x]$ that is periodic under (2.3) of period q, then it follows that

$$R_\Omega^q([x]) = [x] \Leftrightarrow x + q\Omega = x \bmod \mathbb{Z}.$$

In that case necessarily Ω is rational and can be written as a fraction with denominator q and all $[x] \in \mathbb{T}^1$ have this same period q. It follows that if Ω is irrational, no periodic points of (2.3) exist. This case of irrational Ω, is the first case where we speak of *quasi-periodic* evolutions: the translation system (2.3) now is also called quasi-periodic. An important property of such a quasi-periodic evolution is that it is dense in $\mathbb{T}^1$. This is the content of the following lemma.

Lemma 2.4 (Dense orbits). *If $\Omega \in \mathbb{R} \backslash \mathbb{Q}$, then for the translation system (2.3) on $\mathbb{T}^1$ defined by Ω, with time set $\mathbb{Z}_+$, any evolution $\{[m\Omega + x] \mid m \in \mathbb{N}\}$ is dense in $\mathbb{T}^1$. This same conclusion also holds for time set $\mathbb{Z}$.*

Proof. We choose an arbitrary $\varepsilon > 0$ and show that, for any $[x] \in \mathbb{T}^1$ the evolution starting at $[x]$ of the translation system defined by Ω, approximates every point of $\mathbb{T}^1$ to a distance ε.

Consider the positive evolution $[x(m)] = R_\Omega^m([x])$, $m = 1, 2, \ldots$. Periodic points do not occur, therefore the (positive) evolution

$$\{[x(m)] \mid m \in \mathbb{N}\}$$

contains infinitely many different points. By compactness of $\mathbb{T}^1$ this set has an accumulation point, which implies that there exist points $[x(m_1)]$ and $[x(m_2)]$ with

$$d([x(m_1)], [x(m_2)]) < \varepsilon.$$

This implies that the map $R_{|m_1 - m_2|\Omega}$ is a translation over a distance $\tilde{\varepsilon} < \varepsilon$ (to the right or to the left). This means that the successive points

$$X_j = R_{|m_1 - m_2|\Omega}^j([x])$$

are within distance ε and thus that the set

$$\left\{ X_j \mid j = 0, 1, 2, \ldots, \operatorname{int}\left(\frac{1}{\tilde{\varepsilon}}\right) \right\},$$

where $\operatorname{int}(1/\tilde{\varepsilon})$ is the integer part of $1/\tilde{\varepsilon}$, approximates each point of the circle to a distance less than ε. Finally we note that all points X_j belong to the (positive) evolution of $[x]$. This finishes the proof. $\square$

We observe that to some extent all evolutions of such a translation system are equal in the sense that the evolutions that start in $[x]$ and $[y]$ are related by the

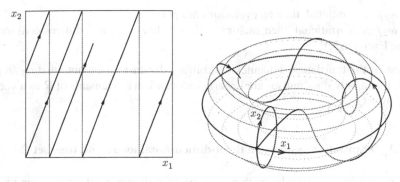

Fig. 2.3 Evolution curve of the translation system generated by a constant vector field on the 2-torus $\mathbb{T}^2$.

translation R_{y-x}. This is the reason why often we only mention properties for the evolution that starts in $[0]$ and not always explicitly say that the same holds for arbitrary other evolutions as well.

Remark. Related to the treatment of the translation dynamics in arbitrary dimension we mention that in the 1-dimensional case Ω is irrational, if and only if the numbers 1 and Ω are linearly independent in $\mathbb{R}$, considered as a vector space over the field $\mathbb{Q}$.

2.2.2.2 Translation systems on the 2-dimensional torus with time set $\mathbb{R}$

A next special case to consider is the 2-dimensional torus $\mathbb{T}^2$ with a translation system with time set $\mathbb{R}$. So the evolution operator is the flow of the vector field $[x_j]' = \omega_j, j = 1, 2$, where $\Omega = (\omega_1, \omega_2)$. Assuming that not both components are equal to zero, without restriction of generality we may assume that $\omega_1 \neq 0$. We can now take a Poincaré map. As a submanifold we take

$$S = \{[x_1, x_2] \in \mathbb{T}^2 \mid x_1 = 0\}.$$

Note that S can be considered as a 1-dimensional torus.[7] Then for the map

$$R_{\omega_1^{-1}(\omega_1, \omega_2)} = R_{(1, \omega_1^{-1}\omega_2)},$$

which is the time ω_1^{-1} map, the set S is mapped onto itself. On S, considered as a 1-dimensional torus, this Poincaré map is a translation over ω_2/ω_1. Combining this with the translation dynamics on a 1-dimensional torus with time set $\mathbb{Z}$, we obtain:

---

[7] Here, and elsewhere in this section, the term 'can be considered as' means that the sets involved are diffeomorphic, and that the diffeomorphism even can be chosen as a linear map in terms of the coordinates used on $\mathbb{R}/\mathbb{Z}$.

1. If ω_2/ω_1 is rational, then all evolutions are periodic.
2. If ω_2/ω_1 is irrational, then each evolution is dense in the 2-dimensional torus. See Exercise 5.1.

In view of the translation dynamics in arbitrary dimension, we note that $\omega_2/\omega_1 \in \mathbb{R}\setminus\mathbb{Q}$, if and only if ω_1 and ω_2 are linearly independent as elements of $\mathbb{R}$ as a vector space over $\mathbb{Q}$.

2.2.2.3 Translation systems on the n-dimensional torus with time set $\mathbb{R}$

We now consider the translation dynamics on an n-dimensional torus, given by the translation vector $\Omega = (\omega_1, \omega_2, \ldots, \omega_n)$ and with time set $\mathbb{R}$, that is, the flow case. An important fact is that whenever $\omega_1, \omega_2, \ldots, \omega_n \in \mathbb{R}$, as a vector space over the rationals, are linearly independent, each evolution of the corresponding translation dynamics is dense in the n-dimensional torus $\mathbb{T}^n$. We do not provide a proof here, inasmuch as this uses theory not included here (Fourier theory, measure theory, and ergodic theory). The curious reader is referred to [59], Appendices 9 and 11, where the analogue for time set $\mathbb{Z}$ is proven. Also see [18, 19, 21]. In cases where $\omega_1, \omega_2, \ldots, \omega_n$ are linearly dependent over $\mathbb{Q}$ one speaks of *resonance*. We return to resonance in §2.2.3 when discussing the driven Van der Pol equation. Still we have a few general remarks.

Lemma 2.5 (Resonant evolutions are not dense). *Suppose that* $\omega_1, \omega_2, \ldots, \omega_n$ *are linearly dependent over* $\mathbb{Q}$. *Then the corresponding translation system* $R_\Omega :$ $\mathbb{T}^n \to \mathbb{T}^n$ *with* $\Omega = (\omega_1, \omega_2, \ldots, \omega_n)$ *and time set* $\mathbb{R}$ *has no dense evolutions.*

Proof. Suppose that $q_1, \ldots, q_n \in \mathbb{Q}$ exist, not all equal to 0, such that $\sum_j q_j \omega_j = 0$. By multiplication of this equality by the product of all denominators we can achieve that all rationals $q_j, 1 \le j \le n$, become integers. Next, by dividing out the common factors, we also achieve that the integers $q_j, 1 \le j \le n$ have no common factors. From algebra it is known that then integers $m_1, \ldots, m_n$ exist such that $\sum_{j=1}^n q_j m_j = 1$; for example, see Exercise 2.5 or compare with van der Waerden [239].

We now consider the function given by

$$f(x_1, \ldots, x_n) = \sum_{j=1}^n q_j x_j.$$

To ensure that f is defined on $\mathbb{T}^n$, we take its values in $\mathbb{R}/\mathbb{Z}$, noting that the coefficients $q_j, 1 \le j \le n$, are integers. Because $\sum_j q_j \omega_j = 0$, the function f is constant on each evolution. For instance on the evolution starting at $[0] = [0, 0, \ldots, 0]$, the function f is everywhere 0. Because f is continuous and not everywhere 0 on all of $\mathbb{T}^n$, it follows that the evolution starting at $[0]$ cannot be dense in $\mathbb{T}^n$. As remarked earlier, this immediately implies that no evolution is dense in $\mathbb{T}^n$. $\square$

We stay for a while in the resonant context of Lemma 2.5, thus with

$$\sum_{j=1}^{n} q_j \omega_j = 0,$$

with relatively prime integers q_j, $1 \leq j \leq n$. We study the function

$$f_{\Omega} : \mathbb{T}^n \to \mathbb{R}/\mathbb{Z}, \quad [x_1, x_2, \ldots, x_n] \mapsto \sum_{j=1}^{n} q_j x_j \bmod \mathbb{Z},$$

as introduced in the above proof, as well as its zero set,

$$N_{f_{\Omega}} = \{ [x_1, x_2, \ldots, x_n] \in \mathbb{T}^n \mid f_{\Omega}[x_1, x_2, \ldots, x_n] = 0 \bmod \mathbb{Z} \}, \tag{2.5}$$

a bit further. The proof asserts that $N_{f_{\Omega}} \subset \mathbb{T}^n$ contains the entire evolution

$$\{ R_t \Omega[0] \mid t \in \mathbb{R} \} \subset \mathbb{T}^n$$

with initial state $[0]$. It can be shown that $N_{f_{\Omega}}$ is diffeomorphic to a torus of dimension $n - 1$. Again, we do not prove this, because the simplest proof we know uses the theory of Lie groups [21]. We, however, do prove that N_f is connected.[8]

Lemma 2.6. *The resonant zero set $N_{f_{\Omega}}$ (2.5) is a connected subset of $\mathbb{T}^n$.*

Proof. Here, as we are dealing with elements $x \in \mathbb{R}^n$ and not with equivalence classes $[x] \in \mathbb{T}^n$, we consider f_{Ω} as a function defined on $\mathbb{R}^n$, with values in $\mathbb{R}$ and not in $\mathbb{R}/\mathbb{Z}$.

Suppose that $[x_1, \ldots, x_n] \in N_f$, or equivalently that $f_{\Omega}(x_1, \ldots, x_n) \in \mathbb{Z}$. The integer coefficients q_j, $1 \leq j \leq n$ have no common factor, thus there is an element $(\bar{x}_1, \ldots, \bar{x}_n) \in \mathbb{R}^n$, such that $\sum_{j=1}^{n} q_j \bar{x}_j = 0$, with

$$(\bar{x}_1, \ldots, \bar{x}_n) \sim (x_1, \ldots, x_n),$$

therefore defining the same element of $\mathbb{T}^n$, such that $f_{\Omega}(\bar{x}_1, \ldots, \bar{x}_n) = 0$. The latter can be seen by being aware that there exist y_j, $1 \leq j \leq n$ such that $f_{\Omega}(x_1, \ldots, x_n) = \sum_{j=1}^{n} q_j y_j$ and then taking $\bar{x}_j = x_j - y_j$, $1 \leq j \leq n$.

Now we can define an arc

$$s \in [0, 1] \mapsto [s\bar{x}_1, s\bar{x}_2, \ldots, s\bar{x}_n] \in N_f,$$

connecting $[0]$ with $[x_1, x_2, \ldots, x_n] \in N_f$, so we proved that N_f is (arcwise) connected. □

---

[8] As we show later on, a similar connectedness does not necessarily hold for the time sets $\mathbb{Z}_+$ or $\mathbb{Z}$.

Although we did not show all the details, we hope the idea may be clear now. When $\omega_1, \ldots, \omega_n$ are linearly dependent over $\mathbb{Q}$, for each initial state there exists a lower-dimensional torus, diffeomorphic to N_{f_Ω}, containing the entire evolution. If in this lower-dimensional torus linear dependencies still exist, the process of taking invariant subtori can be repeated, and so on. Eventually we reach the situation in which the new ωs are rationally independent. If then we have a torus of dimension larger than 1, we speak of quasi-periodic dynamics; if the latter torus has dimension 1, the dynamics is periodic.

2.2.2.4 Translation systems on the n-dimensional torus with time set $\mathbb{Z}$ or $\mathbb{Z}_+$

We next consider the translation dynamics on $\mathbb{T}^n$ with translation vector $\Omega = (\omega_1, \omega_2, \ldots, \omega_n)$, but now with time set $\mathbb{Z}$ (or $\mathbb{Z}_+$). In this case each evolution is dense if and only if $\omega_1, \omega_2, \ldots, \omega_n$ and 1, as elements of $\mathbb{R}$, are linearly independent over $\mathbb{Q}$. For a proof we again refer to [59]. Also here we speak of *resonance* whenever the above independence fails.

The connection with the translation dynamics having time set $\mathbb{R}$ rests on the suspension construction; compare §1.2.2.3. When constructing the suspension of a translation map of $\mathbb{T}^n$, which is given by a translation vector $\Omega = (\omega_1, \omega_2, \ldots, \omega_n)$, as is the case here, we obtain a system the state space of which can be considered as an $(n + 1)$-dimensional torus $\mathbb{T}^{n+1}$, where the dynamics is given by $\tilde{\Phi}^t = R_{t(\omega_1, \ldots, \omega_n, 1)}$. A proof of this assertion is not really hard; also see the exercises. Although this, in principle, reduces the case with time set $\mathbb{Z}$ to that of time set $\mathbb{R}$ by the suspension construction, we point out a few problems. Therefore in the forthcoming considerations we do not use the foregoing analysis with time set $\mathbb{R}$.

Remark (Digression on the resonant case with time set $\mathbb{Z}$ or $\mathbb{Z}_+$). The aim of this excursion is to show why, in the case of time set $\mathbb{Z}$ (or $\mathbb{Z}_+$), considering quasi-periodic evolutions on the disjoint union of a finite number of tori cannot be avoided.

To be precise, we investigate the resonant case when $\omega_1, \ldots, \omega_n$ and 1 are rationally dependent. Analogously to the setting of §2.2.2.3 with time set $\mathbb{R}$, here also there exist integers $q_1, q_2, \ldots, q_n, q$ without a common divisor and for which $\sum_j q_j \omega_j = q$.

1. For $q = 0$, as in the previous case, one can argue that there is an $(n - 1)$-dimensional torus that contains the entire evolution. Again it follows that the evolution is not dense in $\mathbb{T}^n$; compare Lemma 2.5.
2. Now we consider the case where $q \neq 0$. Let r be the greatest common divisor of $q_1, \ldots, q_n$. In the case where $r = 1$, we learn from algebra that integers $m_1, m_2, \ldots, m_n$ exist with $\sum_j q_j m_j = 1$. Here recall from the introduction of §2.2.2 that, because the time set is $\mathbb{Z}$, we may add to each $\omega_j, 1 \leq j \leq n$, an integer. It follows that for $r = 1$, by choosing suitable 'realisations' of $\omega_1, \omega_2, \ldots, \omega_n$, we can achieve that $q = 0$, and we are in the above case with

an invariant $(n-1)$-torus. Next we assume that $r \neq 1$ and that $q \neq 0$. Because $q_1, q_2, \ldots, q_n, q$ have no common divisors, it follows that q and r are coprime. In this setting we define the function $f_\Omega : \mathbb{T}^n \to \mathbb{R}/\mathbb{Z}$ by

$$f_\Omega[x_1, x_2, \ldots, x_n] = \sum_j \frac{q_j}{r} x_j \bmod \mathbb{Z};$$

compare the proof of Lemma 2.5. The integers $q_j/r, 1 \leq j \leq n$, have no common divisors, thus as in the previous case, the resonant zero set $N_{f_\Omega} \subset \mathbb{T}^n$ of f is connected. And again it can be shown that N_{f_Ω} can be considered as a torus of dimension $n-1$.

From the above it follows that

$$f_\Omega(R_\Omega^k([x])) = f_\Omega([x]) + \frac{qk}{r}.$$

Because q and r are coprime, the numbers qk/r as elements of $\mathbb{R}/\mathbb{Z}$, for $k \in \mathbb{Z}$, attain all values $0, 1/r, 2/r, \ldots, (r-1)/r$. We now define

$$\tilde{N}_{f_\Omega} = \left\{ [x] = [x_1, \ldots, x_n] \in \mathbb{T}^n \mid f_\Omega([x]) \in \left\{ 0, \frac{1}{r}, \frac{2}{r}, \ldots, \frac{r-1}{r} \right\} \right\},$$

where $0, 1/r, 2/r, \ldots, (r-1)/r \in \mathbb{R}/\mathbb{Z}$. From this it may be clear that $\tilde{N}_{f_\Omega}$ is the disjoint union of r tori of dimension $n-1$ and that the evolution starting at $[0]$ remains within $\tilde{N}_{f_\Omega}$. A fortiori the evolution is not dense in $\mathbb{T}^n$; again compare Lemma 2.5. Furthermore the translation R_Ω^r maps the torus $N_{f_\Omega} \subset \tilde{N}_{f_\Omega}$ onto itself and the restriction $R_\Omega^r|_{N_{f_\Omega}}$ is a translation. Now if any evolution of the restriction $R_\Omega^r|_{N_{f_\Omega}}$ is dense in $\tilde{N}_{f_\Omega}$, we call the original dynamics, restricted to $\tilde{N}_{f_\Omega}$ quasi-periodic. If the latter is not the case for $R_\Omega^r|_{N_{f_\Omega}}$, dependencies still occur between the new ωs and the reduction can be continued.

2.2.3 General definition of multi- and quasi-periodic evolutions

Before discussing multi- and quasi-periodic evolutions in general, we first define what are multi- and quasi-periodic subsystems; multi- and quasi-periodic evolutions then will be evolutions within such a subsystem. A formal definition follows soon, but, inasmuch as this is quite involved, we first sketch the global idea. A quasi-periodic subsystem is an invariant subset such that the restriction to this subset (compare §1.2.2) is *conjugated* to a translation system. Here a conjugation is understood as a bijection, in this case a diffeomorphism, that maps the evolutions of one system to evolutions of the other system in a time-preserving way.

2.2.3.1 Multi- and quasi-periodic subsystems

Now we come to the formal definition. Consider a dynamical system (M, T, Φ). Here M is a differentiable manifold (where it is not an essential restriction to think of M as a vector space) and Φ a smooth evolution operator, taking for the time set T either $\mathbb{Z}$, $\mathbb{Z}_+$, or $\mathbb{R}$.

Definition 2.7 (Multi- and quasi-periodic subsystem). We say that the system (M, T, Φ) has a *multi-periodic subsystem* if there exists a translation system $(\mathbb{T}^n, T, \Phi_\Omega)$, where

$$\Phi_\Omega([x], t) = R_{t\Omega}([x]) = [x + t\Omega],$$

and a differentiable map $f : \mathbb{T}^n \to M$, such that for all $[x] \in \mathbb{T}^n$ and $t \in T$ the *conjugation equation*

$$f(\Phi_\Omega([x], t)) = \Phi(f([x]), t) \tag{2.6}$$

holds. The multiperiodic subsystem is called *quasi-periodic* if the translation vector Ω contains no resonances. The map f is required to be an embedding, meaning that $f(\mathbb{T}^n) \subset M$ is a submanifold and that the derivative of f at each point has rank n. We call $f(\mathbb{T}^n)$, with the dynamics defined by the restriction of Φ, the multiperiodic subsystem.[9]

$$
\begin{array}{ccc}
\mathbb{T}^n \times T & \xrightarrow{\;\Phi_\Omega\;} & \mathbb{T}^n \\
{\scriptstyle f\times\mathrm{Id}}\big\downarrow & & \big\downarrow{\scriptstyle f} \\
M \times T & \xrightarrow[\;\Phi\;]{} & M
\end{array}
$$

Validity of the conjugation equation (2.6) is expressed by commutativity of the above diagram. We note that (2.6) and the fact that f has to be an embedding imply not only that f is a diffeomorphism from $\mathbb{T}^n$ to the subsystem of (M, T, Φ), but also that f maps evolutions of Φ_Ω to evolutions of Φ, in such a way that the time parametrisation is preserved. This means that f is a *conjugation* between the translation system on $\mathbb{T}^n$ and the restriction of the original system to $f(\mathbb{T}^n)$. We call the translation system *conjugated* with the restriction of the original system to $f(\mathbb{T}^n)$. Conjugations will occur more often; within the 'category' of dynamical systems they play the role of 'isomorphisms'.[10]

In the case of time set $T = \mathbb{Z}$ (or $\mathbb{Z}_+$) we also speak of a multiperiodic subsystem when the conjugation equation (2.6) only holds for values of $t \in rT$ for a fixed integer r. Here it is required that the sets $\Phi(f(\mathbb{T}^n), j)$ are disjoint for $j = 0, \ldots, r - 1$. In that case, the state space of the quasi-periodic subsystem is

---

[9] Arnold [21] instead of 'multiperiodic' uses the term 'conditionally periodic'. In Broer et al. [88] the term 'parallel' is used for this.

[10] Later on we meet conjugations that are only homeomorphic instead of diffeomorphic.

the union $\bigcup_j \Phi(f(\mathbb{T}^n), j)$ of those n-tori. This extension follows earlier remarks concluding §2.2.2.

From Definition 2.7 it follows that a multiperiodic subsystem is quasi-periodic if and only if the orbits of the corresponding translation system are dense in the torus. Let us discuss the case where the subsystem is multi- but not quasi-periodic. We have seen that, if a translation on an n-torus defines translation dynamics, the latter is only multi- (and not quasi-) periodic, if each evolution is contained in a lower-dimensional torus or in a finite union of such subtori. This means that if we have a multi- (and not quasi-) periodic subsystem, by restriction to such a lower-dimensional torus, we get a lower-dimensional subsystem. As in the case of translation systems, we can continue to restrict to lower dimensions until no further reduction is possible. In this way we end up in a periodic evolution or a quasi-periodic evolution. The latter we have defined as an evolution in a quasi-periodic subsystem.

When comparing with asymptotically periodic evolutions (see §2.1.2) it may now be clear what is understood by *asymptotically multi-* or *quasi-periodic* evolutions: these are evolutions that tend to multi- or quasi-periodic evolutions with increasing time. As in the case of periodic evolutions, this can be taken either in a strict sense, or only as convergence in distance. Furthermore, eventually multi- or quasi-periodic evolutions could be defined in a completely analogous way, but we are not aware of any relevant examples where such evolutions play a role.

2.2.3.2 Example: The driven Van der Pol equation

As a concrete example of the above, we consider the driven Van der Pol equation, mainly restricting to the case where the driving force is 'switched off.' Here we illustrate how a multiperiodic subsystem can be constructed. Of course it would be nice to do the same when the driving force is 'switched on.' We go into the problems that are involved here and mention some of the results that research has revealed in this case.

The driven Van der Pol equation is given by

$$x'' = -x - x'(x^2 - 1) + \varepsilon \cos(2\pi \omega t); \qquad (2.7)$$

compare the equation of motion of the driven pendulum (see §1.1.2 and 1.3.1). Our present interest is with multi- and quasi-periodic dynamics of (2.7) as this occurs for small values of $|\varepsilon|$, although our mathematical exploration for now is mainly confined to the case $\varepsilon = 0$. See Figure 1.11 showing the phase portrait of the free Van der Pol oscillator (1.14). However, for as long as possible we work with the full equation (2.7). The present purpose is to construct a multiperiodic subsystem. As in the case of the driven pendulum, we can put this equation in the format of an autonomous system with three state variables, namely x, $y = x'$, and z. Here z takes on values in $\mathbb{R}/\mathbb{Z}$, playing the role of the time t, up to a scaling by ω. The new equations of motion thus are

$$x' = y$$
$$y' = -x - y(x^2 - 1) + \varepsilon \cos(2\pi z) \tag{2.8}$$
$$z' = \omega.$$

In §1.3.1 we already noted that for $\varepsilon = 0$, this system has a unique periodic evolution. All other evolutions, except for the stationary evolution given by $x = y = 0$, are asymptotically periodic and tend to this periodic evolution; again see Figure 1.11. This periodic evolution can be parametrised by a map

$$f_1 : \mathbb{R}/\mathbb{Z} \to \mathbb{R}^2$$

with coordinates x and y as follows. The periodic evolution is given by

$$(x(t), y(t)) = f_1(\alpha t),$$

where α^{-1} is the prime period. This gives a mapping

$$f : \mathbb{T}^2 \to \mathbb{R}^3 = \{(x, y, z)\}, \quad ([w], [z]) \mapsto (f_1(w), z),$$

where both w and z are variables in $\mathbb{R}/\mathbb{Z}$. On the 2-torus $\mathbb{T}^2 = \{w, z\}$ we next consider the translation dynamics corresponding to the system of differential equations

$$w' = \alpha$$
$$z' = \omega,$$

which has the format of the constant vector field (2.4). In other words, we consider the translation dynamics corresponding to the translation vector (α, ω). It is not hard to show that f satisfies the conjugation equation (2.6) and that in this way we have found a multiperiodic subsystem of the driven Van der Pol equation (2.7) for the case $\varepsilon = 0$.

As said before, we would have preferred to deal with the case $\varepsilon \neq 0$, because for $\varepsilon = 0$ the z-coordinate is superfluous and we seem to be dealing with a kind of 'fake multiperiodicity.' The analysis for the case $\varepsilon \neq 0$, even for the case where $|\varepsilon| \ll 1$, belongs to perturbation theory, in particular Kolmogorov–Arnold–Moser (or KAM) theory, and is introduced in Chapter 5; also see the historical remarks and references in §2.2.5 below.

First, because for $\varepsilon = 0$ the torus is normally attracting (in an exponential way), it can be shown that also for $|\varepsilon|$ small there is a normally attracting invariant manifold, both diffeomorphic to and near the original torus. This follows from the theory of normally hyperbolic invariant manifolds; see [25, 143]. The dynamics in the 'perturbed' torus can be considered as a small perturbation of the dynamics that we obtained in the multiperiodic subsystem for $\varepsilon = 0$. It turns out that in the 'perturbed' dynamics it is no longer true that either all evolutions are periodic or quasi-periodic. Compare Appendix C.

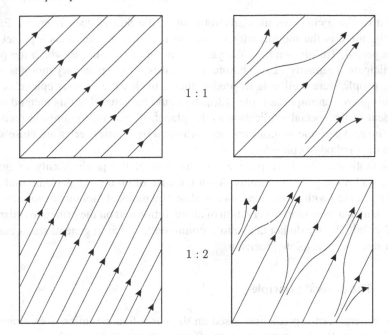

Fig. 2.4 Periodic dynamics on a 2-torus. Left: multiperiodic evolution in resonance. Right: a (not too small) perturbation of the former.

Next, to describe this situation we consider the system (2.8) for a fixed small value of $\varepsilon \neq 0$ and for variable values of ω. The analysis of such systems can be found in [27,37,87,88,140]. For a summary also see Chapter 5 and Appendix B. In fact, it turns out that for many values of ω, to be precise for a set of ω-values of positive Lebesgue measure, the evolutions in the perturbed torus are quasi-periodic. Apart from this, in general there are also many values of ω (to be precise, for ω-values forming an open and dense set) where the perturbed torus contains periodic evolutions, generally only finitely many. In Figure 2.4 we indicated how on a 2-torus multiperiodic dynamics, that happens to be periodic, can be perturbed to dynamics with finitely many periodic evolutions. For the values of ω where such periodic dynamics occurs we have resonance between the period of the driving and the period of the 'free' Van der Pol system; compare §2.2. Here one usually speaks of *phase locking*. Note that without driving, these resonances correspond to values of ω for which the frequency-ratio $\alpha : \omega$ is rational.

2.2.4 The prediction principle l'histoire se répète

Just like periodic evolutions, quasi-periodic evolutions in principle are quite predictable. An important historical example of this is the motion of the planets. As these motions appear to us, at least approximately, three frequencies play a role.

The first of these concerns the daily rotation of the Sun as seen from the Earth. Secondly, there is the motion with respect to the Sun, in which each planet has its own period. Thirdly, we have the yearly motion of the Sun, in which the planets participate 'indirectly'. Long before the 'causes' of the planetary motions were known, people were well able to predict them. In this the famous epicycles and deferents played an important role [136]. It should be noted that this method is independent of the special application to the planetary system: the present methods where (quasi-) periodic motions are approximated by Fourier series are closely related to the method of epicycles.

This motivates the more precise investigation of the predictability of quasi-periodic and more general motions. As in the case of predicting periodic motions, we are concerned with predictions not based on equations of motion or evolution operators, but on a long-term observation of an evolution and on the principle *l'histoire se répète*. Initially we do not take into account the fact that in general states cannot be registered with absolute precision.

2.2.4.1   The general principle

In very general terms, prediction based on the past of an evolution goes along the following lines. We denote the segment of our evolution in the past by

$$\{x(s) \mid 0 \le s \le t\}, \tag{2.9}$$

without loss of generality assuming that we have time set $T = \mathbb{R}$ or $\mathbb{R}_+$. In order to predict the future values of this evolution we look for an instant $\bar{t} \in [0, \frac{1}{2}t]$ in the past, such that the distance $d(x(t), x(\bar{t}))$ between the past and the present state is minimal (the factor $\frac{1}{2}$ is explained later). Because we assume that the evolutions depend continuously on the initial states, it is rational to predict the future states $x(t + s)$ as

$$\hat{x}(t + s) = x(\bar{t} + s), \tag{2.10}$$

which we can safely do until $s = t - \bar{t}$. This is the principle *l'histoire se répète*. We call s the *prediction interval*.

We note that for $s > t - \bar{t}$ in the right-hand side of (2.10) $\bar{t} + s > \bar{t} + t - \bar{t} = t$ from which is follows that

$$\bar{t} + s > t.$$

and $x(\bar{t} + s)$ is not yet known at time t. So the last state we can predict by (2.10) is $x(2t - \bar{t})$ which we predict as

$$\hat{x}(2t - \bar{t}) = x(t).$$

In order to continue our predictions we next look for an instant $\bar{\bar{t}} \in [0, \frac{1}{2}t]$ such that

$$d(x(\bar{\bar{t}}), \hat{x}(2t - \bar{t})) = d(x(\bar{\bar{t}}), x(t))$$

is minimal, but we know already that this means that we have to take $\bar{\bar{t}} = \bar{t}$.

So, following the same procedure, we predict the state $x(2t - \bar{t} + s)$ as

$$\hat{x}(2t - \bar{t} + s) = x(\bar{t} + s), \tag{2.11}$$

for $0 < s \le t - \bar{t}$. Comparison of (2.10) and (2.11) gives that our predictions become periodic with period $t - \bar{t}$, with a discontinuity between each two periods; between the discontinuities the predictions follow a true evolution of the system. Note that the size of the discontinuity is $d(x(t), x(\bar{t}))$.

Remarks.

- It should be clear that large prediction errors will occur if the period $t - \bar{t}$ is small, because then the discontinuities, involving increasing errors, occur relatively often. We took $\bar{t} \in [0, \frac{1}{2}t]$ so as to make the period of prediction at least $\frac{1}{2}t$. It is clear that the factor $\frac{1}{2}$ is arbitrary.
- In general the size of the discontinuities becomes smaller once we know a longer segment of the past states: with more past states in general one can find a better approximation $x(\bar{t})$ of the present state $x(t)$. This is about as much as one can say about predictions of future behaviour of deterministic systems with evolutions that depend continuously on initial states.

2.2.4.2 Application to quasi-periodic evolutions

Next we aim to show that if we apply the above prediction method to a quasi-periodic evolution the prediction error will only grow linearly with the prediction interval s.

We first consider the case of translation dynamics given by Φ_Ω on $\mathbb{T}^n$:

$$\Phi_\Omega(x, t) = x + t\Omega \bmod \mathbb{Z}^n.$$

We recall that, by the definition of quasi-periodicity, $\Omega = (\omega_1, \omega_2, \ldots, \omega_n)$, where $\omega_1, \omega_2, \ldots, \omega_n$ are linearly independent over $\mathbb{Q}$. For any $\varepsilon > 0$ one can find $\mathcal{T}(\varepsilon)$ such that any orbit of length $\mathcal{T}(\varepsilon)$ is ε-dense in $\mathbb{T}^n$, in the sense that its ε-neighbourhood is all of $\mathbb{T}^n$. Then for any orbit segment $\Phi_\Omega(x, [0, t])$ with $t > 2\mathcal{T}(\varepsilon)$ there is some $\bar{t} \in [0, \frac{1}{2}t]$ such that

$$d(\Phi_\Omega(x, t), \Phi_\Omega(x, \bar{t})) < \varepsilon;$$

in fact we may and do take $\bar{t}$ even so that the distance $d(\Phi_\Omega(x, t), \Phi_\Omega(x, \bar{t}))$ is minimal for $\bar{t} \in [0, \frac{1}{2}t]$. Applying the above prediction method, we now predict the continuation as

$$\hat{\Phi}_\Omega(x, t + s) = \Phi_\Omega(x, \bar{t} + s) \quad \text{for } 0 \le s \le t - \bar{t}$$
$$= \Phi_\Omega(x, \bar{t} + s') \quad \text{for } s = m(t - \bar{t}) + s' \text{ for all } m \in \mathbb{Z}_+.$$

From this we see that the prediction error increases at most by ε when increasing s by $t - \bar{t}$. This gives an upper estimate for the prediction error of

$$\left(\left[\frac{s}{t - \bar{t}} \right] + 1 \right) \varepsilon,$$

where [.] takes the integer part. This means that there is a linear growth of prediction errors with time (or rather with the prediction interval s). [11] It should be clear that, by taking a sufficiently long interval of 'past known values,' this linear growth can be made as slow as we wish.

For arbitrary quasi-periodic dynamics in a state space M, by definition we have a subsystem which is differentiably conjugated with translation dynamics in $\mathbb{T}^n$. Such a conjugation is a diffeomorphism

$$f : \mathbb{T}^n \to T \subset M.$$

This means that there is a constant $C \geq 1$ such that for all $x, y \in \mathbb{T}^n$,

$$C^{-1} d(x, y) \leq d_M(f(x), f(y)) \leq C d(x, y),$$

where d_M denotes the distance in M. From this it easily follows that also in this case the same arguments apply and hence the growth of prediction errors is linear in the prediction interval.

Remarks.

– In the discussion of the predictability of the translation dynamics in $\mathbb{T}^n$ we implicitly made use of the fact that the distance $d(\Phi_\Omega(x, s), \Phi_\Omega(\bar{x}, s))$ is independent of s. For general quasi-periodic dynamics this is no longer true, but we do have: for any $\varepsilon > 0$ there is a $\delta > 0$ such that for any $p, q \in T \subset M$ (where T is the closure of the quasi-periodic evolution under consideration) with $d(p, q) < \delta$, we have

$$d_M(\Phi_M(p, s), \Phi_M(q, s)) < \varepsilon$$

for all $s \in \mathbb{R}$, where Φ_M is the evolution operator on M.
– Comparing the above conclusions with those in the periodic case (see §2.1.1) we witness a great similarity. One difference, however, is that in the above it is not yet taken into account that the states can only be observed with a finite precision. (In the periodic case this was essentially the only source of prediction errors.) This is related to the following. Assuming discrete time, in the periodic case the period can be determined with absolute precision after observing at least over one full period (leaving aside the observation errors). In the quasi-periodic case none of the periods can be determined by just observing a finite piece of the evolution,

---

[11] To be more precise, the prediction error is at most ε plus a linear function of s.

because it does not close up and the continuation is not fully determined. In other words, continuation cannot be concluded from the given observation. When taking into account this inaccuracy, the above conclusion about the linear dependence of the prediction error on the prediction interval remains the same, only the growth of the errors becomes larger.

- Linear growth of prediction errors, where the speed of the growth can be made arbitrarily small by carrying out sufficiently long and accurate observations, means that we are dealing with predictability of an exceptionally high quality. In many systems, with the weather as a notorious example, predictability is far more problematic. The common idea about this is that 'random' perturbations always occur, which spoil the predictability. The only exception is formed by planetary motion, that seems to be of an 'unearthly' predictability. As we show hereafter, there are also evolutions of simple *deterministic* systems, the so-called chaotic evolutions, which are essentially less predictable than the (asymptotically) quasi-periodic evolutions met thus far. This being said, we have no wish to deny that in practical circumstances 'randomness' can also play a role.

2.2.5 Historical remarks

For a long time people wondered whether quasi-periodic evolutions, as defined above, do occur in 'natural' dynamical systems other than the special examples invented by mathematicians. Even in the case of the driven Van der Pol equation (compare §2.2.3.1), the proof that quasi-periodic evolutions occur is far from simple: it was only found in the 1950s and 1960s; for an account of this see [27, 37, 87, 88]. The underlying problem has puzzled mathematicians for a long time and a complete treatment of the solution of this problem would carry too far here and we just refer to Chapter 5 and to Appendix B. However, this is a suitable place to present more information on the problem setting, as it is deeply connected to the stability of the solar system. It was for a good reason that our first example of motions that, at least to a very good approximation, are multiperiodic, was given by the planets. We now consider the entire solar system. In this system the motion to a good approximation is multiperiodic. The simplest approximation is obtained by neglecting the mutual attraction of the planets. In this approximation each planet, only attracted by the Sun, carries out a perfectly periodic motion along an ellipse, as was discovered by Kepler and explained by Newton. Now, if the ratios of the periods of revolution of all these planets have no rational dependence (which indeed would be quite unlikely), then the motion of the entire system is quasi-periodic.

The question then is, what are the consequences of the mutual attraction of the planets. From our experience we know that the short-term effects, say over several thousands of years, are quite restricted. The stability problem of the solar system, however, is more involved: the question is whether these effects on an infinitely long time scale are also restricted. Here several scenarios come to mind. To start with, there is the most optimistic scenario, where the mutual interaction does not affect the quasi-periodic character of the global motion, but only position and form of the corresponding invariant torus in the state space are a bit distorted. In terms of the above Definition 2.7 of a quasi-periodic subsystem, this would just mean a small distortion of the map f, which conjugates the translation system to the quasi-periodic subsystem. This is the most orderly way in which the small perturbations keep each other in balance. Another scenario occurs when the small perturbations do not balance on the average, but keep accumulating. In principle this may lead to various interesting calamities: in the not too near future planets could collide or even escape from the solar system. Also the Earth might be captured as a satellite by Jupiter, and so on, and so on. Evidently, these questions are only of interest when looking further ahead than only a few generations. It was King Oskar II of Sweden and Norway,[12] who initiated a large-scale contest where mathematicians[13] were challenged to shed light on this matter. The contest was won by Poincaré,[14] not because he clarified the situation, but because he showed that the complications by far exceeded everyone's expectations. Poincaré's investigations [191] were of tremendous influence on the development of the mathematical theory of dynamical systems.[15] Nevertheless, the original problem concerning the solar system remained. Later, around 1960, some further progress was made in this direction. Due to the joint efforts of Kolmogorov,[16] Arnold,[17] and Moser [18] (one speaks of KAM theory), eventually it was 'approximately' proven that, if the mutual interaction between the planets is sufficiently small, there is a set of initial conditions of positive volume, in the state space which gives rise to quasi-periodic dynamics. Compare [58].[19] The term 'approximately' is used here, because this result has not been elaborated for the full generality of the three-dimensional solar system. Anyway, it was shown that quasi-periodic evolutions fill out a substantial part (i.e., of positive volume) in the state space of systems, the practical relevance of which cannot be denied. KAM theory probably is the most important contribution of the twentieth century to classical mechanics. Further elaboration of the consequences of this discovery is still an important area of research, not only for celestial mechanics, but also for statistical physics (which aims to explain macroscopic properties of matter by the molecular structure, i.e., by molecular interactions) and for the analysis of resonance phenomena.

---

[12] King Oskar II 1829–1907.

[13] Celestial mechanics was already mainly the domain of mathematicians.

[14] Jules Henri Poincaré 1854–1912.

[15] The precise history is quite involved; compare Barrow-Green [64].

[16] Andrey Nikolaevich Kolmogorov 1903–1987.

[17] Vladimir Igorevich Arnold 1937–.

[18] Jürgen K. Moser 1928–1999.

[19] Certain approximate 'strong' resonances in planetary motion, for example, concerning Jupiter and Saturn, were taken into account by the model of Arnold [56].

For more complete information on the historical details we refer to [64,113,208]. For a recent overview from a more general perspective with many references to the literature, also see [37,87]. In Chapter 5 we present an introduction to KAM theory.

2.3 Chaotic evolutions

We just saw in §2.2.4 that the qualitatively good predictability of a (quasi-) periodic evolution $x(t)$ of a dynamical system (M, T, Φ) rests on the facts that

A. For any $\varepsilon > 0$ there exists $T(\varepsilon) \in \mathbb{R}$ such that, for all $0 \leq t \in T$, there is a $\bar{t} \in [0, T(\uparrow)] \cap T$ with $d(x(t), x(\bar{t})) < \varepsilon$.
B. For any $\varepsilon > 0$ there is a $\delta > 0$ such that, if $d(x(t), x(\bar{t})) < \delta$, with $0 \leq t, \bar{t} \in T$, then for all $0 < s \in T$ also $d(x(t + s), x(\bar{t} + s)) < \varepsilon$.

See the first remark at the end of §2.2.4.2.

For the chaotic evolutions, discussed now, property A still holds, but property B is no longer valid. Based on A, it will be possible, given an initial piece of the evolution $[0, T(\varepsilon)]$ of sufficient length, to find states $x(\bar{t})$ with $\bar{t} \in [0, T(\varepsilon)]$, which are very close to the 'present' state $x(t)$, but predictions based on the principle *l'histoire se répète* will be less successful. Before giving a formal definition of what is going to be understood as a chaotic evolution, we first investigate the doubling map (see §1.3.8) for the occurrence of badly predictable, and therefore possibly chaotic, evolutions. Although similar evolutions occur in many of the other examples, only the doubling map is sufficiently simple to be mathematically analysed by rather elementary means.

2.3.1 Badly predictable (chaotic) evolutions of the doubling map

We already saw in §1.3.8 that in the dynamics defined by iteration of the doubling map, where we have the situation in mind with the state space $\mathbb{S}^1 = \mathbb{R}/\mathbb{Z}$, there exist many periodic and eventually periodic evolutions. Evidently these evolutions are not chaotic. We now consider an evolution $x(n)$, $n \in \mathbb{Z}_+$ that is neither (eventually) periodic nor stationary. Such an evolution necessarily has infinitely many different states (i.e., different points on the circle $\mathbb{S}^1$). Indeed, if for certain $n_1 < n_2$ one has that $x(n_1) = x(n_2)$, then the evolution surely from time n_1 on is periodic and contains no more than n_2 different points. So assuming that we are dealing with a not (eventually) periodic evolution, for any $n_1 \neq n_2$ we have $x(n_1) \neq x(n_2)$.

Lemma 2.8. *For any evolution $\{x(n)\}_{n\in\mathbb{Z}_+}$ of the iterated doubling map that is not (eventually) periodic, property A is valid but not property B.*

Proof. Let $\varepsilon > 0$ be chosen. Consider a sequence of nonnegative integers $0 = n_0 < n_1 < n_2 < \cdots$ such that for each $0 \leq j < i$ one has $d(x(n_i), x(n_j)) \geq \varepsilon$. To make the definition of the sequence $\{n_j\}$ unique, we require that for any $n > 0$ that does not occur in the sequence, there exists $n_j < n$, such that $d(x(n), x(n_j)) \leq \varepsilon$. From the definition of the sequence $\{n_j\}$ it now follows that for any two $i \neq j$ we have that $d(x(n_i), x(n_j)) \geq \varepsilon$. The circle $\mathbb{S}^1$ has finite length, thus the sequence $\{n_j\}$ has to be finite. Let $N = n_k$ be the final element of the sequence. (Note that both k and N, as well as the elements $n_1, n_2, \ldots, n_k$ of the sequence, depend on ε.) Now we have that for any $n > 0$ there exists an element n_j of the sequence such that $d(x(n), x(n_j)) < \varepsilon$. From this it clearly follows that property A holds for our evolution: for the given value of ε we can take $T_\varepsilon = N = n_k$.

Next we show that property B does not hold. Fix $\varepsilon < \frac{1}{4}$. For any $\delta > 0$ there are $n \neq m$ such that

$$0 < d(x_n, x_m) < \delta.$$

This is due to the facts that for $n \neq m$ also $x_n \neq x_m$ and that $\mathbb{S}^1$ is compact. Then

$$d(x_{n+s}, x_{m+s}) = 2^s \, d(x_n, x_m) \tag{2.12}$$

as long as $2^s \, d(x_n, x_m) < \frac{1}{2}$. From this it immediately follows that, for some finite s,

$$d(x_{n+s}, x_{m+s}) > \varepsilon.$$

This means that property B does not hold.   □

With this same argument we can also indicate how prediction errors increase. Let us assume the same setup as before where we compare with an initial segment of the evolution $\{x(t) \mid t \in T\}$. Given a 'present' state $x(t)$, if we are able to find a state $x(\bar{t})$ in the initial, 'past' segment, such that $d(x(t), x(\bar{t})) < \delta$, where we use $x(\bar{t} + s)$ as a prediction for the 'future' state $x(t + s)$, then we know that, according to (2.12), the prediction error is $\delta 2^s d(x(t), x(\bar{t}))$, as long as there is no saturation. We see that here we have *exponential growth* of the prediction error as a function of the prediction interval s. Also here we should take into account that for $\bar{t} + s > t$ extra errors will occur, but presently this is of lesser interest, because by the rapid increase of prediction errors the making of predictions far in the future has become useless anyway.

Remark. We note that, even if we have exact knowledge of the evolution operator, a small inaccuracy in our knowledge of the initial state already leads to the same exponential growth of prediction errors. This is contrary to what we saw for (quasi-) periodic evolutions.[20] If, in the dynamics of the doubling map, we know the initial condition up to an error of at most δ, then prediction with the help of the evolution operator and prediction interval s, involves an error of at most $\delta 2^s$.[21] The effect of exponential growth of errors is illustrated in Figure 2.5.

---

[20] This remark is elaborated later: here we are always dealing with one evolution whereas inaccuracy in the initial condition involves many evolutions.

[21] Again leaving alone saturation.

Fig. 2.5 Effect of a small inaccuracy in the initial state in evolutions of the logistic map: the difference in the initial states is approximately 2.10^{-16}. (We used the logistic map rather than the doubling map, because in the latter case numerical artefacts occur as a consequence of the computer using the binary number system. Compare Exercise 2.2.)

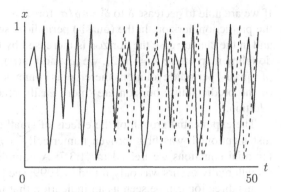

In general we express the lowest upper bound for the prediction error in a chaotic evolution, defined shortly, as a function of the prediction interval s, for large s, by an expression of the form

$$\delta e^{\lambda s}.$$

Here δ can be made small by the amount of information we have over a long interval in the past of the evolution (if our predictions are based on *l'histoire se répète*) or on the initial condition (if we use the exact evolution operator). The constant λ is completely determined by the evolution operator of the system. In the case of the doubling map $\lambda = \ln(2)$.

To clarify the difference in predictability of such chaotic evolutions in comparison with the (quasi-) periodic evolutions mentioned earlier, we investigate how much trouble it takes to enlarge the prediction interval, where we then have to produce useful predictions. Here the notion of the *predictability horizon* comes up. This is the maximal time interval s that as a prediction interval still gives reasonable predictions. Apart from the saturation mentioned before, in the chaotic case the prediction error has an upper bound of $\delta e^{\lambda s}$, whereas in the (quasi-) periodic case we found an upper bound δs. In both cases s is the prediction interval, whereas δ depends on the length of the initial interval of the evolution we used for making our predictions, plus uncertainty in the initial state. Decreasing δ mostly will be difficult and expensive. The predictability horizon now is obtained by introducing a maximal prediction error c and solving the equations

$$\delta e^{\lambda s} = c \quad \text{and} \quad \delta s = c,$$

thus obtaining the prediction horizon

$$s = \begin{cases} \dfrac{1}{\lambda}(\ln(c) - \ln(\delta)) & \text{chaotic case} \\ \dfrac{c}{\delta} & \text{(quasi-) periodic case .} \end{cases}$$

If we are able to decrease δ to $\delta' = \delta/\sigma$, for $\sigma > 1$, this will lead to an increase of the prediction horizon. In the (quasi-) periodic case the horizon is multiplied by σ, and in the chaotic case the horizon is increased by the addition of $\ln(\sigma)/\lambda$. For the doubling map where $\lambda = \ln(2)$, we just add $\ln(\sigma)/\ln(2)$. The better predictability of (quasi-) periodic evolutions therefore is connected with the fact that larger values of the predictability horizon are 'more easily reached by multiplication than by addition.'

The exponential growth of the effects of small deviations in the initial state, as just mentioned, was first 'proven numerically' by Lorenz for the system of differential equations we treated in §1.3.7. A complete proof that such exponential growth really occurs was only found in 1999; see [234]. Already numerical results in this direction can be seen as an indication that in such cases practically no reliable predictions can be made on events farther in the future. Lorenz's system was a strong simplification of the meteorological system that describes the dynamics underlying the weather. In [60] one goes further: only using heuristic, but theoretically based arguments, it is argued that in weather predictions the errors, as a function of the prediction interval s, grow by a daily factor of 1.25. This factor may seem low, given our experience with the Royal Netherlands Meteorological Institute (KNMI), but this can be well explained by the incompleteness of the arguments used. In any case, a better understanding of the predictability of phenomena such as the weather is a fascinating area of research about which the last word has not yet been spoken.

2.3.2 Definition of dispersion exponent and chaos

Here we formalise the concepts, used above in an informal way, in order to arrive at mathematical definitions. In Chapter 6 we return to the problem of quantifying the unpredictability. There we are mainly interested in algorithms for the investigation of experimental data, for which a different approach is required.

We start with property A (see §2.3.1), needed to be able to speak of predictions based on the past, that is, on the principle *l'histoire se répète*. We assume a state space M with a metric $d : M \times M \to \mathbb{R}_+$, assuming that M, under the metric d, is a complete space. This means that every Cauchy sequence has a limit. Also we assume the evolution operator $\Phi : M \times T \to M$ to be continuous, where $T \subset \mathbb{R}$ is the time set as before. Throughout we use the following notation.

1. Instead of $x(0)$ we also just write x.
2. The positive evolution of x, as a subset of M, is denoted by

$$\mathcal{O}^+(x) = \{x(t) \mid t \in T \cap \mathbb{R}_+\}.$$

3. The initial segment of the evolution of x, during time $\mathcal{T}$, is denoted by

$$\mathcal{O}^+(x, \mathcal{T}) = \{x(t) \mid t \in T \cap [0, \mathcal{T}]\}.$$

In this notation property A is equivalent to: for any $\varepsilon > 0$ there exists $T(\varepsilon)$, such that

$$\mathcal{O}^+(x) \subset U_\varepsilon(\mathcal{O}(x, T(\varepsilon))),$$

where U_ε stands for 'ε-neighbourhood of'.

Lemma 2.9. *Let the dynamical system (M, T, Φ) be as above, where M is a complete metric (metrisable) space and Φ continuous. Then, for any positive evolution of $x(t)$ property A holds if and only if the closure $\overline{\mathcal{O}^+(x)}$ is compact.*

Proof. A complete metric space is compact if and only if, for any $\varepsilon > 0$, it allows a cover with finitely many ε-neighbourhoods. Here an ε-neighbourhood is understood to be an ε-neighbourhood of one point. It is not hard to show that this assertion is equivalent to the characterisation of compactness, which says that every infinite sequence has an accumulation point.

We first show that property A implies compactness of $\overline{\mathcal{O}^+(x)}$. So we assume that A holds true. Then, for any $\varepsilon > 0$ a number $T(\frac{1}{2}\varepsilon) > 0$ exists, such that $\mathcal{O}^+(x)$ is contained in the $\frac{1}{2}\varepsilon$-neighbourhood of $\mathcal{O}^+(x, T(\frac{1}{2}\varepsilon))$. Now $\mathcal{O}^+(x, T(\frac{1}{2}\varepsilon))$ is compact, as the image of the compact set $T \cap [0, T(\frac{1}{2}\varepsilon)]$ under the continuous map $x(\ )$. This implies that $\mathcal{O}^+(x, T(\frac{1}{2}\varepsilon))$ can be covered by a finite number of $\frac{1}{2}\varepsilon$-neighbourhoods, say of the points $x(s_1), \ldots, x(s_N)$, with $0 < s_j < T(\frac{1}{2}\varepsilon)$, for $1 \le j \le N$. Then also the ε-neighbourhoods of $x(s_1), \ldots, x(s_N)$ cover a $\frac{1}{2}\varepsilon$-neighbourhood of $\mathcal{O}^+(x, T(\frac{1}{2}\varepsilon))$, and hence of $\overline{\mathcal{O}^+(x)}$. This concludes the first part of our proof.

We now assume that property A does not hold and show that $\overline{\mathcal{O}^+(x)}$ is not compact. Because A does not hold, there exists $\varepsilon > 0$ such that $\mathcal{O}^+(x)$, for any T, is not contained in $U_\varepsilon(\mathcal{O}^+(x, T))$. In turn this implies that there exists an infinite sequence $0 < T_1 < T_2 < \cdots$, such that for each $j > 0$ we have that $x(T_{j+1}) \notin U_\varepsilon(\mathcal{O}^+(x, T_j))$. This means that $d(x(T_i), x(T_j)) \ge \varepsilon$, for all positive $i < j$. Now, if we have a cover of $\overline{\mathcal{O}^+(x)}$ with $\frac{1}{2}\varepsilon$-neighbourhoods, then two different points of the form $x(T_j)$ cannot lie in one same $\frac{1}{2}\varepsilon$-neighbourhood. This means that there is no finite cover of $\overline{\mathcal{O}^+(x)}$ with $\frac{1}{2}\varepsilon$-neighbourhoods. Again using the above characterisation, it follows that $\overline{\mathcal{O}^+(x)}$ is not compact. $\square$

It is wellknown [152] that a subset of a finite-dimensional vector space has a compact closure if and only if the subset is bounded. This implies that for systems with a finite-dimensional state space M property A holds for any positive evolution that is bounded. In infinite-dimensional vector spaces, in particular, in function spaces, this conclusion is not valid. Under certain circumstances, such as for positive evolutions of parabolic partial differential equations, it still is possible to show that positive evolutions do have a compact closure.

Definition 2.10 (Positively compact). Let (M, T, Φ) be a dynamical system as above, with M a topological space and Φ continuous. A positive evolution $x(t)$ of x with property A (i.e., which as a subset $\mathcal{O}^+(x) \subseteq M$ has a compact closure) is called *positively compact*.

When considering the not eventually periodic or stationary evolutions of the doubling map, we mentioned that for an evolution which we like to call 'chaotic,' property B is not valid. Now we introduce a quantity for the 'measure in which B does not hold.' This quantity measures the exponential growth of small errors as discussed before.

Definition 2.11 (Dispersion exponent). Let (M, T, Φ) be a dynamical system. Assume that M is a complete metric space with metric d and that Φ is continuous. Consider an evolution $x(t), t \in T$, that is *positively compact* and not (eventually) periodic.[22]

1. For $s \in T$ with $s > 0$ and for $\varepsilon > 0$ we define

$$E(s, \varepsilon) = \sup_{0 < t \neq t' \in T; d(x(t), x(t')) < \varepsilon} \frac{d(x(t + s), x(t' + s))}{d(x(t), x(t'))}. \tag{2.13}$$

2. For $s \in T$ with $s > 0$ we define

$$E(s) = \lim_{\varepsilon \to 0} E(s, \varepsilon). \tag{2.14}$$

3. Finally we define the *dispersion exponent*

$$E = \lim_{s \to \infty} \frac{1}{s} \ln(E(s)) \tag{2.15}$$

Definition 2.11 needs a few comments. Firstly, the function $E(s, \varepsilon)$ indicates the maximal factor by which the distance between two points of the evolution increases as time increases by s units, assuming that the initial distance between the two points does not exceed ε. This is a measure for the growth of the prediction errors in dependence on the prediction interval s. In the case where we predict with the help of the exact evolution operator, but on the basis of restricted precision of the initial state, ε indicates this lack of precision. In the case where we predict by comparison with the past, ε is the distance within which an arbitrary state is approximated by a 'past state.'

The function $E(s, \varepsilon)$ in (2.13) does not have to be finite. However, it can be proven that for a positively compact evolution of a system with a C^1-evolution operator, $E(s, \varepsilon)$ is finite for all positive values of s and ε. See Exercise 2.9.

Secondly, we already pointed at the phenomenon of saturation: prediction errors generally grow (linearly or exponentially) until saturation takes place. When the initial error decreases, saturation will occur later. Therefore, we can eliminate the effect of saturation by considering a fixed prediction interval s, restricting to smaller initial errors. This motivates taking the limit $E(s)$ in (2.14). The existence of this limit follows from the fact that $E(s, \varepsilon)$, for fixed values of s, is a nondecreasing function of ε.

---

[22] The latter is needed so that for all $\varepsilon > 0$ there exist $t' \neq t \in T$, with $0 < d(x(t), x(t')) < \varepsilon$.

It is not hard to show that for a quasi-periodic evolution the function $E(s)$ is bounded and has a positive infimum. Compare the exercises.

Thirdly, in the above example of a non (eventually) periodic or stationary evolution of the doubling map we have $E(s) = 2^s$, expressing exponential growth. This exponential growth is characteristic for a chaotic evolution. Whenever $E(s)$ increases approximately exponentially with s, we expect that the expression $s^{-1} \ln(E(s))$ does not tend to 0 as $s \to \infty$. In fact it can be proven that the limit E as taken in (2.15), always exists; again see the exercises. In the case of a positively compact evolution, where the evolution operator is of class C^1, the limit E takes on finite values. Also see the exercises. We call E the *dispersion exponent* of the evolution $x(t)$. It indicates how rapidly evolutions that start close together in the state space, may separate, or disperse, by time evolution.

Definition 2.12 (Chaos). Given is a dynamical (M, T, Φ) system as above, with M a complete metric space and Φ continuous. A positively compact evolution $x(t)$ of x is called *chaotic* if the dispersion exponent E (see (2.15)) is positive.

Remarks.

– We have to say that the definitions given here are not standard. Usually one defines chaotic dynamics in a way that uses information on all, or at least a large number of, evolutions. Presently we consider one single evolution, where the notion of predictability, and hence of chaos, should make sense. This is achieved by our definitions.
– Alternative definitions of chaos are named after Devaney and Milnor and others, but are not discussed here. For details see, for example, [4, 8, 168].
– The dispersion exponent is strongly related to the largest Lyapunov exponent; see Chapter 6, in particular §6.9. We note, however, that Lyapunov exponents cannot be defined in terms of one single evolution.

2.3.3 Properties of the dispersion exponent

We discuss a few properties of the dispersion exponent in relation to predictability. By adapting the time scale, the dispersion exponent can be 'normalised'. Indeed, if we scale the time t to $t' = \alpha^{-1} t$, the dispersion exponent changes to $E' = \alpha E$. Note that in the case where the time set is $T = \mathbb{Z}$, we get a new time set $T' = \alpha^{-1} \mathbb{Z}$. In this way we can always achieve that $E \leq 0, E = 1$, or $E = \infty$. For the possibility of $E < 0$ see Exercise 2.16. For chaotic evolutions we therefore find a 'natural' unit of time corresponding to the rate of increase of unpredictability as a function of the prediction interval s. We already noted that for a deterministic system, under a few simple conditions (involving differentiability of the evolution operator), always $E < \infty$. However, the dispersion exponent E also can be defined for nondeterministic motions, such as Brownian motions. In such cases, usually $E = \infty$. Granted some simplification we therefore can say that there are three kinds

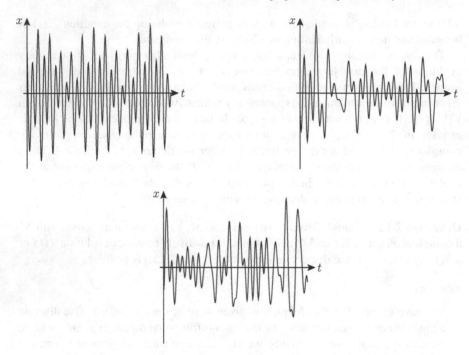

Fig. 2.6 Quasi-periodic evolutions with 3, 8, and 13 frequencies: in all cases a 1-dimensional projection of the evolution is displayed as a function of time.

of motion. The first kind consists of very predictable evolutions, like the stationary, periodic, and quasi-periodic ones, characterised by $E \leq 0$. The second kind is badly predictable, chaotic, but still deterministic and characterised by $0 < E < \infty$. The third kind finally is formed by nondeterministic random motions, characterised by $E = \infty$. In the latter case one also speaks of 'stochastic' motions.

The three classes of motion as just described are not strictly separated, but allow 'continuous transitions' from one class to the other. Without attempting formal mathematical precision, we now discuss a few examples of such 'continuous transitions'.

2.3.3.1 'Transition' from quasi-periodic to stochastic

By simple numerical experiments one can see that quasi-periodic motions with a large number of different frequencies do not look predictable at all. Such motions take place on a torus of large dimension. Compare Figure 2.6, where we displayed a 1-dimensional projection of evolutions with 3, 8, and 13 frequencies. This phenomenon is related to the fact that any reasonable function, on a finite interval, can be approximated by quasi-periodic functions, namely by truncated Fourier series. Compare, for example, [105, 200].

2.3.3.2 'Transition' from periodic to chaotic

There is yet another way to approximate an 'arbitrary' motion by periodic motions, namely by letting the period tend to ∞. When observing a motion for only a finite time interval, we never can conclude that the motion is not periodic. Evidently, it does not make sense to describe a motion that has not been observed to be periodic and for the periodicity of which there are no theoretical arguments, as periodic. This kind of uncertainty on whether the motion is chaotic or periodic occurs in many systems with chaotic evolutions. In the case of the doubling map, the points with a periodic evolution are dense in the state space, thus by a small change of the initial state a periodic evolution can be obtained. When restricting to a finite time interval, the approximation by a periodic initial state can be chosen in such a way that the difference between the original evolution and the periodic approximation is arbitrarily small. Yet, when the initial state is chosen randomly, there is no reason to assume that we will have a periodic evolution: although the periodic points form a dense subset, it has Lebesgue measure 0.

2.3.3.3 'Transition' from chaotic to stochastic

Let us now briefly discuss how a computer simulates systems governed by chance. In its simplest form, a sequence of numbers in $[0, 1]$ is produced that closely resembles the realisation of independent sampling from the uniform distribution. However, a computer is a deterministic system and therefore we have to cheat a bit. In principle this goes as follows. The first number is provided by the programmer, but the subsequent numbers are produced by application of a kind of doubling map, where the multiplication is not by a factor 2, but by a very large number N. So one uses iteration of the map $x \mapsto Nx \mod \mathbb{Z}$. For this deterministic dynamical system, most evolutions have a dispersion exponent $E = \ln(N)$. By taking N larger, we get a better approximation of the nondeterministic case where $E = \infty$. For sufficiently large N, in a graphical representation, the (obvious) lack of independence in these *pseudo-random numbers*, no longer is visible. Indeed, because each pixel of the screen corresponds to an interval of numbers, for sufficiently large N no deterministic relationship exists between the successive pixels representing the pseudo-random numbers.[23]

2.3.4 Chaotic evolutions in the examples of Chapter 1

Unfortunately, it often is difficult to show chaoticity of a given evolution. For the doubling map this was not the case: each point in $\mathbb{R}/\mathbb{Z}$ which is not rational is the initial state of a chaotic evolution.

---

[23] In fact the computer only computes with integers and not with real numbers. This gives rise to further complications, that we do not address here.

Also in the case of the logistic family $\Phi_\mu : x \mapsto \mu x(1-x)$ for $\mu = 4$, it is rather easy to prove that many states exist giving rise to chaotic evolutions. We show this now, illustrated by Figure 2.7. Consider the doubling map on the circle, now of the form [24]

$$\psi \mapsto 2\psi \bmod 2\pi \mathbb{Z}. \qquad (2.16)$$

The circle is considered as a subset of the plane, according to

$$\psi \leftrightarrow (\cos \psi, \sin \psi).$$

As indicated in Figure 2.7, we can project the doubling map dynamics on the interval $[-1, +1]$ of the horizontal axis. Projection amounts to the map

$$\psi \mapsto \cos \psi. \qquad (2.17)$$

The doubling map on the circle maps ψ to 2ψ, which in the horizontal interval corresponds to

$$\cos \psi \mapsto \cos 2\psi = -1 + 2\cos^2 \psi.$$

Expressed in the variable $s = \cos \psi$ this gives the map

$$s \mapsto 2s^2 - 1. \qquad (2.18)$$

The affine coordinate transformation

$$x = \tfrac{1}{2}(1 - s) \quad \text{or} \quad s = 1 - 2x \qquad (2.19)$$

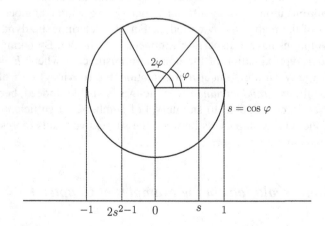

Fig. 2.7 The logistic map (for $\mu = 4$) as a 'projection' of the doubling map on the circle.

---

[24] We use 2π instead of 1 in order to get simpler formulæ.

takes (2.18) into the logistic map $\Phi_4 : x \mapsto 4x(1-x)$, as can be checked by a direct computation.

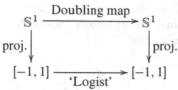

What to conclude from these considerations? First of all the coordinate transformation (2.19) conjugates the map (2.18) into the logistic map Φ_μ for $\mu = 4$. Such a conjugation evidently preserves all dynamical properties. Second the projection map (2.17) is a 2-to-1 map, forming a semiconjugation between the doubling map (2.16) and (2.18), the logistic map in disguise; see the above diagram. It is not hard to show that the semiconjugation preserves chaoticity, as well as the value of the dispersion exponent $E = \ln(2)$ for nonfinite orbits. See Exercise 2.11.

Note that the above argument regarding the logistic system only holds for $\mu = 4$. For other values of μ there is no proof like this, showing the existence of chaotic orbits, for example, see [104].

Regarding the other examples, we observe the following. For the free pendulum as well as for the free Van der Pol oscillation, no chaotic evolutions occur and the only possible evolutions are (asymptotically) stationary and periodic. For the driven pendulum and driven Van der Pol system, chaotic evolutions can occur. Around 1950, this phenomenon has cost mathematicians such as Cartwright, Littlewood, and Levinson a lot of headache. Eventually, during the decades between 1960 and 1990 many methods have been developed to prove that such systems do have chaotic evolutions. This breakthrough owes a lot to the geometrical methods introduced by S. Smale, in an area that for a long time was dominated by analysis. The paper [213] where these ideas have been exposed still is highly worth reading. Much of the later work has been summarised in [5,30,37,87]; also see the various handbooks [33–36].

Also regarding the logistic family, the Hénon family, the Lorenz attractor, and the Rössler attractor, it was clear that for many parameter values the numerical simulations looked chaotic. Still, it has taken a long time, sometimes more than 20 years, before these phenomena could be understood, at least partly, in terms of fully proven theorems. We here just refer to [66,67,114,234] and continue this discussion in Chapter 4, when more mathematical tools have become available.

2.3.5 Chaotic evolutions of the Thom map

We now turn to a new example, the so-called Thom map.[25] Also in this example it is not too hard to prove that many evolutions are chaotic. We include it here, because it is prototypical for a general class of dynamical system, the so-called Anosov

---

[25] Also called Arnold's Cat map; compare [2], page 120, Figure 81.

systems (e.g, [8]). The state space is the 2-dimensional torus $\mathbb{T}^2 = \mathbb{R}^2/\mathbb{Z}^2$ and the time set is $\mathbb{Z}$. So the dynamics is generated by an invertible map $\mathbb{T}^2 \to \mathbb{T}^2$. This map is constructed as follows. We take a 2×2-matrix A with integer coefficients, with $|\det(A)| = 1$ and where the absolute values of the eigenvalues are not equal to 1. As a leading example we take

$$A = \begin{pmatrix} 2 & 1 \\ 1 & 1 \end{pmatrix}. \tag{2.20}$$

The matrix A as in (2.20) defines an invertible transformation of the plane. The coefficients are integers, therefore A induces a map $\varphi : \mathbb{T}^2 \to \mathbb{T}^2$. Because $\det(A) = \pm 1$, the matrix A^{-1} also has integer coefficients. The torus map induced by A^{-1} then is the inverse of φ. We now discuss a number of properties of the map φ, from which it eventually follows that the dynamics contains chaos.

First of all we note that the point $[(0,0)]$ [26] is a fixed point of φ. For the derivative we have

$$d\varphi_{[(0,0)]} = A.$$

A is symmetric, therefore its eigenvalues $\lambda_{1,2}$ are real. Moreover, it easily follows that

$$0 < |\lambda_1| < 1 < |\lambda_2|.$$

This implies that there are stable and unstable separatrices; see §1.3.2. Our main interest is with the unstable separatrix, denoted by W^u.

The fixed point $[(0,0)] \in \mathbb{T}^2$ corresponds to the fixed point $(0,0) \in \mathbb{R}^2$ of the linear map of the plane given by the matrix A. The unstable separatrix of this fixed point is a straight line, thus of the form

$$\{(\alpha_1 s, \alpha_2 s) \mid s \in \mathbb{R}\},$$

where (α_1, α_2) is an eigenvector of A for the eigenvalue λ_2. The unstable separatrix W^u then is the projection of this line to the torus. The successive intersections of W^u with the circle

$$\{[(x_1, x_2)] \mid x_2 = 0\} \subset \mathbb{T}^2,$$

are given by

$$\ldots, \left[\left(-2\frac{\alpha_1}{\alpha_2}, 0\right)\right], \left[\left(-\frac{\alpha_1}{\alpha_2}, 0\right)\right], [(0,0)], \left[\left(\frac{\alpha_1}{\alpha_2}, 0\right)\right], \left[\left(2\frac{\alpha_1}{\alpha_2}, 0\right)\right], \ldots \tag{2.21}$$

We have seen such sequences of points earlier when studying translation systems on the 1-dimensional torus or circle $\mathbb{T}^1 = \mathbb{S}^1$; compare §2.2.2.1. The main point is whether the ratio α_1/α_2 is rational or irrational. We have the following.

---

[26] Again $[-]$ takes the equivalence class mod $\mathbb{Z}^2$; for a while we adopt this notation.

Proposition 2.13 (Thom map has dense separatrix). *The ratio* $\alpha_1 : \alpha_2$ *is irrational and* $\overline{W^u} = \mathbb{T}^2$.

Proof. The case that α_1/α_2 is rational leads to a contradiction as follows. Indeed, the assumption that α_1/α_2 is rational implies that W^u is a closed curve, and hence a circle. Now by the definition of unstable separatrix we have $\varphi(W^u) = W^u$, where the restriction $\varphi|_{W^u}$ magnifies the distance between (nearby) points with a factor $|\lambda_2| > 1$. This contradicts the fact that φ is a diffeomorphism. We conclude that the ratio α_1/α_2 is irrational, whence by Lemma 2.4 it follows that the intersections

$$W^u \cap \{[(x_1, x_2)] \mid x_2 = 0\}$$

(i.e., the set of points (2.21)) are dense in the circle $\{[(x_1, x_2)] \mid x_2 = 0\}$. This directly implies that W^u is dense in $\mathbb{T}^2$. $\qquad\square$

Definition 2.14 (Topologically mixing). We call $f : M \to M$ *topologically mixing* if for any two open (nonempty) subsets $U, V \subset M$ there exists $n_0 \in \mathbb{N}$, such that for all $n > n_0$

$$f^n(U) \cap V \neq \emptyset.$$

Proposition 2.15 (Thom map topologically mixing). *The Thom map* $\varphi : \mathbb{T}^2 \to \mathbb{T}^2$ *is topologically mixing.*

Proof. Let $U, V \subset \mathbb{T}^2$ be given as above. In U we take a segment $\mathcal{L}$ parallel to W^u and choose $\varepsilon > 0$, such that V contains an ε-neighbourhood of one of its points. Application of φ^n to $\mathcal{L}$ yields a segment, still parallel to W^u, but $|\lambda_2|^n$ times as long as $\mathcal{L}$. This segment $\varphi^n(\mathcal{L})$, up to a translation in $\mathbb{T}^2$ is a segment of the same length inside W^u. We know that W^u is dense in $\mathbb{T}^2$. Therefore, a sufficiently long segment is within ε-distance of any point in $\mathbb{T}^2$. (The required length of the segment evidently depends on ε.) This means that $\varphi^n(\mathcal{L})$, for n sufficiently large, also is within ε-distance of any point of $\mathbb{T}^2$. Let $n_0 \in \mathbb{N}$ be such that this property holds for all $n > n_0$. By the choice of ε it now follows that

$$(\varphi^n(U) \cap V) \supset (\varphi^n(\mathcal{L}) \cap V) \neq \emptyset,$$

which proves that the dynamics of φ is topologically mixing. $\qquad\square$

Proposition 2.16 (Chaotic evolutions in Thom map). *The* φ*-dynamics has many* [27] *evolutions that are dense in* $\mathbb{T}^2$. *The dynamics of such evolutions is chaotic, with dispersion exponent* $E = \ln(|\lambda_2|)$.

Proof. We first generalise the argument in the proof of Proposition 2.15 as follows. If $V_1, V_2, \ldots$ is a countable sequence of open nonempty subsets of $\mathbb{T}^2$, then there

---

[27] The corresponding set of initial values is *residual*; see Exercise 2.15 and Appendix A.

exists a sequence of open subsets $U_1 \supset U_2 \supset \cdots$ of $\mathbb{T}^2$ and a sequence of natural numbers $N_1 < N_2 < \cdots$, such that

$$\varphi^{N_j}(U_j) \cap V_j \neq \emptyset \quad \text{and} \quad \overline{U}_{j+1} \subset \varphi^{-N_j}(V_j) \cap U_j,$$

for all $j \in \mathbb{N}$. This follows from the above by induction on $j \in \mathbb{N}$. We note that in these circumstances

$$\bigcap_{j \in \mathbb{N}} \overline{U}_j \neq \emptyset$$

and, writing $Z = \cap_j \overline{U}_j$, we observe that each evolution starting in Z contains at least one point of V_j, for each $j \in \mathbb{N}$.

Now to show that the φ-dynamics has a dense evolution, we take a countable sequence $V_1, V_2, \ldots$ of open sets as follows. Using an enumeration of $\mathbb{Q}^2$, we can arrange that the V_j, $j \in \mathbb{N}$, run through all ε-neighbourhoods of points of $\mathbb{Q}^2$, with $\varepsilon \in \mathbb{Q}$. Following the above construction it easily follows that each element of the set Z is the initial state of an evolution that is dense in $\mathbb{T}^2$.

We next show that any evolution that is dense in $\mathbb{T}^2$ is chaotic with dispersion exponent $E = \ln(|\lambda_2|)$. Here it is convenient not to work with the usual metric on $\mathbb{T}^2$, but to take a metric that fits better to the Euclidean metric of the plane:

$$d(([x_1, y_1], [(x_2, y_2)]) = \min_{n, m \in \mathbb{Z}} \sqrt{(x_1 - x_2 + n)^2 + (y_1 - y_2 + m)^2}.$$

Now, first observe that in the plane, by application of the linear map A, the distance between two points is multiplied by a factor that is at most $|\lambda_2|$. The maximal multiplication factor only occurs if the line connecting the two points lies in the direction of the eigenspace associated with λ_2. This multiplication factor depends continuously on the direction of this connecting line. All these facts follow from linear algebra. What happens in the plane we also witness on $\mathbb{T}^2$, as long as no saturation takes place. To have maximal dispersion we surely have to take segments of the evolution, for which not only the initial points are nearby, but also for which at the end the corresponding points are still close together so that no saturation is in order. Moreover, the initial points of the segments have to be chosen such that the connecting line is parallel, or approximately parallel, to W^u. Because our evolution is dense in $\mathbb{T}^2$, such a choice is possible. In this way we get two segments $x(n), \ldots, x(n + s)$ and $x(m), \ldots, x(m + s)$ of the evolution, where

$$\limsup_{s \to \infty} \frac{d([x(n + s)], [x(m + s)])}{d([x(n)], [x(m)])} \leq |\lambda_2|^s.$$

In fact this expression is less than or equal to, but gets arbitrarily close to $|\lambda_2|^s$. From this it easily follows that $E = \ln(|\lambda_2|)$ and hence that the evolution is chaotic.

$\square$

Remark. Mutatis mutandis the above considerations also go through for any other matrix $A \in \mathrm{GL}(n, \mathbb{Z})$, the group of invertible $n \times n$-matrices with integer entries which are hyperbolic, that is, with the absolute values of the eigenvalues different from 1.

2.4 Exercises

We recall that for all dynamical systems (M, T, Φ) in this chapter it is assumed that the state space M is a topological space and that the evolution operator Φ : $M \times T \to M$ is continuous.

Exercise 2.1 (Negative dispersion). Consider the dynamical system with state space $M = \mathbb{R}$, given by the equation of motion

$$x' = -x, \quad x \in \mathbb{R}.$$

Prove that in this case, for any nonstationary evolution we have $E = -1$. Generalise this to the case of arbitrary linear systems.[28]

Exercise 2.2 (Finitely many digits and doubling map). Consider the symbolic representation of the doubling map. Show that any expansion with finitely many digits is an eventual fixed point.

Exercise 2.3 (A dense set of periodic points). We consider the dynamical system generated by the doubling map $x \mapsto 2x \bmod \mathbb{Z}$, with state space $[0, 1)$. Show that the evolution with initial state x is eventually periodic (or stationary) if and only if x is rational. (Hint: A real number is rational if and only if its decimal (and also its binary) expansion in the long run is periodic.)

Exercise 2.4 (Prime period not defined). We consider a dynamical system, the state space of which is equal to $\mathbb{R}/\mathbb{Q}$, the time set $\mathbb{R}$, and with an evolution operator $\Phi([t], s) = [t + s]$, where $[\,]$ indicates that we take the equivalence class mod $\mathbb{Q}$. Show that in this system each point is periodic, but that for no point the prime period is well defined. See Appendix A for a definition of quotient topology, and show that Φ is continuous.

Exercise 2.5 (Some algebra). In the text we frequently used the following fact from algebra. If $k_1, k_2, \ldots, k_n \in \mathbb{Z}$ are relatively prime, then there exist $m_j \in \mathbb{Z}$, $1 \le j \le n$, such that

$$\sum_{j=1}^{n} m_j k_j = 1.$$

---

[28] A similar result holds in general for hyperbolic point attractors (treated in a later chapter). These are cases where the prediction errors do not increase when the prediction interval s increases.

Prove this. (Hint: Consider the subgroup $G \subset \mathbb{Z}$ generated by $k_1, k_2, \ldots, k_n$. Any such subgroup is of the form $G_K = K\mathbb{Z}$, for some $K \in \mathbb{N}$, thus the aim is to show that $K = 1$. Now use that if $G = G_K$, then K is a common factor of $k_1, k_2, \ldots, k_n$.)

Exercise 2.6 (Resonant subtori). In §2.2.2 the resonance case was described for translation systems on a torus, meaning rational dependence of the components of the translation vector (possibly including 1). It was indicated in principle that the evolution starting at [0] lies in a lower-dimensional torus, or in a (finite) union of such lower-dimensional tori. Moreover, the restriction to this torus (or to the torus containing [0] in the case of several tori) again is a translation system. Below we give three examples of a translation system on the 2-torus. In all cases there is resonance. The problem is to determine the 1-torus (or the union of 1-tori) in which the evolution lies which starts at [0], to prove that the evolution is dense in the 1-torus (or tori) and, finally, to determine the translation system restricted to the 1-torus that contains [0].

  i. The time set is $\mathbb{R}$, the translation vector $\Omega = (\frac{1}{2}, \frac{1}{3})$.
 ii. The time set is $\mathbb{Z}$, the translation vector $\Omega = ((\sqrt{2}/3), (\sqrt{2}/4))$.
iii. The time set is $\mathbb{Z}$, the translation vector $\Omega = ((1/8) + (\sqrt{2}/5), (\sqrt{2}/4))$.

Exercise 2.7 (A conjugation). We mentioned that the suspension of a translation system with time set $\mathbb{Z}$ and translation vector $\Omega = (\omega_1, \ldots, \omega_n)$ 'is the same as' the translation system with time set $\mathbb{R}$ and translation vector $\bar{\Omega} = (\omega_1, \ldots, \omega_n, 1)$. This 'being the same' means that the systems are conjugated; see §2.2.3. In terms of the triple 'state space, time set, evolution operator' we write these two systems as $(S_\Omega(\mathbb{T}^n), \mathbb{R}, S\Phi_\Omega)$ and $(\mathbb{T}^{n+1}, \mathbb{R}, \Phi_{\bar{\Omega}})$. The problem now is to construct a conjugation between these two systems, in other words, a homeomorphism $h : S_\Omega(\mathbb{T}^n) \to \mathbb{T}^{n+1}$ satisfying the conjugation equation

$$h(S\Phi_\Omega(p, t)) = \Phi_{\bar{\Omega}}(h(p), t)$$

for each $p \in S_\Omega(\mathbb{T}^n)$ and $t \in \mathbb{R}$.

Exercise 2.8 (Quasi-periodic is not chaotic). Show that a quasi-periodic evolution has dispersion exponent $E = 0$.

Exercise 2.9 (Finite dispersion exponent). Consider a positively compact evolution $x(t)$ of a dynamical system $(\mathbb{R}^n, T, \Phi)$, where for all $s \in T$ the time s map Φ^s, defined by $\Phi^s(x) = \Phi(x, s)$, is at least of class C^1. Show that in this case the value of $E(s, \varepsilon)$, as defined in §2.3.2, is finite. (Hint: Use $\max_v (\|d\Phi^s(v)\|/\|v\|)$.)

Exercise 2.10 (Dispersion in equivalent metrics). In the definition of the dispersion exponent E we make use of a metric. Show that if another, but equivalent,[29] metric is used, the value of E does not change.

---

[29] For this use of terminology see Appendix A.

Exercise 2.11 (The logistic and the doubling map). Consider the logistic map
$\Phi_4 : x \in [0, 1] \mapsto 4x(1 - x) \in [0, 1]$.

1. Show that Φ_4 is conjugated to the map $s \in [-1, 1] \mapsto 2s^2 - 1 \in [-1, 1]$.
2. From the fact that the latter map is semiconjugated to the doubling map on the circle, prove that nonfinite orbits of the logistic map Φ_4 are chaotic, with dispersion exponent $E = \ln(2)$.

Exercise 2.12 (Existence of a limit). Show that for the function $E(s)$, as defined in §2.3.2, we have

$$E(s_1 + s_2) \le E(s_1) \cdot E(s_2).$$

Using this, show that the limit

$$E = \lim_{s \to \infty} \frac{1}{s} \ln(E(s))$$

exists.

Exercise 2.13 (Conjugation of translation systems). On the 2-torus $\mathbb{T}^2$ consider the translation vectors Ω_1 and Ω_2. Moreover $\varphi : \mathbb{T}^2 \to \mathbb{T}^2$ is a linear diffeomorphism induced by a matrix $A \in GL(2, \mathbb{Z})$ as in §2.3.5. This means that the 2×2-matrix A is invertible, and both A and A^{-1} have integer entries. Show that

$$\varphi \circ R_{\Omega_1} = R_{\Omega_2} \circ \varphi$$

is equivalent to

$$\Omega_2 = A\Omega_1,$$

where R_{Ω_j} is the translation on $\mathbb{T}^2$ determined by Ω_j. Also show that if these equalities hold, then it follows that φ is a conjugation between the translation systems defined by Ω_1 and Ω_2.

NB: The components of a translation vector can be viewed as the frequencies of 'composed periodic motions.' These frequencies are not necessarily preserved under conjugation.

Exercise 2.14 ((∗) Periodic points of Thom map). Show that the set of points in $\mathbb{T}^2$ that are periodic under the Thom map, coincides with the set of points with rational coordinates. For this it is not necessary that the matrix A have the same form as in the example of § 2.3.5. The only point of interest is that A has integer coefficients and that $\det(A) = \pm 1$, where none of the eigenvalues is a root of unity. Also try to formulate and prove this result in higher dimension.

Exercise 2.15 ((∗) Many chaotic initial states of Thom map). Show that for the Thom map A as in (2.20) the set of initial states having a chaotic evolution is residual.[30]

---

[30] For this use of terminology see Appendix A.

Exercise 2.16 ((∗) **Orderly dynamical behaviour for negative dispersion exponent**). Let $x(t)$ be a positively compact evolution of a dynamical system (M, T, Φ), where M is a complete metric space and Φ continuous. Moreover we assume that $x(t)$ is not (eventually) periodic. Show that if the dispersion exponent E of this evolution is negative (i.e., $E < 0$), then the evolution is either asymptotically periodic or stationary. In the case of time set $\mathbb{R}$ the evolution necessarily is asymptotically stationary.

Exercise 2.17 ((∗) **An example of eventual periodicity**). Construct a dynamical system with time set $\mathbb{R}_+$ with an eventually periodic evolution.

Exercise 2.18 ((∗∗) ε-**Dense orbits**). In §2.2.4 we mentioned that in a quasi-periodic translation system, for each positive ε there exists $T(\varepsilon) \in \mathbb{R}$, such that each point of the torus is within ε-distance of the segment $\{x(t)\}_{t \in T \cap [0, T(\varepsilon)]}$ of an evolution. To make the definition unique, we take $T(\varepsilon)$ minimal. Now consider the case of an irrational rotation $[x] \mapsto [x + \omega]$ on the circle $\mathbb{R}/\mathbb{Z}$. The function $T(\varepsilon)$ also depends on ω (and is only defined for irrational ω).

1. Show that for each ω we have $T(\varepsilon) > \frac{1}{2}\varepsilon^{-1}$.
2. Show that there is a ratio ω such that

$$\frac{T(\varepsilon)}{\varepsilon}$$

   is uniformly bounded for all $\varepsilon > 0$.
3. Show that for certain ω

$$\limsup_{\varepsilon \to 0} \frac{T(\varepsilon)}{\varepsilon} = \infty.$$

(Hint: Use continued fractions.)

Chapter 3
Persistence of dynamical properties

When interpreting observed data of a dynamical system, it is of the utmost importance that both the initial state and the evolution law are only known to an acceptable approximation. For this reason it is also good to know how the type of the evolution changes under variation of both. When the type does not change under small variations of the initial state, we speak of *persistence* under such variation and call the corresponding evolution *typical* for the system at hand.

Next to this persistence of properties of individual evolutions, persistence of properties of the entire dynamics of the system is also discussed. Here we speak of persistence under perturbations of the evolution law. Such properties are often also called typical or robust.

In the 1960s and 1970s René Thom [1] and Stephen Smale [2] pointed at the great interest of persistent properties of dynamical system that serve as robust models for concrete systems [213, 231].

3.1 Variation of initial state

To fix our thoughts, we first consider a periodic evolution of the free undamped pendulum; see Chapter 1, in particular §1.1.1, in the case where the pendulum oscillates (i.e., is not turning over). This happens for values of the energy

$$H(x, y) = m\ell^2 \left(\frac{1}{2} y^2 - \omega^2 \cos x \right).$$

in the open interval $(-m\ell^2\omega^2, +m\ell^2\omega^2)$. It should be clear that then also evolutions with an initial state that is sufficiently near this oscillating evolution are periodic in this same way, that is, oscillating and not turning over. Compare Figure 1.2. The dynamical property of being periodic without turning over for this reason is persistent under variations of the initial state. We also say that such evolutions are typical

---

[1] René Thom 1923–2002.

[2] Stephen Smale 1930–.

H.W. Broer and F. Takens, *Dynamical Systems and Chaos*,
Applied Mathematical Sciences 172, DOI 10.1007/978-1-4419-6870-8_3,
© Springer Science+Business Media, LLC 2011

for this dynamical system. We observe, however, that as the energy of the oscillation gets closer to the limit values $\pm m\ell^2\omega^2$, the allowed perturbations of the initial state such that these do not affect the dynamical property at hand, are smaller. (In fact, in §1.1.1 we saw that for the energy $-m\ell^2\omega^2$ the pendulum hangs vertically down and that for the energy $+m\ell^2\omega^2$ the pendulum either points up vertically, or carries out a motion in which for $t \to \pm\infty$ it approaches the vertically upward state.)

Analogous remarks can be made with respect to persistence of the periodic evolutions in which the pendulum goes 'over the top,' either to the right, or to the left. On the other hand, for the remaining evolutions the characterising properties, such as being stationary or of approaching a stationary state, are not persistent under variation of the initial state.

Although a general definition of an attractor is treated in the next chapter, we wish to indicate a connection here between persistence and attractors. This is done for the example of the damped free pendulum

$$x'' = -\omega^2 \sin x - cx',$$

$\omega \neq 0, c > 0$, discussed in §1.1.1. Here the stationary evolution $(x, x') = (0, 0)$, where the pendulum is hanging down, is a point attractor, inasmuch as all evolutions starting sufficiently nearby approach this state. Generally, any attractor $A \subset M$, defined below, is required to have a neighbourhood $U \subset M$, such that all evolutions with initial state in U converge to A. Therefore the convergence to A is a persistent property. For more details see the next chapter.

Definition 3.1 (Persistence). Let (M, T, Φ) be a dynamical system, with M a topological space and Φ continuous. We say that a property P of an evolution $x(t)$ with initial state $x(0)$ is *persistent under variation of initial state* if there is a neighbourhood U of $x(0)$, such that each evolution with initial state in U also has property P.

Apart from the concept of persistence just defined, weaker forms are also known, and described now with the help of examples.

We begin considering the dynamics generated by the doubling map; see §1.3.8. We have seen that this system has both many chaotic and many (eventually) periodic evolutions. To some extent the chaotic evolutions form the majority: from the second proof of the existence of such evolutions (see §1.3.8), it follows that when picking the initial state at random (with respect to Lebesgue measure) on $M = [0, 1)$, the corresponding evolution with probability 1 is chaotic. On the other hand, we know that the initial states of the (eventually) periodic evolutions form a dense subset of $M = [0, 1)$. Therefore we cannot say that it is a persistent property of evolutions of the doubling map dynamics to be chaotic: with an arbitrarily small perturbation of the initial value, we can achieve (eventual) periodicity. Life changes when we decide to neglect events of probability 0. Thus we say that for evolutions of the doubling map chaoticity is *persistent under variation of initial states, in the sense of probability*. Note that for this concept of persistence, we make use of a probability measure. In many cases it is not obvious which probability measure should be used.

However, this is not as bad as it sounds, because we only use the concept of zero measure sets, of *null sets*. Often all obvious probability measures have the same null sets. Compare Appendix A.

Definition 3.2 (Persistence in the sense of probability). Let (M, T, Φ) be a dynamical system with M a topological space with a Borel measure and Φ both continuous and measurable.[3] We say that a property P of an evolution $x(t)$ with initial state $x(0)$ is persistent under variation of initial state *in the sense of probability*, if there is a neighbourhood U of $x(0)$ with a null set $Z \subset U$, such that each evolution with initial state in $U \setminus Z$ also has property P.

The above example, where, due to the abundance of periodic orbits, chaoticity was only persistent in the sense of probability, is not exceptional. However, see Exercise 3.7. An even weaker form of persistence is known for quasi-periodic evolutions. Again we first consider an example, namely the driven undamped pendulum; see §1.1.2. We are dealing with a system with equation of motion

$$x'' = -\omega^2 \sin x + \varepsilon \cos(2\pi\omega t).$$

As in the case of the driven Van der Pol system, dealt with in §1.3.1 to illustrate quasi-periodicity, quasi-periodic evolutions at occur for this system for small values of ε. To explain this further, we turn here to an autonomous system of first order

$$
\begin{aligned}
x' &= y \\
y' &= -\omega^2 \sin x + \varepsilon \cos(2\pi z) \\
z' &= \omega,
\end{aligned}
$$

where both x and z are variables in $\mathbb{R}/\mathbb{Z}$. If in this system we set $\varepsilon = 0$, thereby essentially just considering the free undamped pendulum, we find a 1-parameter family of multiperiodic evolutions. Indeed, there is one for each $(x(0), y(0))$ on an oscillation of the free undamped pendulum, thus corresponding to values of the energy in between $-m\ell\omega^2$ and $+m\ell\omega^2$, and for an arbitrary $z(0)$ the evolution $(x(t), y(t), z(t))$ is multiperiodic; compare Figure 1.2 and the discussion at the beginning of this section. Each of these multiperiodic subsystems has two frequencies: the frequency in the z-direction equals ω and the other frequency decreases with the amplitude. The latter is the frequency of the corresponding free oscillation in the (x, y)-plane.

From KAM theory as briefly discussed in §2.2.5 (also compare Chapter 5), Appendix B, and the reviews [37, 87], it follows that if ε is sufficiently small, there are still many quasi-periodic subsystems present. In fact, they are so many that their union has positive Lebesgue measure. This means that, when randomly choosing the initial condition, we have positive probability to land on a quasi-periodic evolution. In this case we speak of *weak persistence*.

---

[3] We note that continuous does not imply Borel measurable; see Appendix A.

Definition 3.3 (Weak persistence). Let (M, T, Φ) be a dynamical system with M a topological space with a Borel measure and Φ both continuous and measurable.[4] We say that a property P of an evolution $x(t)$ with initial state $x(0)$ is *weakly persistent* under variation of initial state, if for each neighbourhood U of $x(0)$ there exists a subset $Z \subset U$ of positive measure, such that each evolution with initial state in Z also has property P.

To explain that for the driven undamped pendulum many quasi-periodic evolutions occur in a weakly persistent way, we have to resort to the Lebesgue density theorem [131], which we state here for the sake of completeness. This theorem deals with *density points* of a (measurable) subset $Z \subset \mathbb{R}^n$ of positive Lebesgue measure. We say that $x \in Z$ is a density point of Z if

$$\lim_{\varepsilon \to 0} \frac{\lambda(Z \cap D_\varepsilon(x))}{\lambda(D_\varepsilon(x))} = 1,$$

where $D_\varepsilon(x)$ is a ε-neighbourhood of x and where $\lambda(V)$ denotes the Lebesgue measure of V.

Theorem 3.4 (Lebesgue density theorem [131]). *If $Z \subset \mathbb{R}^n$ is a Lebesgue measurable set with $\lambda(Z) \neq 0$, then almost every point of Z, in the sense of Lebesgue measure, is a density point of Z.*

The conclusion of Theorem 3.4 implies that the property 'belonging to Z' for almost every point of Z is weakly persistent. Also see Appendix A.

3.2 Variation of parameters

In many of the examples of dynamical systems we met in Chapter 1 several constants occur. In several variations of the pendulum we saw ε, c, ω, and Ω, for the logistic system this was μ and for the Hénon system a and b. The questions we presently ask are of the form: which properties of the dynamics are persistent under (small) variation of these parameters. Again persistence may hold in various stronger or weaker forms and again we demonstrate this with the help of examples.

As a first example we consider the damped free pendulum; see §1.1.1. Recall that this system is governed by the following equation of motion,

$$x'' = -\omega^2 \sin x - cx',$$

where $\omega \neq 0$ and $c > 0$. For each choice of these parameters ω and c, within the given restrictions, the typical evolution is asymptotically stationary to the lower equilibrium (which is unique once we take the variable x in $\mathbb{R}/(2\pi\mathbb{Z})$). Within this setting it therefore is a persistent property of the dynamics to have exactly one point attractor.

---

[4] See Appendix A.

A warning is in order here. It is important to know exactly what the setting is in which persistence holds or does not hold. Indeed, for the undamped free pendulum, with equation of motion

$$x'' = -\omega^2 \sin x,$$

it is a persistent property that the typical evolution is periodic: this holds for all values of $\omega \neq 0$. However, if we would consider the latter system as a special case of the damped free pendulum occurring for $c = 0$, then the property of typical evolutions being periodic would not be persistent, inasmuch as the least damping would put an end to this.

An analogous remark goes with respect to quasi-periodicity, discussed in the previous subsection as an example of a weakly persistent property under variation of initial state. Indeed, quasi-periodicity only occurs in the driven pendulum without damping.[5] For a proof of this, based on a general relation between divergence and contraction of volume, see Exercise 3.2.

An example of weak persistence under variation of parameters we find with respect to quasi-periodic dynamics in the driven Van der Pol system

$$x'' = -x - x'(x^2 - 1) + \varepsilon \cos(2\pi\omega t);$$

see §2.2.3. As mentioned there, for all values of ε, with $|\varepsilon|$ sufficiently small, there is a set of ω-values of positive Lebesgue measure, where this system has a quasi-periodic attractor. With the help of the Fubini theorem [131], it then follows that in the (ω, ε)-plane, the parameter values for which the driven Van der Pol system has a quasi-periodic attractor, has positive Lebesgue measure. This leads to weak persistence, under variation of the parameter(s), of the property that the system has a quasi-periodic attractor. Also see Chapter 5 and Appendix B.

We next discuss persistence (i.e., persistent occurrence) of chaotic evolutions in a number of the examples discussed in Chapter 1. In these cases, in general, persistence is hard to prove and we confine ourselves to giving references to the literature. First consider the logistic system given by

$$\Phi_\mu(x) = \mu x (1 - x)$$

(see §1.3.3), with state variable $x \in [0, 1]$ and where $\mu \in [2.75, 4]$ is a parameter. It has been proven [148], that for a random choice of both x and μ, the evolution is chaotic with positive probability. As a consequence of the Lebesgue density theorem 3.4, we have here an example of weak persistence of chaoticity. However, the evolution is also asymptotically periodic or stationary with positive probability, in which case it is not chaotic. Persistence of asymptotically stationary evolutions is easy to prove for $\mu \in [0, 3]$. The bifurcation diagram of the logistic system, as sketched in

---

[5] At least if negative damping as in the Van der Pol oscillator is excluded.

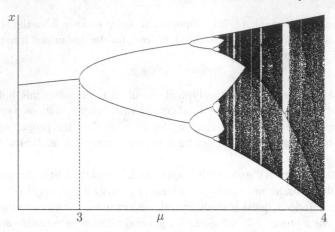

Fig. 3.1 Bifurcation diagram of the logistic system; see §1.3.3. For the various (horizontal) values of $\mu \in [2.75, 4]$, asymptotic x-values are plotted (vertically) as these occur after many iterations (and after deleting transient values). For the μ-values with only a finite number of x-values, we have asymptotic periodicity. For the μ-values for which in the x-direction whole intervals seem to occur, we probably are dealing with chaotic dynamics.

Figure 3.1, illustrates the extremely complex and chaotic dynamics well. Also for the Hénon system (see §1.3.2) given by

$$\Phi_{a,b}(x, y) = (1 - ax^2 + y, bx),$$

one has been able to prove that the occurrence of chaotic evolutions is weakly persistent under variation of the parameters a and b. Again it is true that for a 'random choice' of a, b, x, and y, the evolution generated by $\Phi_{a,b}$ with initial state (x, y) is chaotic with positive probability; the same statement also holds when 'chaotic' is replaced by 'asymptotically periodic.' For a proof, which for this case is much more complicated than for the logistic system, we refer to [67, 114]. There is, however, an important perturbative connection between the Hénon and the logistic system, in which the latter occurs as a special case for $b = 0$. As a consequence, the set of (a, b)-values for which the Hénon system is known to be chaotic, is contained in a small strip around the line $\{b = 0\}$. In Chapter 4, §4.2 we give more details on this perturbative setting.

Remark. The Hénon attractor as introduced in §1.3.2 occurs for $a = 1.4$ and $b = 0.3$. This value of b probably is far too large to be covered by the perturbative approach of [67, 114]. So it is not clear whether the attractor shown in Figure 1.13 is really chaotic or just a periodic attractor of extremely large period.

Finally, we briefly mention the Rössler attractor; see §1.3.7. We here expect the same situation as for the logistic and the Hénon system. Although there are good grounds for such a conjecture, no proof is known yet.

3.3 Persistence of stationary and periodic evolutions

We show here that stationary and periodic evolutions which satisfy a simple extra condition, are persistent, in the sense that such evolutions persist under small perturbations of the evolution law. We note, however, that the position in the state space, as well as the period, may vary.

3.3.1 Persistence of stationary evolutions

We start with stationary evolutions of dynamical systems with discrete time. Such evolutions correspond exactly to the fixed points of the map that determines the time evolution. We first give an example from which we see that not all fixed points persist. This problem is directly related to the implicit function theorem.

Example 3.1 (A 'saddle-node bifurcation'). An example of a nonpersistent fixed point we find in the map

$$\varphi_\mu : \mathbb{R} \to \mathbb{R}, \quad \text{with} \quad x \mapsto x + x^2 + \mu,$$

where $\mu \in \mathbb{R}$ is a parameter. For $\mu < 0$, the map φ_μ has two fixed points, for $\mu = 0$ one fixed point, and for $\mu > 0$ no fixed point; compare Figure 3.2.

We conclude that the fixed point $x = 0$ of φ_0 is not persistent. It turns out that in this and in similar examples, in such a nonpersistent fixed point the derivative of the map at the fixed point has an eigenvalue equal to 1. Indeed, in the example we see that $d\varphi_0(0) = 1$.

Remark. In situations such as Example 3.1, where for a special parameter value the number of stationary points changes, we speak of a *bifurcation*, in this case it is a saddle-node bifurcation. For more details see the Appendices C and E.

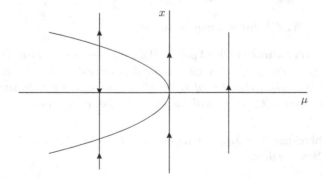

Fig. 3.2 Bifurcation diagram of the 'saddle node bifurcation' $x \mapsto x + x^2 + \mu$. The μ-axis is horizontal and the x-axis vertical. The stationary points are indicated by a solid line in the (μ, x)-plane. The arrows indicate the direction in which the nonstationary evolutions move. Compare Appendix C.

Theorem 3.5 (Persistence of fixed points I). *Let* $\varphi_\mu : \mathbb{R}^n \to \mathbb{R}^n$ *be a family of maps, depending on a parameter* $\mu \in \mathbb{R}^p$, *such that* $\varphi_\mu(x)$ *is differentiable in* (μ, x). *Let* x_0 *be a fixed point of* φ_{μ_0} (*i.e., with* $\varphi_{\mu_0}(x_0) = x_0$). *Then, if the derivative* $\mathrm{d}\varphi_{\mu_0}(x_0)$ *has no eigenvalue* 1, *there exist neighbourhoods* U *of* $\mu_0 \in \mathbb{R}^p$ *and* V *of* $x_0 \in \mathbb{R}^n$, *as well as a differentiable map* $g : U \to V$ *with* $g(\mu_0) = x_0$, *such that* $x = g(\mu)$, *for each* $\mu \in U$, *is a fixed point of* φ_μ, *that is, such that* $\varphi_\mu(g(\mu)) = g(\mu)$. *Moreover, for each* $\mu \in U$, *the map* φ_μ *in* V *has no other fixed points.*

Remark. Theorem 3.5 precisely states in which sense fixed points are persistent in parametrised systems. Indeed, under 'small' variation of the parameter μ (within U) the position of the fixed point $x = g(\mu)$ changes only 'slightly' (within V), because the map g is smooth and therefore surely continuous.

Proof. Define the map $F : \mathbb{R}^{p+n} \to \mathbb{R}^n$ by $F(\mu, x) = \varphi_\mu(x) - x$. Then fixed points of φ_μ correspond to zeroes of $F(\mu, x)$. Moreover, the derivative $\partial_x F(\mu_0, x_0)$, as a function of x, (i.e., for the fixed value $\mu = \mu_0$), does not have an eigenvalue 0 and therefore is an invertible $n \times n$-matrix. Then Theorem 3.5 immediately follows from the implicit function theorem (compare, e.g., [105]). $\qquad\qquad\square$

A slightly different formulation of the persistence of fixed points is possible (again without the derivative having an eigenvalue 1). Here we wish to speak of arbitrary perturbations of a map φ. For a precise formulation of such a persistence result, we have to know which perturbations of φ should be considered 'small'. We do not enter into the question of which kinds of topologies can be put on the space of differentiable maps; again see Appendix A. For our present purposes it suffices to define the 'C^1-norm' $\| - \|_1$ as follows.

Definition 3.6 (C^1-distance). For differentiable maps $f, g : \mathbb{R}^n \to \mathbb{R}^m$ we define the C^1-distance by

$$\|f - g\|_1 = \max_{1 \le i \le n, 1 \le j \le m} \sup_{x \in \mathbb{R}^n} \{|f^j(x) - g^j(x)|, |\partial_i(f^j - g^j)(x)|\}.$$

We note that such a C^1-distance may be infinite.

Theorem 3.7 (Persistence of fixed points II). *Let* $\varphi : \mathbb{R}^n \to \mathbb{R}^n$ *be a differentiable map with a fixed point* $x_0 \in \mathbb{R}^n$, *such that the derivative* $\mathrm{d}\varphi_{x_0}$ *has no eigenvalue* 1. *Then there is a neighbourhood* V *of* x_0 *in* $\mathbb{R}^n$ *and a number* $\varepsilon > 0$, *such that any differentiable map* $\tilde{\varphi} : \mathbb{R}^n \to \mathbb{R}^n$ *with* $\|\varphi - \tilde{\varphi}\|_1 < \varepsilon$ *has exactly one fixed point in* V.

Proof. We abbreviate $A = \mathrm{d}\varphi_{x_0}$ and define $B = (A - \mathrm{Id})^{-1}$, where Id is the $n \times n$ unit matrix. Next we define

$$N_\varphi(x) = x - B(\varphi(x) - x).$$

It is not hard to verify the following properties:

- $N_\varphi(x_0) = x_0$;
- $d(N_\varphi)_{x_0} = 0$;
- x is a fixed point of φ if and only if x is a fixed point of N_φ.

We specify the neighourhood $V \subset \mathbb{R}^n$ of x_0 as a δ-neighbourhood, defined coordinate-wise; that is,

$$x \in V \Leftrightarrow |x_j - x_{0,j}| < \delta, \quad j = 1, 2, \ldots, n.$$

We take δ sufficiently small to ensure that

$$\left| \frac{\partial N_\varphi}{\partial x_j}(x) \right| < \frac{1}{4n}, \quad \text{for } j = 1, 2, \ldots, n$$

for all $x \in V$. As a consequence the restricted map $N_\varphi|_V$ is a contraction, in the sense that it contracts the distances between the points of V by at least a factor 4. Here we use the distance defined by taking the maximum of the absolute values of the differences of corresponding coordinates, called the ℓ_∞-distance. This can be seen as follows.

For all $x, y \in V$ we have

$$N_\varphi(x) - N_\varphi(y) = \left(\int_0^1 d(N_\varphi)_{(1-t)x+ty} dt \right) (x - y).$$

The elements of the matrix $M = (\int_0^1 d(N_\varphi)_{(1-t)x+ty} dt)$ all are less than $1/(4n)$. This implies that for any $x \in \mathbb{R}^n$ we have

$$|M(x, y)| \leq \frac{1}{4}|x - y|.$$

This implies that

$$N_\varphi(U_\delta(x_0)) \subset U_{1/4\delta}(x_0).$$

By the contraction principle (see Appendix A), this implies that x_0 is the unique fixed point of N_φ in V. (Compare, e.g., [25].)

For arbitrary smooth maps $\tilde{\varphi} : \mathbb{R}^n \to \mathbb{R}^n$ we define $N_{\tilde{\varphi}}(x) = x - B(\tilde{\varphi}(x) - x)$. Again it follows that $x \in \mathbb{R}^n$ is a fixed point of $\tilde{\varphi}$ if and only if x is a fixed point of $N_{\tilde{\varphi}}$. We now choose $\varepsilon > 0$ small enough to ensure that, if $\|\tilde{\varphi} - \varphi\|_1 < \varepsilon$, we have for all $x \in V$:

- $\|N_\varphi(x) - N_{\tilde{\varphi}}(x)\| < \frac{1}{3}\delta$,
- $\left| \frac{\partial N_{\tilde{\varphi}}}{\partial x_j}(x) \right| < \frac{1}{2n}$, for $j = 1, 2, \ldots, n$.

Given this, $N_{\tilde{\varphi}}$ still maps the closure of the δ-neighbourhood V of x_0 within itself and also the restriction $N_{\tilde{\varphi}}|_V$ is a contraction. This implies that also $N_{\tilde{\varphi}}$ within V has a unique fixed point. $\qquad \square$

Remark. We like to point to the connection with the Newton method to find zeroes, because fixed points of a map φ exactly are the zeroes of $\varphi - \mathrm{Id}$. According to §1.3.4, the corresponding Newton operator is

$$N(x) = x - (\mathrm{d}\varphi_x - \mathrm{Id})^{-1}(\varphi(x) - x),$$

which is almost identical to the operator N_φ we used in the above proof. The only difference is that we replaced $\mathrm{d}\varphi_x$ by $\mathrm{d}\varphi_{x_0}$. This was done to simplify the argument.

3.3.2 Persistence of periodic evolutions

The persistence theory of periodic evolutions of dynamical systems with discrete time directly follows from the previous subsection. If such a system is generated by the smooth map φ, and if the point x lies on a periodic evolution with (prime) period k, then x is a fixed point of φ^k. Now, if the derivative $\mathrm{d}(\varphi^k)_x$ has no eigenvalue 1, then x is persistent as a fixed point of φ^k, and hence also as a periodic point of φ.

For dynamical systems with continuous time, given by (a system of) ordinary differential equations

$$x' = f(x), \quad x \in \mathbb{R}^n, \tag{3.1}$$

the situation is somewhat different. A stationary evolution corresponds to (3.1) to a zero of the associated map $f : \mathbb{R}^n \to \mathbb{R}^n$, say of class C^1, such that the derivative $\mathrm{d}f$ at a zero has no eigenvalue 0. The proof of persistence of such a zero is completely analogous to that of the persistence of fixed points of maps (cf. the proofs of Theorems 3.5 and 3.7).

Regarding periodic evolutions of dynamical systems with continuous time, in §1.2.2 we have seen how a (local) Poincaré map can be constructed. Such a periodic evolution then corresponds to a fixed point of the Poincaré map. Moreover, it is known from the theory of ordinary differential equations (e.g., see [25]) that for C^1-small perturbations of the map f the corresponding evolution operator Φ, when restricted to a compact domain, also undergoes only C^1-small perturbations, which thus lead to C^1-small perturbations of the corresponding Poincaré map. From this it follows that for a dynamical system defined by a system (3.1) a periodic solution, such that the derivative of the corresponding Poincaré map at its fixed point has no eigenvalue 1, is persistent under C^1-small perturbations of f. Also compare Exercise 3.9.

Remarks.

- It should be said that the above discussion is not complete. For instance, we have only defined the C^1-distance for maps $\mathbb{R}^n \to \mathbb{R}^m$ and not for maps between arbitrary smooth manifolds. Although here there are no essential difficulties, many details are needed to complete all the arguments. In Appendix A we present some of these; more information can be found in [142].

– In general it is not so easy to get analytical information on Poincaré maps. For an example of this see Exercise 1.21.

3.4 Persistence for the doubling map

We already observed that persistent occurrence of chaotic evolutions often is hard to prove. In case of the doubling map, however, the analysis is rather simple. We show here that in this case not only chaoticity is persistent, but that even the whole 'topology' of the system persists in the sense of *structural stability*, defined below.

3.4.1 Perturbations of the doubling map: Persistent chaoticity

We return to the doubling map from the circle $\mathbb{S}^1 = \mathbb{R}/\mathbb{Z}$ onto itself; see §1.3.8. The doubling map φ is given by $\varphi([x]) = [2x]$, where $[x]$ denotes the equivalence class of $x \in \mathbb{R}$ modulo $\mathbb{Z}$. In general such a map $\varphi : \mathbb{S}^1 \to \mathbb{S}^1$ can be lifted to a map $\varphi : \mathbb{R} \to \mathbb{R}$,[6] such that for all $x \in \mathbb{R}$ one has

$$\varphi(x + 1) - \varphi(x) = k,$$

for an integer k, called the *degree* of φ; compare Exercise 3.11 and also see Appendix A. In case of the doubling map φ we have $k = 2$. We consider perturbations $\tilde{\varphi}$ of the doubling map φ with the same degree 2. Moreover we only consider perturbations $\tilde{\varphi}$ of φ, the derivative of which is larger than 1 everywhere. By compactness of $[0, 1]$ (and of $\mathbb{S}^1 = \mathbb{R}/\mathbb{Z}$) it then follows that a constant $a > 1$ exists, such that

$$\tilde{\varphi}'(x) > a. \tag{3.2}$$

It may be clear that all smooth maps $\tilde{\varphi}$ that are C^1-near φ and defined on $\mathbb{S}^1 = \mathbb{R}/\mathbb{Z}$, meet with the above requirements.

Proposition 3.8 (Perturbations of the doubling map I). *For a smooth perturbation $\tilde{\varphi}$ of the doubling map φ, which is sufficiently C^1-near φ, we have:*

1. *As a map of $\mathbb{S}^1 = \mathbb{R}/\mathbb{Z}$ onto itself, $\tilde{\varphi}$ has exactly one fixed point,*
2. *Modulo a conjugating translation, the above fixed point of $\tilde{\varphi}$ equals $[0]$,*
3. *As a map of $\mathbb{S}^1 = \mathbb{R}/\mathbb{Z}$ onto itself, $\tilde{\varphi}$ has exactly $2^n - 1$ periodic points of period n,*
4. *For the map $\tilde{\varphi}$, as a map of $\mathbb{S}^1 = \mathbb{R}/\mathbb{Z}$ onto itself, the inverse image of each point in $\mathbb{R}/\mathbb{Z}$ consists of two points.*

---

[6] For simplicity the lift of φ is given the same name.

Proof. We prove the successive items of the proposition.

1. The map $x \in \mathbb{R} \mapsto \tilde{\varphi}(x) - x \in \mathbb{R}$ is monotonically increasing and maps the interval $[0, 1)$ onto the interval $[\tilde{\varphi}(0), \tilde{\varphi}(0) + 1)$. Therefore, there is exactly one point in $\bar{x} \in [0, 1)$ that is mapped on an integer. The equivalence class of $\bar{x}$ mod $\mathbb{Z}$ is the unique fixed point of the map $\tilde{\varphi}$ in $\mathbb{S}^1 = \mathbb{R}/\mathbb{Z}$.

2. Let $\bar{x}$ be the fixed point, then by a 'translation of the origin' we can turn to a conjugated system, given by the map $\bar{\varphi}$ given by

$$\bar{\varphi}(x) = \tilde{\varphi}(x + \bar{x}) - \bar{x};$$

compare §2.2.3. Seen as a map within $\mathbb{S}^1 = \mathbb{R}/\mathbb{Z}$, the fixed point of $\bar{\varphi}$ equals $[0]$. As we saw earlier, conjugated systems have 'the same' dynamics. Therefore it is no restriction of generality to assume that $[0]$ is the unique fixed point in $\mathbb{S}^1 = \mathbb{R}/\mathbb{Z}$.

3. This property can be checked most easily by considering $\tilde{\varphi}$ as a map $[0, 1) \to [0, 1)$ with fixed point 0. In this format $\tilde{\varphi}^n$ exactly has $2^n - 1$ points of discontinuity that divide the interval $[0, 1)$ into 2^n intervals of continuity. This can be proven by induction. Each of the latter intervals is mapped monotonically over the full interval $[0, 1)$. Therefore in each such interval of continuity we find one fixed point of $\tilde{\varphi}^n$, except in the last one, because 1 does not belong to $[0, 1)$. This proves that $\tilde{\varphi}^n$ indeed has $2^n - 1$ fixed points.

4. The last property can be immediately concluded from the graph of $\tilde{\varphi}$, regarded as a map $[0, 1) \to [0, 1)$; compare Figure 3.3.

□

Remark. From Proposition 3.8 it follows that the total number of periodic and eventually periodic points is countable. In fact, as we show later in this chapter when dealing with structural stability, the whole structure of the (eventually) periodic

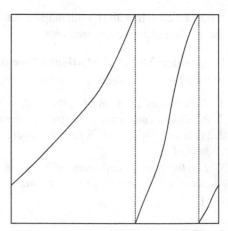

Fig. 3.3 Possible graph of $\tilde{\varphi}$.

points, up to conjugation, is equal to that of the original map φ. For our present purposes, namely showing that the occurrence of chaotic evolutions for the doubling map is persistent, it suffices to know that the set of (eventually) periodic points is only countable. The importance is that this implies that many points exist that are not (eventually) periodic.

Proposition 3.9 (Perturbations of the doubling map II). *Under the conditions of Proposition 3.8, if $[x] \in \mathbb{R}/\mathbb{Z}$ is neither a periodic nor an eventually periodic point of $\tilde{\varphi}$, then the $\tilde{\varphi}$-evolution of $[x]$ is chaotic.*

Proof. The proof of this Proposition is almost the same as the corresponding proof of chaoticity of an evolution of the doubling map itself which is not (eventually) periodic; compare §2.3.1. We briefly indicate which adaptations are needed. To start with, as for the case of the doubling map, each positive evolution that is not (eventually) periodic has infinitely many different points. Because $\mathbb{S}^1 = \mathbb{R}/\mathbb{Z}$ is compact, such an evolution also contains points that are arbitrarily close to each other (without coinciding).

The most important difference between φ and $\tilde{\varphi}$ is that in the latter map the derivative is not constant. However, there exist constants $1 < a < b$, such that for all $x \in [0, 1)$ one has

$$\tilde{\varphi}'(x) \in [a, b].$$

From this it directly follows that in $\mathbb{R}/\mathbb{Z}$ an interval of length $\ell < b^{-1}$ by $\tilde{\varphi}$ is mapped to an interval at least of length $a\ell$. Furthermore, two points at a distance $d < \frac{1}{2}b^{-1}$ by $\tilde{\varphi}$ are mapped to points at least of distance ad.

By n-fold iteration of $\tilde{\varphi}$ the distance between two nearby points therefore grows at least by a factor a^n (implying exponential growth till saturation takes place). From this it follows that the dispersion exponent (see §2.3.2) is at least $\ln(a) > 0$. This implies that Definition 2.12 applies to such evolutions, implying that they are chaotic.   □

In summary we now state the following theorem.

Theorem 3.10 (Persistence of chaoticity in the doubling map). *If $\tilde{\varphi}$ is a C^1-perturbation of the doubling map $\varphi([x]) = [2x]$, such that*

1. *$\tilde{\varphi}$ as a map on $\mathbb{R}$ has the property $\tilde{\varphi}(x + 1) - \tilde{\varphi}(x) = 2$, and*
2. *For each $x \in [0, 1)$ one has $\tilde{\varphi}'(x) > 1$,*

then $\tilde{\varphi}$, as a map of the circle $\mathbb{S}^1 = \mathbb{R}/\mathbb{Z}$, has chaotic evolutions.

3.4.2   Structural stability

Below we revisit and generalise the notion of two dynamical systems being conjugated. Roughly speaking this means that the systems coincide up to a change of coordinates, where in many important cases the corresponding conjugations are just homeomorphisms. It may be clear that conjugated systems share all their dynamical properties in a qualitative way. When speaking of dynamical properties that are persistent, we are presently interested in systems that are conjugated to all nearby systems. Such a system is called 'structurally stable.' It follows that all dynamical properties, defined in terms of the topology of the state space, of a structurally stable system are persistent in a qualitative way.

In §3.4.3 when dealing with an example concerning fair and false coins, we encounter interesting paradoxes. The notion of structural stability was strongly propagated by Stephen Smale and René Thom [213,231] in the 1960s and 1970s. In fact, the ensuing persistence properties make structurally stable dynamical systems very suitable for modelling purposes.

Definition 3.11 (Topological conjugation). Let two dynamical systems (M, T, Φ) and (N, T, Ψ) be given with M and N topological spaces and with continuous evolution operators Φ and Ψ. We say that (M, T, Φ) and (N, T, Ψ), are *topologically conjugated* whenever there is a homeomorphism $h : M \to N$ such that for each $t \in T$ and $x \in M$ one has

$$h(\Phi(x,t)) = \Psi(h(x),t). \tag{3.3}$$

The map h is called a *topological conjugation*.

The conjugation relation (3.3) is equivalent to commutativity of the following diagram.

$$
\begin{array}{ccc}
M \times T & \xrightarrow{\ \Phi\ } & M \\
{\scriptstyle h \times \mathrm{Id}} \downarrow & & \downarrow {\scriptstyle h} \\
N \times T & \xrightarrow[\ \Psi\ ]{} & N
\end{array}
$$

Compare §2.2.3, in particular with the conjugation equation (2.6).

Remarks.

– If $T = \mathbb{Z}$ or $\mathbb{Z}_+$ and the evolution operators therefore are given by maps $\varphi = \Phi^1$ and $\psi = \Psi^1$, then the equation (3.3) is equivalent to

$$h \circ \varphi = \psi \circ h.$$

– The conjugation h takes the dynamics in M to that in N, in the sense that each evolution $t \mapsto \Phi(x_0, t)$ by h is mapped to the evolution $t \mapsto \Psi(h(x_0), t) = h(\Phi(x_0, t))$. Thus it also turns out that two dynamical systems cannot be conjugated when their time sets do not coincide.
– When defining quasi-periodic evolutions, (see Definition 2.7), we used differentiable conjugations. We do not do this here for reasons that are explained in Chapter 5 and Appendix B. Also compare Exercise 3.8.

- In cases where $T = \mathbb{R}$, the notion of *topological equivalence* also exists. A homeomorphism $h : M \rightarrow N$ between the state spaces of two dynamical systems is a topological equivalence if for each $x \in M$ there exists a monotonically increasing function $\tau_x : \mathbb{R} \rightarrow \mathbb{R}$, such that

$$h(\Phi(x, t)) = \Psi(h(x), \tau(t)).$$

This condition means that h does map evolutions of the former system to evolutions of the latter, but that the time parametrisation does not have to be preserved, although it is required that the time direction is preserved. This concept is of importance when considering periodic evolutions. Indeed, under a conjugation the periods of the corresponding periodic evolutions do have to coincide, which is not the case under a toplogical equivalence.

Definition 3.12 (Structural Stability). A dynamical system (M, T, Φ) where $T = \mathbb{Z}$ or $\mathbb{Z}_+$ is called C^1 structurally stable if any dynamical system (M, T, Φ') with Φ' sufficiently C^1-near Φ, is topologically conjugated to (M, T, Φ). For dynamical systems with $T = \mathbb{R}$ the same definition holds, when we replace 'conjugation' by 'equivalence'.

Remarks.

- We have not yet given a general definition of 'C^1-near'. For such definitions on manifolds we refer to [142]. For a brief discussion also see Appendix A. Note that here we only use the concept of structural stability for the doubling map, in which case, by Definition 3.6, we already know what is meant by C^1-small. In fact, a C^1-small perturbation of the doubling map φ is a map $\tilde{\varphi}$, which as a map within $\mathbb{R}$ is C^1-near φ, while always observing the relation $\tilde{\varphi}(x+1) - \tilde{\varphi}(x) = 2$.
- More generally we speak of C^k-structural stability when such a conjugation or equivalence exists whenever Φ and Φ' are C^k-nearby; again see [142] and Appendix A. Below by 'structural stability' we always mean C^1-structural stability. Theoretically one also may define C^0-structural stability, but such a concept would be not very relevant. Compare Exercise 3.12.
- The Russian terminology is slightly different (compare, e.g., [2]). In fact, a topological conjugation is called 'orbital equivalence,' whereas structurally stable is called 'rough', and so on.

In this section we aim to show the following.

Theorem 3.13 (Structural stability of the doubling map). *The doubling map is (C^1)-structurally stable.*

Proof. In fact, we show that any perturbation $\tilde{\varphi}$ as met in the previous section, hence determined by the properties that $\tilde{\varphi}(x + 1) - \tilde{\varphi}(x) = 2$ and $\tilde{\varphi}' > 1$, is conjugated with the original doubling map φ. To do this, we investigate the set of eventually stationary evolutions of both the doubling map φ and its perturbation $\tilde{\varphi}$ in greater detail.

We view φ and $\tilde{\varphi}$ as maps on the interval $[0, 1)$ with a stationary point at 0. We label the inverse images and the iterated inverse images of this point 0, in such a way that for the actual doubling map φ the labelling is exactly that of their binary expansions. This is now explained as follows for both φ and $\tilde{\varphi}$ at the same time.

The stationary point 0 first gets the label '0'. The inverse image of 0 consists of two points: one in the 'left half' of the interval and the other in the 'right half'. The former of these, that coincides with the original stationary point now gets the label '00' and the latter '10'. The point with the label 10 from now on is called the first point in the 'right half' of the interval. This process we repeat inductively. Each label consists of a finite sequence of digits 0 and 1 and at each repetition the sequences get longer by one digit. The point in the left half gets a label consisting of a 0 followed by the sequence of the point on which it is mapped and in the right half we get a label starting with 1 followed by the sequence of the point in which it is mapped. In the diagram below we indicate the result. On the nth row we wrote the labels of the elements that are mapped to the stationary point in n iterations, starting with $n = 0$. Labels in one column belong to the same point. The order of the points in the interval is equal to the horizontal order of the labels in the diagram.

$$
\begin{array}{llllllll}
n = 0 & 0 \\
n = 1 & 00 & & & 10 \\
n = 2 & 000 & & 010 & & 100 & & 110 \\
n = 3 & 0000 & 0010 & 0100 & 0110 & 1000 & 1010 & 1100 & 1110
\end{array}
$$

We define the sets $A_n = \varphi^{-n}(0)$. The labels of the points in A_n are the sequences in the diagram of length $n + 1$. For $\tilde{\varphi}$ we denote the corresponding set by $\tilde{A}_n$. It now follows from the definition of this coding that the map $h_n : A_n \to \tilde{A}_n$ with the following properties preserves the coding.

- h_n is an order-preserving bijection; that is, if $p < q \in A_n$, then also $h_n(p) < h_n(q) \in \tilde{A}_n$,
- For $p \in A_n$ the conjugation equation $h_n(\varphi(p)) = \tilde{\varphi}(h_n(p))$ holds,
- h_{n-1} is the restriction $h_n|_{A_{n-1}}$.

This statement can be proven by induction.

In the above we take the limit as $n \to \infty$. To this end we define the sets $A_\infty = \bigcup_n A_n$ and $\tilde{A}_\infty = \bigcup_n \tilde{A}_n$. It is clear that A_∞ is dense in $[0, 1)$: any point of $[0, 1)$ is within distance 2^{-n} of A_n. Also $\tilde{A}_\infty$ is dense in $[0, 1)$. To see this we recall (3.2), saying that for all $x \in [0, 1)$ one has $\tilde{\varphi}' > a > 1$. This implies that any point of $[0, 1)$ is within a^{-n}-distance of $\tilde{A}_n$.

From the above properties of h_n it first of all follows that there exists exactly one map $h : A_\infty \to \tilde{A}_\infty$, such that for each n we have $h_n = h|_{A_n}$. Secondly it follows that h is an order-preserving bijection from A_∞ onto $\tilde{A}_\infty$, where both of these sets are dense in $[0, 1)$. This means that h has a unique extension to a homeomorphism $h : [0, 1) \to [0, 1)$. The proof of this is an exercise in topology; see Exercise 3.14. Moreover, for all $p \in A_\infty$ the conjugation equation $h(\varphi(p)) = \tilde{\varphi}(h(p))$ is satisfied.

Now, because A_∞ is dense in $[0, 1)$ and because h is continuous, the conjugation equation also holds on all of $[0, 1)$. This means that h is a conjugation from the system generated by φ to the system generated by $\tilde{\varphi}$. This proves structural stability of the doubling map. $\square$

Remarks.

- The map h, as constructed in the above proof, in general will not be differentiable. To see this we consider a periodic point p of φ of (prime) period k. We define the *exponent* of φ at such a periodic point p as

$$\lambda_p = \frac{1}{k} \ln(d\varphi_p^k).$$

From the definition of the doubling map it follows that for the exponent at each periodic point one has $\lambda_p = \ln(2)$.
For $\tilde{\varphi}$ the point $h(p)$ also is a periodic point of (prime) period k and

$$\tilde{\lambda}_{h(p)} = \frac{1}{k} \ln \left(d\tilde{\varphi}_{h(p)}^k\right).$$

The conjugation equation $h \circ \varphi = \tilde{\varphi} \circ h$ implies that

$$h \circ \varphi^k = \tilde{\varphi}^k \circ h.$$

If we would assume differentiability of h, then by the chain rule it would follow that

$$\lambda_p = \tilde{\lambda}_{h(p)};$$

compare Exercise 3.8. This would lead to a contradiction, because it is easy to produce perturbation $\tilde{\varphi}$ for which not all exponents equal $\ln(2)$. Thus for such perturbations the conjugation h cannot be differentiable.
- From the above it should be clear that if in Definition 3.12 of structural stability, we would require differentiability of the conjugation, in general there would practically be no structurally stable dynamical systems, in which case the concept would be meaningless; see the Exercises 3.8 and 3.15.

3.4.3   The doubling map modelling a (fair) coin

From the above one may have obtained the impression that when two dynamical systems are conjugated, all their 'most important properties' coincide. This is, however, not altogether true. We show here that conjugated systems can have entirely different 'statistical properties.' In Exercise 3.17 we show that dispersion exponents also do not have to be preserved by conjugations.

We may view the doubling map as a mathematical model for throwing a (fair) coin. Again, we consider the map as being defined on $[0, 1)$. The left half $[0, \frac{1}{2})$ we now consider as 'heads' and the right half $[\frac{1}{2}, 1)$ as 'tails'. For a 'random' choice of initial state $q \in [0, 1)$, the corresponding sequence of events regarding the coin is $s_0, s_1, s_2, \ldots$, where

$$s_j = \begin{cases} \text{'heads' when } \varphi^j(q) \in [0, \frac{1}{2}), \\ \text{'tails' when } \varphi^j(q) \in [\frac{1}{2}, 1). \end{cases}$$

A choice of the initial state q thus is equivalent with the choice of an infinite heads–tails sequence. If the choice is random in the sense of Lebesgue measure, then indeed for each j the probabilities to get heads and tails are equal, and the events are independent for different values of j. In this sense we can speak of the behaviour of a *fair* coin.

These assertions can be proven in various ways. Below we give a heuristic argument that can be made rigourous with the help of probability theory. It also can be adapted to certain perturbations φ_p of φ dealt with below.

We first deal with the independence of different coin throwings. Under a random choice of $q \in [0, 1)$ with respect to Lebesgue measure, q lands in an interval I of length ℓ with probability ℓ. A similar remark goes for the probability that q lands in a disjoint union of a number of intervals. With the help of this it can be simply shown that, under random choice of q, the probability that the point $\varphi(q)$ lies in such a union of intervals I, also is equal to the length of I. Moreover, this event is stochastically independent of the fact whether $q \in [0, \frac{1}{2})$ or $[\frac{1}{2}, 1)$. This is due to the fact that

$$\ell\left(\varphi^{-1}(I)\right) = \ell(I) \quad \text{and} \quad \ell\left(\varphi^{-1}(I)\right) = 2\ell\left(\varphi^{-1}(I) \cap \left[0, \frac{1}{2}\right]\right).$$

This means the following. Apart from the fact that the probability[7]

$$P\{s_0 = 0\} = P\{s_0 = 1\} = \frac{1}{2}$$

also

$$P\{s_1 = 0\} = P\{s_1 = 1\} = \frac{1}{2},$$

and the events regarding s_0 and s_1 are stochastically independent. This means that

$$P\{s_0 = 0 \,\&\, s_1 = 0\} = P\{s_0 = 0 \,\&\, s_1 = 1\}$$
$$= P\{s_0 = 1 \,\&\, s_1 = 0\} = P\{s_0 = 1 \,\&\, s_1 = 1\} = \frac{1}{4}.$$

By induction we get a similar statement for s_i and s_j, $i \neq j$. This accounts for the independence stated above. To be explicit, it follows that for any sequence

---

[7] By $P\{s_0 = 0\}$ we denote the probability that $s_0 = 0$, and so on.

$(s_0, s_1, \ldots, s_n)$ of zeroes and ones, the set of points in $[0, 1]$ which are the initial states leading to a coding that starts in this way, is an interval of length, hence of 'probability', $2^{-(n+1)}$. In this situation, according to the law of large numbers [196], the set $B_{1/2}$ defined by

$$B_{\frac{1}{2}} = \left\{ x \in [0, 1) \mid \lim_{n \to \infty} \frac{\#\{j \in \{0, 1, 2, \ldots, n-1\} \mid \varphi^j(x) \in [0, \frac{1}{2})\}}{n} = \frac{1}{2} \right\},$$

where # denotes the number of elements, has Lebesgue measure 1.

We next consider a modified doubling map $\varphi_p : [0, 1) \to [0, 1)$ defined by

$$\varphi_p(x) = \begin{cases} \frac{1}{p}x & \text{as } 0 \le x < p, \\ \frac{1}{1-p}(x - p) & \text{as } p \le x < 1, \end{cases} \tag{3.4}$$

for a parameter $p \in (0, 1)$. Before investigating the statistics of φ_p, we first address a number of dynamical properties.

1. The maps φ_p, for $p \ne \frac{1}{2}$, are not smooth as maps within $\mathbb{R}/\mathbb{Z}$: at the points 0 and p the derivative jumps from $1/(1 - p)$ to $1/p$, and vice versa.
2. Yet all maps φ_p, $0 < p < 1$, are conjugated to $\varphi = \varphi_{1/2}$. The proof as given in the previous section, also works here. The conjugation h_p that takes the system defined by φ to that defined by φ_p, maps 0 to 0 and $\frac{1}{2}$ to p; this is an immediate consequence of the conjugation equation.

We now turn to the statistics of φ_p. In this case we let the interval $[0, p)$ coincide with 'heads' and the interval $[p, 1)$ with 'tails'. It is clear that, when choosing an initial state $q \in [0, 1)$ at random with respect to Lebesgue measure, the probability to land in $[0, p)$ equals p. So if we view iteration of φ_p as a mathematical model for tossing a coin, the probability of 'heads' equals p, which implies that for $p \ne \frac{1}{2}$, the coin is no longer fair. As for the case $p = \frac{1}{2}$, however, the events for different tossings of the coin are independent.

We conclude that the dynamics describing a fair coin is conjugated to that describing an unfair one. This has strange consequences. First of all the conjugating homeomorphisms h_p are highly nondifferentiable, especially on the dense set of periodic points. See the above remarks. Secondly, such homeomorphisms send subsets of Lebesgue measure 0 to subsets of Lebesgue measure 1 and vice versa. The latter can be seen as follows. For each $p \in (0, 1)$, again by the law of large numbers [196], the set

$$B_p = \left\{ x \in [0, 1) \mid \lim_{n \to \infty} \frac{\#\{j \in \{0, 1, 2, \ldots, n-1\} \mid \varphi_p^j(x) \in [0, p)\}}{n} = p \right\},$$

has Lebesgue measure 1. However, for $p \ne \frac{1}{2}$, the map h_p maps the set $B_{1/2}$ in the complement of B_p, which is a set of Lebesgue measure 0. Conversely, the map h_p^{-1}

also maps the set B_p of Lebesgue measure 1 to a set of Lebesgue measure 0 in the complement of $B_{1/2}$.

Observe that this shows that measure theoretical persistence is not preserved under homeomorphisms.

Remarks.

- A final comment concerns the special choice we made of perturbations $\tilde{\varphi} = \varphi_p$ of the doubling map φ, which is inspired by the fact that it is easy to deal with their statistical properties. A more general perturbation $\tilde{\varphi}$, as considered earlier, does not only provide another model for an unfair coin, but in general the successive tossings will be no longer stochastically independent.
- This probabilistic analysis of (deterministic) systems has grown to an extensive part of ergodic theory. An introductory treatment of this can be found in [204]. In [8, 18] this subject is dealt with in greater generality.

3.5 Exercises

Exercise 3.1 (Typical Van der Pol evolution). What types of evolution occur persistently in the dynamical system given by the free Van der Pol oscillator?

Exercise 3.2 (A set of positive measure). For the driven undamped pendulum (see §1.1.2) with $A = 0$, it is known from KAM theory that in particular the quasi-periodic subsystems, the frequency ratios of which are badly approximated by rationals, persist for (small) values of ε. Such frequency ratios typically contain a set such as

$$F_{\tau,\gamma} = \{x \in \mathbb{R} \mid \left| x - \frac{p}{q} \right| > \frac{\gamma}{q^\tau}, \text{ for all } p, q \in \mathbb{Z}, q \neq 0\}.$$

Show that, for $\tau > 2$ and $\gamma > 0$ sufficiently small, the set $F_{\tau,\gamma}$ has positive Lebesgue measure.

Exercise 3.3 (Divergence and contraction of volume). For a vector field X on $\mathbb{R}^n$, with components $X_1, X_2, \ldots, X_n$, the *divergence* is defined as

$$\text{div}(X) = \sum_j \partial_j X_j.$$

By $\Phi_X : \mathbb{R}^n \times \mathbb{R} \to \mathbb{R}^n$ we denote the evolution operator associated with X, assuming that it is globally defined. For a subset $A \subset \mathbb{R}^n$ we let $\text{vol}_A(t)$ be the Lebesgue measure of $\Phi^t_X(A)$, where $\Phi^t_X(x) = \Phi_X(x, t)$.

1. Show that $d/dt \, \mathrm{vol}_A(0) = \int_A \mathrm{div}(X)$.

Our aim is to show, with the help of this result, that the driven damped pendulum cannot have quasi-periodic evolutions. To this end we first determine the relevant divergence. The 3-dimensional vector field X associated with the driven pendulum has components $(y, -\omega^2 \sin x - cy + \varepsilon \cos(\Omega z), 1)$, with $c > 0$.

2. Show that $\mathrm{div}(X) = -c$.

We now consider an arbitrary dynamical system with state space $\mathbb{R}^3$ (where possibly one or more coordinates have to be taken modulo $\mathbb{Z}$ or $2\pi\mathbb{Z}$) and continuous time, the dynamics of which is generated by a vector field X. We assume that there is a quasi-periodic subsystem, each evolution of which densely fills the differentiable image T of the 2-torus $\mathbb{T}^2$, where T is the boundary of a 3-dimensional domain W.

3. Show that in the above situation $\int_W \mathrm{div}(X) = 0$.
4. Show that the driven pendulum, with positive damping, cannot have any quasi-periodic evolutions.

Exercise 3.4 (Conservative flow preserves volume). On $\mathbb{R}^{2n}$ with coordinates $p_1, p_2, \ldots, p_n, q_1, q_2, \ldots, q_n$ consider a Hamiltonian system

$$p'_j = -\frac{\partial H}{\partial q_j} \tag{3.5}$$

$$q'_j = \frac{\partial H}{\partial p_j}.$$

$j = 1, 2, \ldots, n$, where

$$H : \mathbb{R}^{2n} \to \mathbb{R}$$

is a smooth function, the *Hamiltonian* of the vector field (3.5). Show that the flow of this vector field preserves the Lebesque measure on $\mathbb{R}^{2n}$. (Hint: Use Exercise 3.3.)

Exercise 3.5 (Poincaré map of the forced pendulum). Consider the equation of motion of a periodically driven pendulum

$$\ddot{y} + \alpha^2 \sin y = \varepsilon \cos(\omega t), \tag{3.6}$$

which can be written as a vector field

$$\dot{x} = \omega$$
$$\dot{y} = z \tag{3.7}$$
$$\dot{z} = -\alpha^2 \sin y + \varepsilon \cos x$$

on the phase space $\mathbb{T}^1 \times \mathbb{R}^2 = \{x, (y, z)\}$. Show that the Poincaré map $P = P_\varepsilon$ of the vector field (3.7) is area-preserving. (Hint: Use the Gauß[8] divergence theorem.)

---

[8] Carl Friedrich Gauß 1777–1855.

Exercise 3.6 (Poincaré recurrence). Let M be a bounded region in $\mathbb{R}^n$ and φ : $M \to M$ a homeomorphism that preserves volume. Show that for any point $x \in M$ and any neighborhood U of x, there is a point $y \in U$ such that $\varphi^n(y) \in U$ for some $n > 0$.

Exercise 3.7 (Product of irrational rotation and doubling map). Consider the map $\varphi : \mathbb{T}^1 \times [0, 1) \to \mathbb{T}^1 \times [0, 1)$ which is the product of an irrational rotation and the doubling map. Show the existence of chaotic and nonchaotic orbits.

Exercise 3.8 (Invariance of eigenvalues). Given are two smooth maps f, g : $\mathbb{R}^n \to \mathbb{R}^n$ with $f(0) = 0$ and $g(0) = 0$. Let $A = d_0 f$ and $B = d_0 g$. Assume that there is a smooth map $h : \mathbb{R}^n \to \mathbb{R}^n$ with $h(0) = 0$ (possibly only defined on a neighbourhood of 0), such that the conjugation equation

$$h \circ f = g \circ h$$

is satisfied. Show that the matrices A and B are similar (i.e., conjugated by an invertible linear map). What can you conclude about the eigenvalues of A and B?

Exercise 3.9 (Eigenvalue derivatives of Poincaré maps). Given is a vector field X on a manifold M with a closed orbit γ. For any section Σ that is transversal to γ, we can define a local Poincaré map $P : \Sigma \to \Sigma$ with a fixed point $p \in \Sigma \cap \gamma$. Show that the eigenvalues of the derivative dP_p do not depend on the choice of the section Σ. (Hint: Use Exercise 3.8.)

Exercise 3.10 (Smooth conjugations for flows). Let two vector fields $\dot{x} = F(x)$ and $\dot{y} = G(y)$, $x, y \in \mathbb{R}^m$, be given. Consider a diffeomorphism $y = \Phi(x)$ of $\mathbb{R}^m$. Show that Φ takes evolutions of the former to the latter vector field in a time-preserving way if and only if

$$D_x \Phi \cdot F(x) \equiv G(\Phi(x)). \tag{3.8}$$

(Hint: Use the chain rule and the existence and uniqueness theorem [7, 144] for solutions of ordinary differential equations.)

Remark. In tensorial shorthand we often rewrite (3.8) as $G = \Phi_* F$; compare [74, 215].

Exercise 3.11 (Lifts and degree of circle maps). From the definition of $\mathbb{R}/\mathbb{Z}$ it follows that a map $F : \mathbb{R} \to \mathbb{R}$ induces a map within $\mathbb{R}/\mathbb{Z}$ if and only if for each $x \in \mathbb{R}$ one has that $F(x + 1) - F(x) \in \mathbb{Z}$. When F is continuous, it follows that $F(x + 1) - F(x)$ is a constant (i.e., independent of x); it is an integer called the *degree* of F.

1. Show that for each continuous map $f : \mathbb{R}/\mathbb{Z} \to \mathbb{R}/\mathbb{Z}$, there exists a continuous map $F : \mathbb{R} \to \mathbb{R}$, such that $f \circ \pi = \pi \circ F$, where $\pi : \mathbb{R} \to \mathbb{R}/\mathbb{Z}$ is the canonical projection.

2. Show that for a given f as in the previous item, the continuous map F is unique up to translation by an integer (i.e., when F and $\tilde{F}$ are two such maps, then $F(x) - \tilde{F}(x)$ is an integer which is independent of x.)

Exercise 3.12 (No C^0-structural stability). We define the C^0-distance between two maps $f, g : \mathbb{R}^n \to \mathbb{R}^m$ by

$$\|f - g\|_0 = \sup_{x \in \mathbb{R}^n} |f(x) - g(x)|.$$

Using this, it also will be clear what is understood by C^0-small perturbations of a map within the circle $\mathbb{R}/\mathbb{Z}$.

Show that any map of the circle that has at least one fixed point, cannot be C^0-*structurally stable*. (NB: It is even possible to prove this statement without assuming the existence of fixed points: for maps within the circle, C^0-structural stability does not occur.)

Exercise 3.13 (Chaos doubling map persistent in sense of probability). Consider the doubling map $\varphi : [0, 1) \to [0, 1)$.

1. Show that the evolutions of φ are either periodic or chaotic,
2. Show that the set of eventually periodic orbits of φ is countable,
3. Show that the chaoticity of the doubling map is persistent in the sense of probability.

Exercise 3.14 (Extension of a homeomorphism from a dense set). Suppose that $Z_1, Z_2 \subset \mathbb{R}$ are dense subsets and that $h : Z_1 \to Z_2$ is a bijection, such that $p < q \in Z_1$ implies that $h(p) < h(q)$.

1. Show that h can be extended in a unique way to a bijection $\bar{h} : \mathbb{R} \to \mathbb{R}$, such that $p < q \in \mathbb{R}$ implies that $\bar{h}(p) < \bar{h}(q)$.
2. Show that this extension $\bar{h}$ is a homeomorphism. (Here it is important that the topology of $\mathbb{R}$ can be defined in terms of the order $(<)$: the sets $U_{a,b} = \{x \in \mathbb{R} \mid a < x < b\}$ form a basis of this topology.)
3. Prove the analogue of the above two statements for the case where $\mathbb{R}$ is replaced by $[0, 1)$.

Exercise 3.15 (Structural stability under smooth conjugation). Show that any (nontrivial) translation on $\mathbb{R}$ is structurally stable under smooth conjugation.

Exercise 3.16 ($(*)$ A structurally stable map of the circle). Show that the map within the circle $\mathbb{R}/\mathbb{Z}$, induced by the map $F(x) = x + 1/10 \sin(2\pi x)$, is $(C^1$-$)$ structurally stable.

Exercise 3.17 ($(*)$ Dispersion exponent of the modified doubling map). Consider the modified doubling map φ_p, for $p \in (0, 1)$, as defined by (3.4). Compute its maximal dispersion exponent E_p, showing that $E_p > \ln 2$ as $p \neq \frac{1}{2}$. (Hint: The possible value of E_p depends on the initial state $x \in \mathbb{R}/\mathbb{Z}$ of the orbit. Here take the maximal value over all possible values of x.)

Exercise 3.18 ((∗) **On skew tent maps**). Consider the skew tent map

$$\varphi_p(x) = \begin{cases} \frac{1}{p}x & \text{as } 0 \le x < p, \\ \frac{1}{1-p}(1-x) & \text{as } p \le x < 1, \end{cases}$$

Develop a theory for the dynamics generated by iteration of φ_p analogous to that of the modified doubling map (3.4).

Chapter 4
Global structure of dynamical systems

Till now we have dealt mainly with the properties of individual evolutions of dynamical systems. In this chapter we consider aspects that in principle concern all evolutions of a given dynamical system. We already saw some of this when dealing with structural stability (see §3.4.2) which means that under C^1-small perturbations the 'topology' of the collection of all evolutions does not change.

We first deal with attractors from a general perspective. Recall that we have already met several examples of these, such as the Hénon attractor. These 'configurations' in the state space were obtained by taking a rather arbitrary initial state, iterating a large number of times, and then plotting the corresponding states in the plane with deletion of a rather short initial (transient) part. Compare Figures 1.13 and C.7. The fact that these 'configurations' are independent of the choice of initial state indicates that we are dealing here with attractors. Wherever one starts, always the evolution tends to the attractor, and in the long run such an evolution also approximates each point of it. Our first problem now is to formalise this intuitive idea of attractor.

A dynamical system can have more than one attractor (this situation is often called multistability). Then it is of interest to know which domains of the state space contain the initial states of the evolutions that converge to the various attractors. Such a domain is called the basin of attraction of the corresponding attractor. Evolutions starting at the boundary of two basins of attraction will never converge to any attractor. In simple cases these basin boundaries are formed by stable separatrices of saddle points, but we show that such boundaries in general can be far more complicated. For an example, see the Newton algorithm applied to the complex equation $z^3 = 1$ in §1.3.4.4. We now turn to formally defining attractors and their basins.

4.1 Definitions

Let (M, T, Φ) be a dynamical system, where M is a topological space and Φ is continuous. Assume that for each $x \in M$ the positive evolution $\mathcal{O}^+(x) = \{\Phi(x,t) \mid T \ni t \geq 0\}$ is relatively compact, which means that it has a compact

H.W. Broer and F. Takens, *Dynamical Systems and Chaos*,
Applied Mathematical Sciences 172, DOI 10.1007/978-1-4419-6870-8_4,
© Springer Science+Business Media, LLC 2011

closure.[1] According to Definition 2.10 (see §2.3.2), this evolution is called *positively compact*.

Definition 4.1 (ω-**Limit set**). For $x \in M$ we define the ω-*limit set* by

$$\omega(x) = \{y \in M \mid \exists \{t_j\}_{j \in \mathbb{N}} \subset T, \text{ such that } t_j \to \infty \text{ and } \Phi(x, t_j) \to y\}.$$

For a few simple consequences of Definition 4.1 we refer to Exercise 4.1. The ω-limit set $\omega(x)$, and the dynamics therein, determine the dynamics starting at x, insofar as this occurs in the far future. To be more precise:

Lemma 4.2 (ω-**Limit attracts evolution**). *For any neighbourhood U of $\omega(x)$ in M there exists $t_U \in \mathbb{R}$, such that for each $t_U < t \in T$ one has that $\Phi(x, t) \in U$.*

Proof. Suppose, by contradiction, that for a given U, such a t_U does not exist. Then, there is a sequence $\{t_j\}_{j \in \mathbb{N}} \subset T$, such that $t_j \to \infty$ but where $\Phi(x, t_j) \notin U$. Because we assumed all positive evolutions to be relatively compact, the sequence $\{\Phi(x, t_j)\}_{j \in \mathbb{N}}$ must have an accumulation point, say p. Such an accumulation point p cannot be in the interior of U, but on the other hand, by Definition 4.1, necessarily $p \in \omega(x)$. This provides a contradiction. □

From Lemma 4.2 it follows that the evolution starting in x in the long run tends to $\omega(x)$. The fact that the evolution operator Φ is continuous then implies that, as the evolution starting at x approaches $\omega(x)$ more closely, it can be approximated more precisely by evolutions within $\omega(x)$, at least when we restrict ourselves to finite segments of evolutions.

Remark. In an implicit way, the ω-limit set was already discussed when speaking of *asymptotically* stationary, periodic, or multiperiodic evolutions. Indeed, for such evolutions, the ω-limit set consists of a stationary, a periodic, or a multiperiodic evolution, viewed as a subset of state space. For chaotic evolutions we expect more complicated ω-limit sets.

Already the ω-limit set looks like what one might call an 'attractor'. Indeed, one may say that $\omega(x)$ is the 'attractor' of the evolution starting at x. Yet, for real attractors we need an extra property, namely that the set of initial states for which the ω-limit set is the same, is 'large' to some extent. The choice of a good definition of the concept of attractor is not so simple. In contrast with ω-limit sets, on the definition of which everyone seems to agree, there are a number of definitions of attractor in use. For a careful discussion we refer to [168]. Here we give a definition that connects well to the concepts introduced earlier, but we do not claim that our choice is the only 'natural' one.

We might put forward that an ω-limit set $\omega(x)$ is an attractor, whenever there is a neighbourhood U of x in M such that for all $y \in U$ one has $\omega(y) = \omega(x)$.

---

[1] In $\mathbb{R}^n$ this just means that the set is bounded.

This attempt at a definition looks satisfactory for the cases in which the ω-limit set is stationary, periodic, or quasi-periodic. However, we note two problems with this. The first of these we illustrate with an example.

Example 4.1 (Saddle-node 'bifurcation' on the circle). On the circle $\mathbb{R}/(2\pi\mathbb{Z})$ consider the dynamical system generated by the diffeomorphism

$$f(x) = x + \frac{1}{2}(1 - \cos x)\mathrm{mod}\ 2\pi\mathbb{Z}.$$

It may be clear that under iteration of f each point of the circle eventually converges to the point $[0]$. However, for an initial state corresponding to a very small positive value of x, the evolution first will move away from $[0]$, only to return to $[0]$ from the other side.

Cases such as Example 4.1 are excluded by requiring that an attractor has arbitrarily small neighbourhoods such that positive evolutions starting there also remain inside. The second problem then has to do with what we already noted in §3.1, namely that the initial states of chaotic evolutions very often can be approximated by initial states of (asymptotically) periodic evolutions. Based on this, we expect that the ω-limit set Λ of a chaotic evolution will also contain periodic evolutions. The points on these periodic evolutions then are contained in any neighbourhood of Λ, but they have a much smaller ω-limit set themselves. This discussion leads to the following.

Definition 4.3 (Attractor). We say that the ω-limit set $\omega(x)$ is an *attractor*, whenever there exist arbitrarily small neighbourhoods U of $\omega(x)$ such that

$$\Phi(U \times \{t\}) \subset U$$

for all $0 < t \in T$ and such that for all $y \in U$ one has

$$\omega(y) \subset \omega(x).$$

Remarks.

- Referring to Exercise 4.1 we mention that, without changing our concept of attractor, in the above definition we may replace 'such that for all $y \in U$ one has $\omega(y) \subset \omega(x)$' by 'such that $\bigcap_{0 \leq t \in T} \Phi^t(U) = \omega(x)$'.
- In other definitions of attractor [168] often use is made of probability measures or classes of such measures.
- A definition that is more common than Definition 4.3 and that also does not use measure-theoretical notions, is the following alternative.
  A closed subset $A \subset M$ is an attractor whenever A contains an evolution that is dense in A, and if A contains arbitrarily small neighbourhoods U in M, such that $\Phi(U \times \{t\}) \subset U$ for all $0 < t \in T$ and such that for all $y \in U$ one has $\omega(y) \subset A$.
  Here the requirement that the attractor is an ω-limit set has been replaced by the stronger requirement that it is the ω-limit set of a point *inside* A. It is possible to give examples of attractors according to Definition 4.3 that do not satisfy

the alternative definition we just gave. Such examples are somewhat exceptional; compare Exercise 4.2. Regarding the considerations that follow, it makes no difference which of the two definions one prefers to use.

- In Definition 4.3 (and in its alternative) it is not excluded that $A = M$. This is the case for the doubling map, for the Thom map (compare §2.3.5) as well as for any other system where the whole state space occurs as the ω-limit set of one state. (NB: Because we require throughout that evolutions are positively compact, it follows that now M also has to be compact.)
- There are many dynamical systems without attractors. A simple example is obtained by taking $\Phi(x, t) = x$, for all $x \in M$ and $t \in T$.

Definition 4.4 (Basin of Attraction). When $A \subset M$ is an attractor then the *basin of attraction* of A is the set $B(A) = \{y \in M \mid \omega(y) \subset A\}$.

Remarks.

- It should be clear that a basin of attraction always is an open set. Often the union of the basins of all attractors is an open and dense subset of M. However, examples exist where this is not the case. A simple example again is given by the last item of the remarks following Definition 4.3, where the time evolution is the identity.
- Finally we mention that less exceptional examples also exist, such as the logistic system for the parameter value that is the limit of the parameter values of the successive period doublings, the so-called Feigenbaum attractor. This attractor does not attract an open set, but it does attract a set of positive measure; for more information see [104].

4.2 Examples of attractors

We start reviewing the simplest types of attractor already met before:

- The stationary attractor or point attractor, for example, in the dynamics of the free damped pendulum; see §1.1.1.
- The periodic attractor, occurring for the free Van der Pol equation; see §2.2.3.
- The multi- or quasi-periodic attractor, occurring for the driven Van der Pol equation; see §§2.2.3, 2.2.5.

Remark. Here we note that multi-, but not quasi-, periodic attractors are mathematically still problematic.

We also met a number of 'attractors' with chaotic dynamics. Except for the doubling map (§§1.3.8, 2.3.1, 2.3.4, also see below) and for the Thom map (§2.3.5), we discussed the Hénon attractor (§§1.3.2, 2.3.4), the Rössler attractor (§§1.3.7, 2.3.4), the Lorenz attractor (§§1.3.6, 2.3.4), and the attractors that occur in the logistic system (§§1.3.3, 2.3.4). We already said that not in all these examples has it been

proven that we are dealing with an attractor as just defined. In some of the cases
a mathematical treatment was possible after adaptation of the definitions (as in the
case of the 'Feigenbaum attractor,' mentioned just after Definition 4.4). There is a
class of attractors, the so-called *hyperbolic* attractors, that to some extent generalise
the point attractors mentioned above. On the other hand, hyperbolic attractors may
contain chaotic dynamics and, notwithstanding this, allow a detailed mathematical
description. More generally speaking, hyperbolic attractors are persistent under C^1-
small perturbations, and the dynamics inside is structurally stable. A proof of this
assertion is sketched later. Because hyperbolic attractors have a certain similarity
with the state space of the doubling map, the description below of these attractors is
based on the case of the doubling map as far as possible.

4.2.1 The doubling map and hyperbolic attractors

The attractor of the doubling map on the circle S^1 is hyperbolic; we define this
notion below. We now give two modifications of this, one on the plane and the other
in 3-space. The latter one, the solenoid, is a hyperbolic attractor according to the
definition given below. Such hyperbolic examples also exist in higher dimensions.

4.2.1.1 The doubling map on the plane

We first present an example of a dynamical system, the state space of which is $\mathbb{R}^2$,
that contains an attractor the dynamics of which coincides with that of the doubling
map. For this purpose, on the plane we take polar coordinates (r, φ), as usual defined by

$$x = r \cos \varphi, \quad y = r \sin \varphi.$$

We now define a map $f(r, \varphi) = (R(r), 2\varphi)$, where $R : \mathbb{R} \to \mathbb{R}$ is a diffeomorphism, such that the following hold.

– $R(-r) = -R(r)$ for all $r \in \mathbb{R}$.
– For $r \in (0, 1)$ we have $R(r) > r$ whereas for $r \in (1, \infty)$ we have $R(r) < r$.
– $R'(1) \in (0, 1)$.

See Figure 4.1. The dynamics generated by iteration of the function R has $r = 1$
as an attracting fixed point; compare §1.3.3. It should now be clear that for each
point $p \in \mathbb{R}^2 \setminus \{0\}$, the iterates $f^j(p)$ tend to the unit circle $S^1 = \{r = 1\}$ as
$j \to \infty$. On this circle, which is invariant under the map f, we exactly have the
doubling map.

This definition made the unit circle with doubling map dynamics into a 'real'
attractor, in the sense that it is smaller than the state space $\mathbb{R}^2$. We now argue
that this attractor is not at all persistent under small perturbations of the map f.
To begin with, consider the saddle point p of f, in polar coordinates given by
$(r, \varphi) = (1, 0)$.

Fig. 4.1 Graph of the function R.

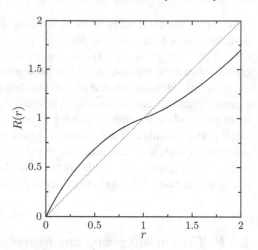

The eigenvalues of the derivative of f at p are as follows. The eigenvalue in the r-direction belongs to $(0, 1)$ and the other, in the φ-direction, equals 2. By the inverse function theorem, the restriction of f to a small neighbourhood of p, is invertible. This means that there exist local stable and unstable separatrices of p as discussed in §1.3.2. Then the global stable and unstable separatrix at p are obtained by iteration of f^{-1}, respectively f. Because f is not invertible, here we restrict to the global unstable separatrix $W^u_f(p)$. It is easy to see that

$$W^u_f(p) = \{r = 1\} = \mathbb{S}^1. \tag{4.1}$$

What happens to this global picture when perturbing f? Let us perturb f to $\tilde{f}$ where $\tilde{f}$ still coincides with f in the sector $-\alpha < \varphi < \alpha$; then the unstable separatrix $W^u_{\tilde{f}}(p)$ coincides with $W^u_f(p)$ in the sector $-2\alpha < \varphi < 2\alpha$. We apply this idea for $\alpha \in (\frac{1}{3}\pi, \frac{1}{2}\pi)$. So in that case the segment

$$S_\alpha = \{r = 1, \varphi \in (-2\alpha, 2\alpha)\} \subset \mathbb{S}^1$$

of the unit circle is a subset of $W^u_{\tilde{f}}(p)$. Now we really make a change from f to $\tilde{f}$ in the sector $\{|\varphi| \in (\alpha, 2\alpha)\}$. This can be done in such a way that, for instance,

$$\tilde{f}(S_\alpha) \cap S_\alpha$$

contains a nontangent intersection. This creates a situation that completely differs from the unperturbed case as described by (4.1). Compare Figure 4.2.

Remark.

– We emphasise that f is not invertible, noting that for a diffeomorphism global separatrices cannot have self-intersections.

Fig. 4.2 Unstable manifold
of the saddle point of the
perturbed doubling map $\tilde{f}$ on
the plane.

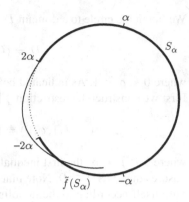

- From §3.3 recall that a saddle point p of a map f is persistent in the sense that for any other C^1-small perturbation $\hat{f}$ of f we get a saddle point $\hat{p}$ that may be slightly shifted.
- It also can be shown, although this is harder [30, 172], that a C^1-small perturbation $\tilde{f}$ of f still has an attractor and that both the saddle point $\tilde{p}$ and its unstable separatrix $W^u_{\tilde{f}}(\tilde{p})$ are part of this. This implies that the original attractor (i.e., the unit circle $\mathbb{S}^1$) can undergo transitions to attractors with a lot more complexity. The present example of the doubling map in the plane therefore has little 'persistence' in the sense of Chapter 3. However, in dimension 3 it turns out to be possible to create a persistent attractor based on the doubling map, but this attractor will have a lot more topological complexity than a circle.
- Although the attractor of the planar doubling map is not persistent in the usual sense, we show later in this chapter, that it has persistent features regarding its 'stable foliation'.
- We like to mention here that the doubling map also occurs when considering the polynomial map $f : \mathbb{C} \to \mathbb{C}, z \mapsto z^2$, namely as the restriction to the complex circle $\mathbb{S}^1 \subset \mathbb{C}$, which is a repellor. Compare Exercise 1.20. Also within the holomorphic setting, this situation is not very persistent. Compare, for example, [4, 169] and with several contributions to [36].

4.2.1.2 The doubling map in 3-space: The solenoid

The idea is now in 3-space to map a 'fattened' circle without self-intersections onto itself in such a way that the image runs around twice as compared to the original. We make this more precise with pictures and formulæ.

Next to the Cartesian coordinates x, y, and z, on $\mathbb{R}^3$ we also use cylindrical coordinates r, φ, and z. As before we then have $x = r \cos \varphi$ and $y = r \sin \varphi$. We are dealing with the circle

$$\mathbb{S}^1 = \{r = 1, z = 0\}.$$

We fatten the circle to a domain D, given by

$$D = \{(r-1)^2 + z^2 \le \sigma^2\},$$

where $0 < \sigma < 1$. As indicated before, the map f is going to map D within itself. First we construct the restriction $f|_{S^1}$ by

$$f(1, \varphi, 0) = (1 + \mu \cos \varphi, 2\varphi, \mu \sin \varphi),$$

where $0 < \mu < \sigma$; the first inequality avoids self-intersections, and the second one ensures that $f(S^1) \subset D$. Note that in the limit case $\mu = 0$, the circle S^1 is mapped onto itself according to the doubling map. Next we define f on all of the fattened circle D by

$$f(1 + \tilde{r}, \varphi, z) = (1 + \mu \cos \varphi + \lambda \tilde{r}, 2\varphi, \mu \sin \varphi + \lambda z). \qquad (4.2)$$

Here $\lambda > 0$ has to be chosen small enough to have $\mu + \lambda\sigma < \sigma$, which ensures that $f(D) \subset D$, and to have $\lambda\sigma < \mu$, which ensures that $f|_D$ also has no self-intersections. In the φ-direction a doubling map takes place: each disc

$$D_{\varphi_0} = \{(r, \varphi, z) \mid (z-1)^2 + z^2 \le \sigma^2, \varphi = \varphi_0\}$$

is mapped within the disc $D_{2\varphi_0}$, where it is contracted by a factor λ; compare Figure 4.3.

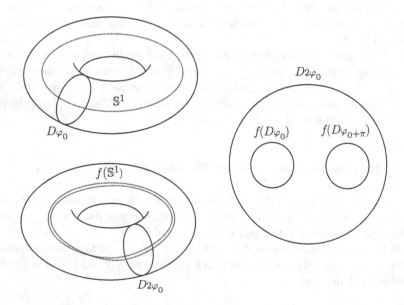

Fig. 4.3 Left: the circle S^1, the fattening D, and $f(S^1) \subset D$. Right: the disc $D_{2\varphi_0}$ with inside $f(D_{\varphi_0})$ and $f(D_{\varphi_0+\pi})$.

The map $f : D \to D$ generates a dynamical system, called the solenoid system. This dynamical system has an attractor $A \subset D$. To describe A, we define the sets $K_j = f^j(D)$. Because $f(D) \subset D$, also $K_i \subset K_j$ whenever $i > j$. We then define

$$A = \bigcap_{j \in \mathbb{N}} K_j.$$

Before showing that A indeed is an attractor of f, we first investigate its geometry.

Theorem 4.5 (Cantor set transverse to solenoid). $A \cap D_{\varphi_0}$ is a Cantor set.

Proof. A Cantor set is characterised as a (nonempty) compact metric space, that is totally disconnected and perfect; for details see Appendix A. Perfect means that no isolated points exist. The fact that $A \cap D_{\varphi_0}$ is totally disconnected means that each point $a \in A \cap D_{\varphi_0}$ has arbitrarily small neighbourhoods with empty boundary.

We just defined $D_{\varphi_0} = D \cap \{\varphi = \varphi_0\}$. From the definitions it now follows that $K_j \cap D_{\varphi_0}$ consists of 2^j discs, each of diameter $\sigma \lambda^j$, and such that each disc $K_{j-1} \cap D_{\varphi_0}$ contains two discs of $K_j \cap D_{\varphi_0}$ in its interior. The discs of $K_j \cap D_{\varphi_0}$ can be coded by the integers between 0 and $2^j - 1$ as follows. The disc with index ℓ then is given by

$$f^j \left(D_{2^{-j}(\varphi_0 + 2\pi \ell)} \right)$$

Two discs with indices $\ell < \ell'$ that differ by 2^{j-1} lie in the same disc of $K_{j-1} \cap D_{\varphi_0}$; that particular disc in $K_{j-1} \cap D_{\varphi_0}$ then has index ℓ.

We now can simplify by expressing these integers in the binary system. The discs of $K_j \cap D_{\varphi_0}$ correspond to integers that in the binary system have length j (where such a binary number may start with a few digits 0). Discs contained in the same disc of $K_i \cap D_{\varphi_0}$, with $i < j$, correspond to numbers of which the final i digits are identical. In this way, the elements of $A \cap D_{\varphi_0}$ correspond to infinite binary expressions of the form

$$(\dots, \alpha_2, \alpha_1, \alpha_0),$$

with $\alpha_j = 0$ or 1. From algebra this set is known as the 2-adic closure of the integers: integers then are closer together whenever their difference contains a larger power of 2. Compare Exercise 4.6.

From the above construction it should be clear that $A \cap D_{\varphi_0}$ has the following properties.

1. It is the limit of a decreasing sequence of nonempty compact sets, $K_i \cap D_{\varphi_0}$, and hence it is also compact and nonempty;
2. $A \cap D_{\varphi_0}$ is totally disconnected, meaning that each point $a \in A \cap D_{\varphi_0}$ has arbitrarily small neighbourhoods in $A \cap D_{\varphi_0}$ with an empty boundary. Indeed, if

$$a \leftrightarrow (\dots, \alpha_2, \alpha_1, \alpha_0),$$

then, as neighbourhoods one can take sets U_j of points that correspond to such expressions, the final j elements of which are given by $(\alpha_{j-1}, \ldots, \alpha_1, \alpha_0)$, in other words all points in one and the same disc of $K_j \cap D_{\varphi_0}$.

3. Any point $a \in A \cap D_{\varphi_0}$ is an accumulation point of $A \cap D_{\varphi_0}$; indeed, the sets U_j we just defined, apart from a, also contain other points.

From these three properties is follows that $A \cap D_{\varphi_0}$ is a Cantor set; again see Appendix A.   □

Remark. We note, however, that A is not homeomorphic to the product of the circle and Cantor set, which is exactly why it is called a solenoid attractor. This is related to the fact that our description is 'discontinuous' at D_0 in the sense that the coding of the points of $A \cap D_0$ differs from that of $A \cap D_{2\pi} (= \lim_{\varphi \uparrow 2\pi} A \cap D_\varphi)$. Also note that $A \setminus (D_0 \cap A)$ is homeomorphic with the product of $(0, 1)$ and the Cantor set.[2] See Appendix A and [142]. In Exercise 4.6 we enter deeper into the algebraic description of the solenoid and the corresponding dynamics.

Theorem 4.6 (Solenoid as attractor). *A is an attractor of the dynamics generated by the solenoid map f; see (4.2).*

Proof. We show that A is an ω-limit set and thus an attractor. The problem is to find a suitable point $p \in \mathbb{R}^3$, such that $\omega(p) = A$. For p we take an element of D, such that the projection of its orbit on $\mathbb{S}^1$ is a dense orbit of the doubling map.

Now let $q \in A$ be an arbitrary point with angular coordinate $\varphi(q)$ and let the binary expression of q as an element of $D_{\varphi(q)}$ be given by

$$(\ldots, \alpha_2(q), \alpha_1(q), \alpha_0(q)).$$

We now have to show that $q \in \omega(p)$, meaning that the positive evolution $\{f^j(p)\}_{j \geq 0}$ has elements that are arbitrarily near q. So let V be an (arbitrarily) small neighbourhood of q. Then V contains a neighbourhood that can be obtained by restricting the angular coordinate to an ε-neighbourhood of $\varphi(q)$ and by restricting to the elements of the corresponding binary expressions, that for a suitable j end like

$$\alpha_{j-1}(q), \ldots, \alpha_0(q).$$

(Here we assume that the angular coordinate has no 'discontinuity' in the ε-neighbourhood of $\varphi(q)$.) Such a neighbourhood also can be written as

$$f^j \left(\left\{ \frac{\varphi_0 - \varepsilon}{2^j} < \varphi < \frac{\varphi_0 + \varepsilon}{2^j} \right\} \right),$$

---

[2] Due to the gluing the solenoid is not such a product; it is a nontrival Cantor set bundle over the circle.

for some φ_0. Because the φ coordinates of the positive evolution of p are dense in $\mathbb{S}^1$, and therefore also contain points of the interval

$$\left(\frac{\varphi_0 - \varepsilon}{2^j}, \frac{\varphi_0 + \varepsilon}{2^j}\right),$$

the positive evolution of p also contains points of V. This concludes the proof. □

4.2.1.3 Digression on hyperbolicity

As indicated earlier, the solenoid is an example of an important class of attractors. This class of *hyperbolic attractors* is defined here, and we also show that the solenoid is one of these. Also we discuss a few properties that characterise hyperbolicity. For a manifold M and $p \in M$ we denote the tangent space in p at M by $T_p(M)$; see Appendix A.

We now deal with hyperbolicity a bit more generally for compact invariant subsets $A \subseteq M$.

Definition 4.7 (Hyperbolicity, discrete time). We call a compact invariant set $A \subseteq M$ for the dynamics generated by a map $f : M \to M$ *hyperbolic* if f, restricted to a neighbourhood of A, is a diffeomorphism and if for each point $p \in A$ there exists a splitting $T_p(M) = E^u(p) \oplus E^s(p)$ with the following properties:

1. The linear subspaces $E^u(p)$ and $E^s(p)$ depend continuously on p, their dimensions being independent of p.
2. The splitting is invariant under the derivative of f :

$$\mathrm{d}f_p(E^u(p)) = E^u(f(p)) \quad \text{and} \quad \mathrm{d}f_p(E^s(p)) = E^s(f(p)).$$

3. The nonzero vectors $v \in E^u(p)$ (respectively, $E^s(p)$) increase (respectively, decrease) exponentially under repeated application of $\mathrm{d}f$ in the following sense. There exist constants $C \geq 1$ and $\lambda > 1$ such that for any $n \in \mathbb{N}$, $p \in A$, $v \in E^u(p)$, and $w \in E^s(p)$ we have

$$\mid \mathrm{d}f_p^n(v) \mid \geq C^{-1}\lambda^n|v| \quad \text{and} \quad \mid \mathrm{d}f_p^n(w) \mid \leq C\lambda^{-n}|w|,$$

where $|-|$ is the norm of tangent vectors with respect to some Riemannian metric.

We call the linear subspace $E^u(p)$, respectively, $E^s(p)$, of $T_p(M)$ the space of *unstable* (respectively, *stable*) tangent vectors at p.

Remarks.

- It is not hard to show that, if such a hyperbolic splitting exists, then it is' necessarily unique. Indeed, a vector in $T_p(M)$ only belongs to $E^s(p)$ when it shrinks exponentially under iteration of $\mathrm{d}f$. Similarly a vector in $T_p(M)$ only belongs to $E^u(p)$ when it shrinks exponentially under iteration of $\mathrm{d}f^{-1}$.

– For the above definition it is not important to know which Riemannian metric on M has been chosen for the norms. Indeed, using the compactness of A, one can prove that whenever such estimates hold for one Riemannian metric they also hold for any other choice. Here it may be necessary to adapt the constant C. Moreover, it can be shown that a Riemannian metric always exists where $C = 1$.

– Hyperbolic invariant subsets can also be defined for dynamical systems with continuous time, given by a vector field, say X. In that case the definition runs differently. To begin with, in this case the attractor should not contain a stationary evolution. Also, if Φ^t denotes the flow of X over time t, then for each $p \in M$ and for each $t \in \mathbb{R}$ we have

$$d(\Phi^t)X(p) = X(\Phi^t(p)).$$

Therefore, if the positive evolution of p is relatively compact, $d(\Phi^t)X(p)$ cannot grow nor shrink exponentially for $t \to \infty$ or $t \to -\infty$. This means that vectors in the X-direction cannot participate in the 'hyperbolic behaviour,' which leads to the following definition.

Definition 4.8 (Hyperbolicity, continuous time). An attractor $A \subset M$ of a dynamical system, defined by a smooth vector field X with evolution operator Φ, is called *hyperbolic* if for each $p \in A$ there exists a splitting $T_p(M) = E^u(p) \oplus E^c(p) \oplus E^s(p)$ with the following properties:

1. The linear subspaces $E^u(p)$, $E^c(p)$, and $E^s(p)$ depend continuously on p, whereas their dimensions are independent of p.
2. For any $p \in A$, the linear subspace $E^c(p)$ is 1-dimensional and generated by $X(p)$.[3]
3. The splitting is invariant under the derivative $d\Phi^t$ in the sense that for each $p \in A$ and $t \in \mathbb{R}$:

$$d\Phi^t(E^u(k)) = E^u(\Phi^t(k)) \quad and \quad d\Phi^t(E^c(k)) = E^c(\Phi^t(k))$$

$$and \quad d\Phi^t(E^s(k)) = E^s(\Phi^t(k)).$$

4. The vectors $v \in E^u(p)$ (respectively, $E^s(p)$), increase (respectively, decrease) exponentially under application of $d\Phi^t$ as a function of t, in the following sense. Constants $C \geq 1$ and $\lambda > 1$ exist, such that for all $0 < t \in \mathbb{R}$, $p \in A$, $v \in E^u(p)$, and $w \in E^s(p)$, we have:

$$\mid d\Phi^t(v) \mid \geq C^{-1}\lambda^t |v| \quad and \quad \mid d\Phi^t(w) \mid \leq C\lambda^{-t}|w|,$$

where $| - |$ is the norm of tangent vectors with respect to a Riemannian metric.

Remark. With respect to the Lorenz attractor mentioned before (§§1.3.6, 2.3.4) which is discussed again later in this chapter, we note that a hyperbolic attractor of a

---

[3] In particular there can be no stationary points in A.

system with continuous time (and defined by a vector field) cannot have equilibrium points (i.e., zeroes of the vector field). This implies in particular that the Lorenz attractor cannot be hyperbolic. Moreover, from [164, 186] it follows that a structurally stable attractor is either a point attractor or hyperbolic, from which it follows that the Lorenz attractor is not structurally stable.

4.2.1.4   The solenoid as a hyperbolic attractor

We now turn to the solenoid system given by the map f in (4.2). In Theorem 4.6 we already showed that this system has an attractor A. The solenoid geometry of A was identified in Theorem 4.5 as a (nontrivial) bundle of Cantor sets over the circle $\mathbb{S}^1$.

Theorem 4.9 (Hyperbolicity of the solenoid). *The solenoid attractor A, as described in Theorems 4.6 and 4.5, is hyperbolic.*

Proof. We need to identify the splitting of the tangent bundle $T_p(M) = E^u(p) \oplus E^s(p)$ according to Definition 4.7.

The choice of the stable linear subspaces $E^s(p)$ is straightforward: these are just the tangent spaces to the discs D_φ. The corresponding estimates of Definition 4.7 then can be directly checked.

Regarding the unstable subspaces $E^u(p)$, one would expect that these lie approximately in the φ-direction, but the existence of this component with the prescribed properties, is not completely obvious. There exists, however, a method to show that the unstable subspaces $E^u(p)$ are well defined. Here we start off with approximate 1-dimensional choices of the unstable subspaces $\mathcal{E}_0(p)$ which lie exactly in the φ-direction. In other words, $\mathcal{E}_0(p)$ is tangent to the circle through p, defined by taking both r and z constant. Now it is possible to improve this approximation in an iterative way, which converges and where the limit is exactly the desired subspace $E^u(k)$. To describe this process of improvement, we need some terminology.

A map $p \in A \mapsto \mathcal{E}(p) \subset T_p(M)$ that assigns to each point a linear subspace of the tangent space of M, is called a *distribution*, defined on A. Compare [215]. We assume that such distributions have the property that $\dim \mathcal{E}(p)$ is constant and that $\mathcal{E}$ is continuous. On the set of all such distributions we define a transformation $\mathcal{T}_f$ by

$$\mathcal{T}_f(\mathcal{E})(p) = \mathrm{d}f_{f^{-1}(p)}\mathcal{E}(f^{-1}(p)).$$

It can be generally shown that any distribution $\mathcal{E}$, defined on A and sufficiently near $\mathcal{E}_0$, under repeated application of $\mathcal{T}_f$ will converge to the distribution E^u of unstable vectors

$$E^u = \lim_{n \to \infty} (\mathcal{T}_f)^n(\mathcal{E}). \tag{4.3}$$

Exercise 4.7 deals with the convergence of this limit process for the solenoid attractor.   □

Remark. For more details regarding this and other limit procedures associated with hyperbolicity, we refer to [143] and to the first appendix of [30]. We do mention here that, if in a general setting we wish to get a similar convergence to the stable distribution E^s, we need to use the transformation $T_{f^{-1}}$.

4.2.1.5 Properties of hyperbolic attractors

We mention a number of properties which generally hold for hyperbolic attractors and which we already partly saw for the solenoid. For proofs of the properties discussed here, we refer to the literature, in particular to [13, 143] and to the first appendix of [30]. The main aim of this section is to indicate why hyperbolic attractors are persistent under C^1-small perturbations, and the dynamics inside is structurally stable.

In the following we assume to deal with a dynamical system as before, with a state space M, time t map Φ^t, and a hyperbolic attractor A. Thus, there exists an open neighbourhood $U \supset A$, such that $\Phi^t(U) \subset U$ for all $t > 0$ and $\bigcap_{T \ni t > 0}(U) = A$. In the case of time set $\mathbb{N}$, we also assume that the restriction $\Phi^1|_U$ is a diffeomorphism onto its image.

One of the tools used for proving persistence of hyperbolic attractors is the 'stable foliation'. First we give a general definition of a foliation of a manifold V, which is a decomposition of V in lower-dimensional manifolds, called the leaves of the foliation.

Definition 4.10 (Foliation). Let V be a C^∞-manifold. A *foliation* of V is a map

$$\mathcal{F} : v \in V \mapsto F_v,$$

where F_v is an injectively immersed submanifold of V containing v, in such a way that for certain integers k, ℓ, and h one has

1. If $w \in F_v$, then $F_v = F_w$.
2. For each $v \in V$ the manifold F_v is of class C^k.
3. For any $v \in V$ there exist C^ℓ-coordinates $x = (x_1, \ldots, x_m)$, defined on a neighbourhood $V_v \subset V$ of v, such that for any $w \in V_v$, the connected component of $F_w \cap V_v$ has the form

$$\{x_{h+1} = x_{h+1}(w), \ldots, x_m = x_m(w)\}.$$

It should be clear that h is the dimension of the leaves of the foliation and that all leaves are C^k-manifolds, where always $k \geq \ell$. We call ℓ the differentiability of the foliation. Here it is not excluded that $\ell = 0$, but in the foliations that follow, always $k \geq 1$.

Definition 4.11 (Stable equivalence). For a hyperbolic attractor A of a dynamical system with time t maps Φ^t and a neighbourhood $U \supset A$ as above, that is, such that $\Phi^t(U) \subset U$ for $0 < t \in T$ and $\cap_{0 < t \in T} \Phi^t(U) = A$, two points $v, w \in U$ are called *stably equivalent* if

$$\lim_{t \to \infty} \rho(\Phi^t(v), \Phi^t(w)) = 0,$$

where ρ is the distance function with respect to a Riemannian metric on M.

For a hyperbolic attractor it can be shown that the connected components of the equivalence classes of this stable equivalence are leaves of a foliation in U. This is called the 'stable foliation,' denoted by $\mathcal{F}^s$.

Theorem 4.12 (Stable foliation). *Let $U \supset A$ be as in Definition* 4.11. *Then for the stable foliation $\mathcal{F}^s$ one has*

1. *The leaves of $\mathcal{F}^s$ are manifolds of dimension equal to* $\dim E^s(p)$, $p \in A$, *and of differentiability equal to that of Φ^t.*
2. *At each point $p \in A$ one has that*

$$T_p \mathcal{F}^s(p) = E^s(p).$$

3. *$\mathcal{F}^s$ is of class C^0.*
4. *For any $p \in U$ and $T \ni t > 0$ one has that*

$$\Phi^t(\mathcal{F}^s(p)) \subset \mathcal{F}^s(\Phi^t(p));$$

that is, the foliation is invariant *under the forward dynamics.*

In the case where the leaves have codimension 1 in M, and in a few exceptional cases, it can be shown that $\mathcal{F}^s$ is of class C^1. For proofs of Theorem 4.12, see the references given before. We now explain the meaning of Theorem 4.12 in a couple of remarks.

Remarks.

- In the case of the solenoid, it should be clear what the stable foliation amounts to: its leaves are just the 2-dimensional manifolds D_φ on which the angle φ is constant.
- Although this is perhaps not the most interesting case, it may be worthwhile to observe that Theorem 4.12 also applies in the case of a hyperbolic point attractor. Indeed, if $A = \{p\}$ is such a point attractor, then $E^s(p) = T_p(M)$ and the foliation $\mathcal{F}^s$ only has one leaf, which is an open neighbourhood of p in M.
- On a neighbourhood $U \supset A$ as in Theorem 4.12 we can define an induced dynamics on the set of leaves of $\mathcal{F}^s$. In the case of the solenoid, this set of leaves in a natural way is a 1-dimensional manifold, namely the circle $\mathbb{S}^1$, on which the induced dynamics is conjugated to the doubling map.

  In general it may be expected that such a construction leads to a manifold of dimension $\dim E^u(p)$, with an induced dynamics on this manifold. In reality, the situation is more complicated than this, but in the case where $\dim E^u(p) = 1$, this idea has been fruitful [247].

− Even in cases where no hyperbolic attractor exists, it may occur that a stable foliation exists with certain persistence properties. In the example of the doubling map on the plane (see §4.2.1.1), we can define a stable foliation $\mathcal{F}_f^s$ on a neighbourhood $U = \{\frac{1}{2} < r < 2\}$ of the circular attractor $\mathbb{S}^1$. The leaves of this foliation then are the line segments on which φ is constant and on which $r \in (\frac{1}{2}, 2)$.

This foliation is even persistent, in the sense that for any map $\tilde{f}$ which is sufficiently C^1-near f there is a stable foliation $\mathcal{F}_{\tilde{f}}^s$ in the sense of Theorem 4.12.

In this case it turns out that the set of leaves naturally gets the structure of a circle, on which the induced dynamics is conjugated to the doubling map.

In this respect we can speak of semiconjugations as follows. If (M_1, Φ_1, T) and (M_2, Φ_2, T) are dynamical systems, then we call a continuous and surjective map $h : M_1 \to M_2$ a *semiconjugation* whenever for all $t \in T$ one has

$$h \circ \Phi_1^t = \Phi_2^t \circ h.$$

We now see that the property that the doubling map on the plane, near its attractor, is semiconjugated to the doubling map on the circle, is persistent. See §4.2.1.1. By Theorem 3.13 we also know that the presence of chaotic evolutions is persistent for the doubling map on the circle. From all this, a similar persistence result follows for the occurrence of chaos in the doubling map on the plane.

The proof of the existence of stable foliations for this kind of attractors in endomorphisms is a standard application of the methods of [143]. Yet, it is not known to the authors where this proof and the corresponding persistent existence of a semiconjugation, were mentioned earlier in the literature.

− Finally we mention that recently it has been shown that for the Lorenz attractor, another nonhyperbolic attractor, a well-defined stable foliation exists. We come back to this later in this chapter.

We now come to the main result of this section, dealing with persistence and structural stability of hyperbolic attractors under C^1-small perturbations.

Theorem 4.13 (Persistence of hyperbolic attractors). *Let Φ define a dynamical system with time t maps Φ^t, and let A be a hyperbolic attractor of this system. Then there exist neighbourhoods $\mathcal{U}$ of Φ in the C^1-topology and $U \subset M$ of A, such that $A = \cap_{T \ni t > 0} \Phi^t(U)$ and $\Phi^t(U) \subset U$, for $0 < t \in T$, and such that for any $\tilde{\Phi} \in \mathcal{U}$:*

1. *$\tilde{A} = \bigcap_{T \ni t > 0} \tilde{\Phi}^t(U)$ is a hyperbolic attractor of $\tilde{\Phi}$.*
2. *There exists a homeomorphism $h : A \to \tilde{A}$, which defines a conjugation (or an equivalence in the case $T = \mathbb{R}$), between the restrictions $\Phi|_A$ and $\tilde{\Phi}|_{\tilde{A}}$.*

Remarks.

− Theorem 4.13 for the solenoid implies, among other things, that the complicated Cantor-like structure is preserved when we perturb the map that generates the dynamics, at least when this perturbation is sufficiently small in the C^1-topology.
− Notice that because the Thom map is hyperbolic (see Exercise 4.5), it is also structurally stable.

4.2.2　Nonhyperbolic attractors

In §3.2 we already briefly discussed the attractors met in Chapter 1, including the chaotic ones. Most of the chaotic attractors discussed there and met in concrete examples are nonhyperbolic. Below we continue this discussion, where we distinguish between Hénon-like attractors and the case of the Lorenz attractor. One aspect of interest in all cases is the persistence. First of all, no cases are known to us where one has plain persistence of chaoticity under variation of the initial state. Indeed, as far as known, in all cases the set of (unstable) periodic evolutions is dense in a persistent attractor. In quite a number of cases it turns out that persistence of chaoticity holds in the sense of probability or in the weak sense; see Definitions 3.2 and 3.3.

Secondly, regarding persistence under variation of parameters, we already mentioned in §3.2 that the logistic family and the Hénon family only show weak persistence of chaoticity; see Definition 3.3. Compare [4, 66, 67, 114, 148, 167]. Indeed, in both these examples, there exist subsets of the parameter space with nonempty interior where the corresponding dynamics is asymptotically periodic.

Remark. Plain persistence of chaoticity under variation of parameters is a property of hyperbolic attractors, such as the solenoid. However, as we show below, the Lorenz attractor, although nonhyperbolic, is also C^1-persistent. In both cases, the persistence proof uses stable foliations.

4.2.2.1　Hénon-like attractors

A number of dynamical systems we met in Chapter 1 have nonhyperbolic attractors. To start with, think of the logistic system with parameter $\mu = 4$. It has even been generally shown that in the case of a 1-dimensional state space no 'interesting' hyperbolic attractors can exist: in the case of hyperbolicity for the logistic system we have exactly one periodic attractor; compare Exercise 4.15. Nonhyperbolic attractors in a 1-dimensional state space often are not persistent in the sense that small variations of parameters may lead to a much smaller (periodic) attractor. In this sense the doubling map is an exception. On the other hand, as we mentioned already, for the logistic family the set of parameter values with a nonhyperbolic attractor has positive measure [148].

Meanwhile the research on more general dynamical systems with a 1-dimensional state space has grown to an important area within the theory of dynamical systems. A good introduction to this area is [4]; for more information see [167].

In Chapter 1, in particular in Exercise 1.13, we saw how the Hénon family, for small values of $|b|$, can be approximated by the logistic family. In general we can fatten 1-dimensional endomorphisms to 2-dimensional diffeomorphisms as follows. Given an endomorphism $f : \mathbb{R} \to \mathbb{R}$, the diffeomorphism can be defined as

$$F : \mathbb{R}^2 \to \mathbb{R}^2, \quad F(x, y) = (f(x) + y, bx),$$

where b is a real parameter. For $b = 0$ the x-axis is invariant, and the corresponding restriction of F reads

$$F(x, 0) = (f(x), 0),$$

so we retrieve the original endomorphism f. Following the principles of perturbation theory, it is reasonable to try to extend the theory as available for the logistic system,[4] to these 2-dimensional diffeomorphisms.

Remark. Also compare the construction of the solenoid map from the doubling map in §4.2.

A first step in this direction was the theory of Benedicks and Carleson [67] for the Hénon system. As a result it became known that the Hénon system, for small values of $|b|$ and for many values of the parameter a, has complicated and non-hyperbolic attractors. As already mentioned in §3.2, it seems doubtful whether these considerations also include the 'standard' Hénon attractor which occurs for $a = 1.4$ and $b = 0.3$.

This theory, although originally only directed at the Hénon system, was extended considerably to more general systems by Viana and others; for reviews of results in this direction see [114, 236, 237].

The corresponding theory is rapidly developing, and here we only briefly sketch a number of properties as these seem to emerge. In a similar perturbative setting (see [30, 67, 114, 236, 237]) it has been proven that there is a chaotic attractor A which is the closure of the unstable manifold

$$A = \overline{W^u(p)}, \tag{4.4}$$

where p is a periodic (or fixed) point of saddle type. Colloquially, such attractors are called *Hénon-like*. Also in a number of cases an invariant measure on A exists, such that the dynamics, restricted to A, has certain mixing properties; compare Appendix A. For continuation of the periodic attractors as these occur in the 1-dimensional case, compare [229]. For partial results in higher dimension, see [172, 242] and also see [223].

Remarks.

– Hénon-like attractors seem to occur quite a lot numerically in examples and applications. For further comments and references see Appendix C.
– The numerical simulations on the Rössler system, as discussed in §1.3.7, suggest the presence of a Poincaré map that can be viewed as a 2-dimensional diffeomorphism, which is approximated by a 1-dimensional endomorphism.

4.2.2.2 The Lorenz attractor

As a final example of a nonhyperbolic attractor we treat the Lorenz attractor as introduced in §1.3.6. We are able to discuss this example to greater detail, because a

---

[4] As well as for more general 1-dimensional systems.

geometrical description of the dynamics in this system is based on concepts treated before, such as the stable foliation; see §4.2.1.4, a Poincaré map, and modified doubling maps. This geometric presentation originally was proposed by Williams and Guckenheimer [129,248] (also see Sparrow [214]) and Tucker [234] in 1999 proved that this description is valid for the Lorenz attractor with the common parameter values. Although we have tried to make the description below as convincing as possible, the mathematical details have not been worked out by far. Often we use properties suggested by numerical simulations. As noted before, everything can be founded mathematically, but this is beyond the present scope.

We recall the equations of motion for the system with the Lorenz attractor, that is, the system (D.11), with the appropriate choice of parameters:

$$x' = 10(y - x)$$
$$y' = 28x - y - xz$$
$$z' = -\frac{8}{3}z + xy. \tag{4.5}$$

We start the analysis of (4.5) by studying the equilibrium point at the origin. Linearisation gives the matrix

$$\begin{pmatrix} -10 & 10 & 0 \\ 28 & -1 & 0 \\ 0 & 0 & -\frac{8}{3} \end{pmatrix}.$$

The eigenvalues of this matrix are

$$-\frac{8}{3} \ (z\text{-direction}) \quad \text{and} \quad -\frac{11}{2} \pm \frac{1}{2}\sqrt{1201}, \tag{4.6}$$

the latter two being approximately equal to 12 and -23.

Following [216], by a nonlinear change of coordinates in a neighbourhood of the origin, we linearise the system (4.5). This means that coordinates ξ, η, ζ exist in a neighbourhood of the origin, such that (4.5) in that neighbourhood gets the form

$$\xi' = \lambda\xi$$
$$\eta' = \mu\eta$$
$$\zeta' = \sigma\zeta,$$

where λ, μ, σ are the eigenvalues (4.6), respectively 12, $-8/3$, and -23 (approximately). Around the origin, within the domain of definition of (ξ, η, ζ) we now choose a domain D as indicated in Figure 4.4. Moreover we define

$$D_+ = D \cap \{\eta \geq 0\}, \quad U_\pm = \partial D_+ \cap \{\xi = \pm\xi_0\} \quad \text{and} \quad V = \partial D \cap \{\eta = \eta_0\}.$$

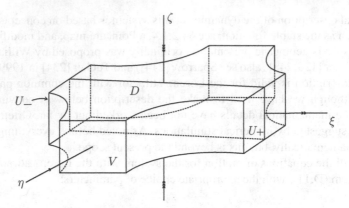

Fig. 4.4 Neighbourhood D of the origin: via the boundary planes $\eta = \pm\eta_0$ and $\zeta = \pm\zeta_0$ the evolution curves enter D and by $\xi = \pm\xi_0$ they exit again. The curved parts of the boundary ∂D are formed by segments of evolution curves.

Fig. 4.5 Typical evolution of the Lorenz system (4.5), projected on the (x, z)-plane; compare Figure 1.23.

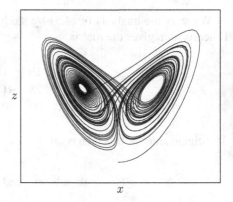

The neighbourhood D moreover can be chosen in such a way that the evolution curves leaving D through $U_\pm$ return again to D by V.

To get a good impression of the way in which the evolution curves return, it is useful to study the numerical simulation of Figure 4.5. In Figure 4.6 this same simulation is represented, where the set D_+ is also shown.

We note that our system is symmetric: under the reflection

$$(x, y, z) \mapsto (-x, -y, z),$$

evolution curves turn into evolution curves. Near $(x, y, z) = (0, 0, 0)$ linearising coordinates (ξ, η, ζ)-coordinates can be chosen such that in these coordinates this reflection is given by

$$(\xi, \eta, \zeta) \mapsto (-\xi, \eta, -\zeta).$$

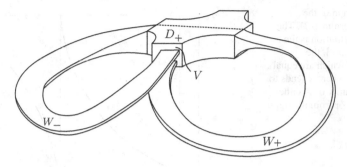

Fig. 4.6 Sketch of the sets V, D_+, W_+, and W_-.

By this symmetry, the evolution curves that leave D via U_+ turn into curves that leave D via U_-. The regions where these evolution curves re-enter into D by V, are symmetric with respect to the η-axis. We define the sets $W_\pm$ as the unions of segments of evolution curves starting at $U_\pm$ that end at V; again compare Figures 4.5 and 4.6.

It may be clear that the set $D_+ \cup W_- \cup W_+$ is 'positively invariant' in the sense that the evolution curves enter it, but never leave it again. Another term for such a set is 'trapping region.' The Lorenz attractor lies inside this trapping region, where it is given by

$$\bigcap_{t>0} \Phi^t (D_+ \cup W_- \cup W_+) \subset D_+ \cup W_- \cup W_+.$$

Here Φ^t is the time t map of (4.5). This means that the origin 0 belongs to the attractor, which is an equilibrium point of (4.5). Therefore the Lorenz attractor cannot be hyperbolic.

The fact that the evolutions that leave D_+ by $U_\pm$ return through V is suggested by numerical simulations. In principle this can also be confirmed by explicit estimates. The final element we need for our description is the existence of a stable foliation,[5] as for hyperbolic attractors. The proof for the existence of these foliations for the Lorenz system (4.5) is extremely involved and in the end only was possible in a computer-assisted way, that is, by numerical computations with a rigorous book-keeping of the maximally possible errors. The result [234] is that for this attractor, though nonhyperbolic, there yet exists a stable foliation $\mathcal{F}^s$. The leaves of this foliation are 1-dimensional. Without restriction of the generality we may assume that V consists of a union of leaves of $\mathcal{F}^s$. Now the global dynamics generated by (4.5) can be described in terms of a Poincaré return map, defined on the leaves of $\mathcal{F}^s$ in V. We next describe this Poincaré map.

We can parametrise the leaves of $\mathcal{F}^s$ in V by the 1-dimensional set $V \cap \{\zeta = 0\}$. When following these leaves by the time t evolution, we see that one of the leaves

---

[5] This is a foliation with 1-dimensional leaves and it is only tangent to the directions of strongest contactions (so at the stationary point to the eigenvector with eigenvalue -23 and not $-8/3$).

Fig. 4.7 Graph of the
Poincaré return map $\mathcal{P}$. The
domain of definition is the
interval $[o_-, o_+]$, the
endpoints of which are equal
to $\lim \mathcal{P}(x)$, where x tends to
the discontinuity of $\mathcal{P}$ either
from the left or from the right.

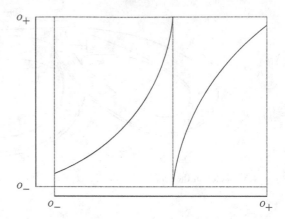

(in the middle) converges to the origin. The points on the left enter W_- by U_-;
the points on the right similarly enter W_+ by U_+; compare Figures 4.4 and 4.6.
After this the leaves again pass through V. Because, by invariance of the stable
foliation (compare Theorem 4.12), each leaf enters within one leaf, it follows that
the Poincaré return map $\mathcal{P}$ on the set of leaves is well defined. From the Figures 4.4
and 4.6 it now should be clear that the graph of $\mathcal{P}$ looks like Figure 4.7. Here we
also use that eigenvalues exist that cause expansion.

The graph of $\mathcal{P}$ looks like that of the doubling map, as defined on the interval
$[0, 1)$; in the middle there is a point of discontinuity and on both the left and on the
right interval the function is monotonically increasing. An important difference with
the doubling map is that the endpoints of the interval $[o_-, o_+]$ of the definition are
not fixed points of $\mathcal{P}$. As for the doubling map, however, we can also describe here
the evolutions of $\mathcal{P}$ by symbolic dynamics. In fact, for each point $x \in [o_-, o_+]$ we
define a sequence of symbols

$$s_0(x), s_1(x), s_2(x), \dots,$$

where

$$s_j(x) = \begin{cases} 0 & \text{if } \mathcal{P}^j(x) \text{ is on the left of the discontinuity,} \\ 1 & \text{if } \mathcal{P}^j(x) \text{ is on the right of the discontinuity.} \end{cases}$$

(For the case where $\mathcal{P}^j(x)$ for some $j \in \mathbb{N}$ exactly is the point of discontinuity, we
refer to Exercise 4.11.) If, except in the discontinuity, $\mathcal{P}$ has everywhere a derivative
larger than 1, the point $x \in [o_-, o_+]$ is uniquely determined by its symbol sequence
$s(x)$. However, as a consequence of the fact that $\mathcal{P}(o_-) > o_-$ and $\mathcal{P}(o_+) < o_+$, in
this case it is not true that any symbol sequence of 0s and 1s has a corresponding
initial point x. For a deeper investigation of this kind of Poincaré return maps, we
refer to Exercise 4.11, where we classify such maps up to conjugation. For more
information on the symbolic dynamics of this map, we refer to [190].

The topology of the Lorenz attractor, up to equivalence, is completely determined by the Poincaré return map $\mathcal{P}$. This means the following. If we consider a perturbation of the Lorenz system, for instance by a slight change of parameters, then the perturbed system, by persistence, again will have a stable foliation, and hence there will be a perturbed Poincaré map. Then $\mathcal{P}$ and the perturbed Poincaré map will be conjugated if and only if the dynamics on the attractor of (4.5) and its perturbation are equivalent.

4.3 Chaotic systems

In Chapter 2 we defined a chaotic evolution by positivity of the dispersion exponent. In this section we address chaoticity of a whole system or of an attractor. For simplicity we only consider dynamical systems (M, T, Φ) where M is a manifold and where Φ is at least continuous. In general we call a dynamical system chaotic if the set of initial states with a chaotic evolution has positive measure. For a good notion of measure and measure class of the Lebesgue measure we refer to Appendix A. This notion of a chaotic system is closely related to the definition that requires positive Lyapunov exponents; compare Chapter 6.

In the case where Φ has an attractor, as defined earlier in Chapter 4, we speak of a *chaotic attractor* if the initial states in its basin with a chaotic evolution have positive measure. In that case the system itself is also called chaotic, although there may be coexistence of attractors, some of which are not chaotic, for examples see [91].

In Exercise 3.4 we met Hamiltonian systems on $\mathbb{R}^{2n}$, with coordinates p_1, $p_2, \ldots, p_n$ and $q_1, q_2, \ldots q_n$, of the form (3.5)

$$p'_j = -\frac{\partial H}{\partial q_j}$$

$$q'_j = \frac{\partial H}{\partial p_j},$$

with Hamiltonian (or energy) $H : \mathbb{R}^{2n} \to \mathbb{R}.$[6] It is easily seen that the function H is constant along its evolutions and that the solution flow preserves the Lebesgue measure on $\mathbb{R}^{2n}$. This means that these systems cannot have 'proper' attractors, that is, attractors that are smaller than a connected component of an energy level. These systems can certainly have chaotic evolutions. For a numerical example in this spirit see Appendix C. It is often very difficult to decide whether the union of chaotic evolutions has positive measure, that is, whether the system is chaotic. In particular this is the case when there are both chaotic evolutions and also a set of positive

---

[6] For instance, think of $q'' = -\text{grad } V(q), q \in \mathbb{R}^n$, where $p = q'$ and $H(p, q) = \frac{1}{2}|p|^2 + V(q)$.

measure exists which is a union of quasi-periodic evolutions. However, there does exist a class of Hamiltonian systems, namely the geodesic flow on compact manifolds with negative curvature, where the flow on each energy level is chaotic and even hyperbolic [8, 43].

4.4   Basin boundaries and the horseshoe map

We conclude this chapter on the global structure of dynamical systems with a discussion of the boundaries of the basins of attraction. We first treat a class of systems, namely gradient systems, where these boundaries have a simple structure. After this more complicated situations are considered.

4.4.1   Gradient systems

A gradient system is a dynamical system generated by a differential equation of the form

$$x' = -\mathrm{grad}\, V(x). \tag{4.7}$$

The state space is a manifold M, where $V : M \to \mathbb{R}$ is a smooth function, and $\mathrm{grad}\, V$ is the gradient vector field of V with respect to a Riemannian metric on M. In the simplest case $M = \mathbb{R}^m$, and the Riemannian metric is given by the standard Euclidean one, in which case the differential equation (4.7) gets the form

$$
\begin{aligned}
x_1' &= -\partial_1 V(x_1, x_2, \ldots, x_m) \\
x_2' &= -\partial_2 V(x_1, x_2, \ldots, x_m) \\
&\ \ \vdots \\
x_m' &= -\partial_m V(x_1, x_2, \ldots, x_m).
\end{aligned}
$$

In the case where $m = 2$, one can think of V as a level function and of the evolution of this system as the orbits of particles that go down as steeply as possible and where the speed increases when the function gets steeper. We now, in the case where $m = 2$, investigate what the basins of attraction and their boundaries look like. To exclude exceptional cases we still have to make an assumption on the functions V considered here. In fact, we require that at each critical point of V (i.e., whenever $dV(x) = 0$), for the Hessian matrix of second derivatives one has

$$\det\begin{pmatrix} \partial_{11} V(x) & \partial_{12} V(x) \\ \partial_{21} V(x) & \partial_{22} V(x) \end{pmatrix} \neq 0.$$

Such critical points are called *nondegenerate*. The corresponding equilibrium of the gradient vector field $\operatorname{grad} V(x)$ is hyperbolic. In the present 2-dimensional case there are three types of hyperbolic equilibria:

- A point attractor.
- A saddle point.
- A repelling point (when inverting time, this is exactly the above case of a point attractor).

Moreover, for simplicity we assume that the evolutions of our systems are positively compact (i.e., bounded). This can, for instance, be achieved by requiring that

$$\lim_{|x| \to \infty} V(x) = \infty.$$

In this situation (compare Exercise 4.14), each point has an ω-limit that consists exactly of one critical point of V. For a point attractor p, the set of points

$$\{y \in M \mid \omega(y) = \{p\}\}$$

is its basin of attraction, which is open. The other ω-limit sets are given by the saddle-points and the point repellors. Assuming that the unstable separatrices tend to point attractors, this implies that the boundaries of the basins have to consist of the stable separatrices of the saddle points (and the repellors). In particular these basin boundaries are piecewise smooth curves and points.

Remarks.

- We return to the geographical level sets we spoke of before. The repellors then correspond to mountaintops and the saddle points to mountain passes. The point attractors are just minima in the bottoms of the seas and lakes. A basin boundary then corresponds to what is called a watershed. This is the boundary between areas where the rainwater flows to different rivers (this is not entirely correct, because two of such rivers may unite later on). These watersheds generally are the mountain combs that connect the tops with the passes.
- Note that basin boundaries consist of branches of stable manifolds of saddle points as well as of the repellors which are in their closures.
- It is possible to create more complicated basin boundaries in dynamical systems with a 2-dimensional state space and continuous time. For this it is helpful if M is a torus, or another closed surface. If we increase the dimension by 1 or turn to discrete time, it is simple to create basin boundaries with a more complicated structure.

4.4.2 The horseshoe map

We already saw that general hyperbolic attractors to some extent are a generalisation of hyperbolic point attractors. A similar generalisation of saddle-points also exists.

Fig. 4.8 By the map f the vertices $1, 2, 3, 4$ of the rectangle D are turned into the vertices $1', 2', 3', 4'$ of $f(D)$.

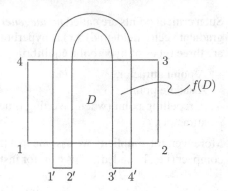

The latter generalisation yields far more complicated boundaries between basins of attraction than in the above example of gradient systems. Historically speaking, the first example of such a 'generalised saddle-point' was given by Smale [212], as the horseshoe map discussed now. In this example we cannot yet speak of the boundary *between* basins of attraction. In fact, as we show, there is only one attractor, which does have a complicated basin boundary. However, the horseshoe map provides the simplest example of such 'generalised saddle points' and plays an important role in the so-called homoclinic dynamics.

A horseshoe map generally is a diffeomorphism $f : \mathbb{R}^2 \to \mathbb{R}^2$ that maps a rectangle D over itself in the form of a horseshoe, as indicated in Figure 4.8. We further assume that in the components of $D \cap f^{-1}(D)$, the map f is affine, and also the horizontal and vertical directions are preserved. On these components f is an expansion in the vertical direction and a contraction in the horizontal direction. In Figure 4.9 we sketched the positioning of D, $f(D)$, $f^2(D)$, and $f^{-1}(D)$, where it was used that the pairs $(f^{-1}(D), D)$, $(D, f(D))$, and $(f(D), f^2(D))$ are diffeomorphic in the sense that each pair is carried to the next by f.

Remark. Sometimes it is useful to extend the horseshoe map as a diffeomorphism of the sphere $\mathbb{S}^2 \approx \mathbb{R}^2 \cup \{\infty\}$, where an appropriate sink is added and where the point ∞ is a source; compare Smale [213].

From the Figures 4.8 and 4.9 it may be clear that $D \cap f(D)$ consists of two components, each of which contains two components of $D \cap f(D) \cap f^2(D)$. By induction it can be easily shown that

$$D_0^n = \bigcap_{j=0}^{n} f^j(D)$$

consists of 2^n vertical strips. Here each component of D_0^{n-1} contains two strips of D_0^n. We can give any of these strips an address in the binary system; compare the construction for the solenoid in §4.2.1.2. The vertical strip L in D_0^n has an address

$$\alpha_{-(n-1)}, \ldots, \alpha_{-2}, \alpha_{-1}, \alpha_0$$

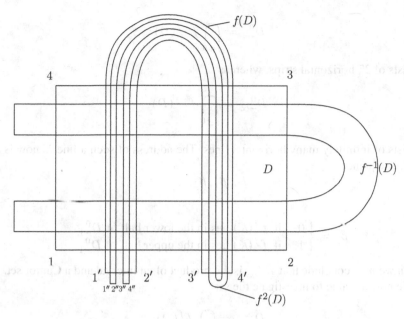

Fig. 4.9 The positioning of $f^{-1}(D)$, D, $f(D)$, and $f^2(D)$.

of length n, where $\alpha_{-j} = 0$ or 1, for all j, assigned as follows.

$$\alpha_{-j} = \begin{cases} 0 & \text{if } f^{-j}(K) \text{ lies in the left half of } D_0^1 \\ 1 & \text{if } f^{-j}(K) \text{ lies in the right half of } D_0^1, \end{cases}$$

where $D_0^1 = D \cap f(D)$. It now may be clear that the set

$$D_0^\infty = \bigcap_{j \geq 0} f^j(D)$$

consists of an infinite number of vertical lines. Any such line $\mathcal{L}$ in this way gets an infinite address

$$\ldots, \alpha_{-2}, \alpha_{-1}, \alpha_0$$

according to the same assignment principle as above. Observe that two of these lines $\mathcal{L}$ and $\mathcal{L}'$ are nearby whenever for the corresponding addresses $\ldots, \alpha_{-2}, \alpha_{-1}, \alpha_0$ and $\ldots, \alpha'_{-2}, \alpha'_{-1}, \alpha'_0$ for a certain j_0 one has

$$\alpha'_{-j} = \alpha_{-j}, \quad \text{for all } j \leq j_0.$$

As before we may conclude that D_0^∞ is the product of a Cantor set and an interval.

From Figures 4.8 and 4.9 is should be clear that the inverse of a horseshoe map f again is a horseshoe map; we only have to rotate the picture over $90°$. In this way we see that

$$D_{-n}^0 = \bigcap_{j=0}^{n} f^{-j}(D)$$

consists of 2^n horizontal strips, whereas

$$D_{-\infty}^0 = \bigcap_{j \geq 0} f^{-j}(D)$$

consists of infinitely many horizontal lines. The address of such a line $\mathcal{K}$ now is an infinite sequence

$$\beta_0, \beta_1, \beta_2, \ldots$$

and

$$\beta_j = \begin{cases} 0 & \text{if } f^j(\mathcal{K}) \text{ lies in the lower half of } D_{-1}^0 \\ 1 & \text{if } f^j(\mathcal{K}) \text{ lies in the upper half of } D_{-1}^0. \end{cases}$$

Again we now conclude that $D_{-\infty}^0$ is the product of an interval and a Cantor set.

We now are able to investigate the set

$$D_{-\infty}^\infty = \bigcap_{j \in \mathbb{Z}} f^j(D),$$

which by definition is invariant under forward and backward iteration of f. It follows that

$$D_{-\infty}^\infty = D_0^\infty \cap D_{-\infty}^0$$

thus it is the product of two Cantor sets: each point is the intersection of a vertical line in D_0^∞ and a horizontal line in $D_{-\infty}^0$. Therefore we can give such a point both the addresses of these lines. In fact, we proceed a little differently, using the fact that the lower component of $f^{-1}(D) \cap D$ by f is mapped to the left component of $D \cap f(D)$ (in a similar way the upper component in the former is mapped to the right component in the latter). This assertion can be easily concluded by inspection of Figures 4.8 and 4.9. Now a point $p \in D_{-\infty}^\infty$ gets the address

$$\ldots, \alpha_{-1}, \alpha_0, \alpha_1, \alpha_2, \ldots,$$

where $\alpha_j = \beta_{j-1}$, for $j \geq 1$, and according to

$$\alpha_j = \begin{cases} 0 & \text{if } f^j(p) \text{ is in the left half of } D_0^1 \\ 1 & \text{if } f^j(p) \text{ is in the right half of } D_0^1, \end{cases}$$

for all $j \in \mathbb{Z}$. By application of the map f a point with address $\ldots, \alpha_{-1}, \alpha_0, \alpha_1, \alpha_2, \ldots$ turns into a point with address $\ldots, \alpha'_{-1}, \alpha'_0, \alpha'_1, \alpha'_2, \ldots$ where

$$\alpha'_j = \alpha_{j+1}.$$

As in §1.3.8.4, we also speak here of a *shift map* on the space of symbol sequences.

4.4.2.1 Symbolic dynamics

The above assignment of addresses defines a homeomorphism

$$\alpha : D_{-\infty}^{\infty} \to \mathbb{Z}_2^{\mathbb{Z}},$$

where the restricted map $f|_{D_{-\infty}^{\infty}}$ corresponds to the shift map

$$\sigma : \mathbb{Z}_2^{\mathbb{Z}} \to \mathbb{Z}_2^{\mathbb{Z}}.$$

For the topology of $\mathbb{Z}_2^{\mathbb{Z}}$ we refer to Chapter 1, in particular §1.3.8. For topological background also see Appendix A. As in the case of the doubling map (see §1.3.8.4) we speak of *symbolic dynamics*, which can help a lot in describing the dynamics generated by f.

First of all it follows that there are exactly two fixed points of f, corresponding to symbol sequences that only consist of 0s denoted by $\{0\}$, or of 1s denoted $\{1\}$. Both are saddle points. The addresses consisting of the sequences

$$\ldots, \alpha_{-1}, \alpha_0, \alpha_1, \ldots,$$

such that $\alpha_j = 0$, for all $j \geq j_0$, for some $j_0 \in \mathbb{Z}$, that is, where all addresses sufficiently far in the future are 0, belong to the stable separatrix $W^s\{0\}$. The unstable separatrix $W^u\{0\}$ contains similar sequences where now $\alpha_j = 0$, for all $j \leq j_0$, for some $j_0 \in \mathbb{Z}$, that is, where all addresses sufficiently far in the past are 0. Analogous remarks hold for the saddle point $\{1\}$.

Again as in the case of the doubling map (see §1.3.8.4) we can also use symbolic dynamics to get information on the periodic points. For instance, the number of periodic points with a given prime period is determined in Exercise 4.9. Also it is easy to see that the set of periodic points is dense in $D_{-\infty}^{\infty}$.

Another subject concerns the *homoclinic points*. We say that a point q is homoclinic with respect to the saddle point p if

$$q \in W^u(p) \cap W^s(p), \quad q \neq p.$$

It now follows that the set of points q that are homoclinic to the saddle point $\{0\}$ also is dense in $D_{-\infty}^{\infty}$. To see this, observe that any sequence $\ldots, \alpha_{-1}, \alpha_0, \alpha_1, \ldots$ can be approximated (arbitrarily) by a sequence $\ldots, \alpha'_{-1}, \alpha'_0, \alpha'_1, \ldots,$ where for a (large) value $j_0 \in \mathbb{Z}$ it holds that

$$\alpha'_j = \alpha_j \quad \text{for } j < j_0 \quad \text{and} \quad \alpha'_j = 0 \quad \text{for } |j| \geq j_0.$$

A similar remark goes for the set of points homoclinic to $\{1\}$.

Apart from homoclinic points, we also distinguish *heteroclinic points* with respect to two saddle points p_1 and p_2. These are the points inside

$$W^u(p_1) \cap W^s(p_2) \quad \text{or} \quad W^s(p_1) \cap W^u(p_2).$$

We leave it to the reader to show that the points that are heteroclinic with respect to $\{0\}$ and $\{1\}$ form a dense subset of $D^\infty_{-\infty}$.

We can continue for a while in this spirit. Any of the periodic points of period k are a saddle fixed point of f^k, and hence have a stable and unstable separatrix and homoclinic points. Heteroclinic points between any two of such periodic points (also of different periods) can also be obtained. Following the above ideas (also compare §1.3.8.4) it can be shown that, for any periodic point p, the set of homoclinic points with respect to p is dense in $D^\infty_{-\infty}$. Similarly, for any two periodic points p_1 and p_2, the set of corresponding heteroclinic points is dense in $D^\infty_{-\infty}$.

Finally, again as in §1.3.8.4, it follows that if we choose a sequence of 0s and 1s by independent tosses of a fair coin, with probability 1 we get a symbol sequence corresponding to a point with a chaotic evolution, which lies densely in $D^\infty_{-\infty}$.

All the facts mentioned up to now rest on methods very similar to those used for the doubling map in §1.3.8. There are, however, a number of differences as well. For the horseshoe map we have bi-infinite sequences, corresponding to invertibility of this map. Also, the assignment map α is a homeomorphism, which it was not in the case of the doubling map. This has to do with the fact that the domain of definition of the doubling map is connected, whereas the symbol space Σ_2^+ is totally disconnected. Presently this difference does not occur: the sets $D^\infty_{-\infty}$ and $\mathbb{Z}_2^{\mathbb{Z}}$ both are totally disconnected; in fact, they are both Cantor sets (see Appendix A).

4.4.2.2   Structural stability

The set $D^\infty_{-\infty}$ is a hyperbolic set for the horseshoe map f. Although here we are not dealing with an attractor, hyperbolicity is defined exactly in the same way as for an attractor; compare Definition 4.7. We do not go into details here, but just mention that this hyperbolicity here is also the reason that the dynamics of the horseshoe map is structurally stable, in the following sense.

Theorem 4.14 (Structural stability of the horseshoe map). *There exists a neighbourhood $\mathcal{U}$ of the horseshoe map f in the C^1-topology, such that for each diffeomorphism $\tilde{f} \in \mathcal{U}$ there exists a homeomorphism*

$$h : \bigcap_{j \in \mathbb{Z}} f^j(D) \to \bigcap_{j \in \mathbb{Z}} \tilde{f}^j(D)$$

such that for the corresponding restrictions one has

$$\left(\tilde{f}|_{\bigcap_j \tilde{f}^j(D)} \right) \circ h = h \circ \left(f|_{\bigcap_j f^j(D)} \right).$$

The description of the homeomorphism in Theorem 4.14 is quite simple. For f and σ we just saw that the following diagram commutes.

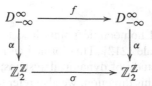

where, as before, $D_{-\infty}^{\infty} = \bigcap_j f^j(D)$. Now it turns out that a counterpart diagram for $\tilde{f}$, when sufficiently C^1-close to f, commutes just as well. This gives an assignment map

$$\tilde{\alpha} : \bigcap_j \tilde{f}^j(D) \to \mathbb{Z}_2^{\mathbb{Z}}.$$

The fact that the symbolic descriptions for f and $\tilde{f}$ match then provides the homeomorphism $h = \tilde{\alpha}^{-1} \circ \alpha$. For more details on the proof of Theorem 4.14 and on similar results regarding hyperbolic sets, see [13, 210].

We just saw that the horseshoe map contains many homoclinic points. We also like to point to a kind of converse of this. Indeed, let φ be a diffeomorphism with a hyperbolic saddle point p and let $q \neq p$ be a transversal, that is, nontangent, point of intersection

$$q \in W^u(p) \bigcap W^s(p),$$

implying that q is homoclinic to p. Then there exists an integer $n > 0$ and a φ^n-invariant set Z (i.e., with $\varphi^n(Z) = Z$) such that the dynamics of $\varphi^n|_Z$ is conjugated to that of the horseshoe map f restricted to $D_{-\infty}^{\infty} = \bigcap_j f^j(D)$. For this statement it is not necessary that the manifold on which φ is defined have dimension 2. In Figure 4.10 we sketched how a transversal intersection of stable and unstable separatrix of a saddle point can give rise to a map that, up to conjugation, can be shown to be a horseshoe map.

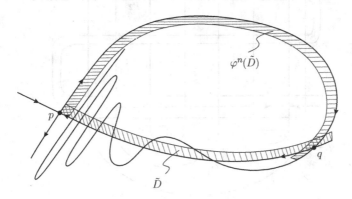

Fig. 4.10 Two-dimensional diffeomorphism φ with a saddle point p and a transversal homoclinic intersection q. The set $\tilde{D}$, homeomorphic to the unit square $[0, 1] \times [0, 1]$, by a well-chosen iterate φ^n is 'mapped over $\tilde{D}$' in a horseshoelike way.

Remarks.

- The fact that transversal homoclinic points lead to horseshoe dynamics, was already observed by Smale [212]. This formed an important basis for his program on the geometric study of dynamical systems; see [213]. For proofs and further developments in this direction, we also refer to [30].
- In general, structural stability implies persistence of all qualitative properties under C^1-small perturbations of the system, thus in particular under variation of parameters.
  In particular the structural stability of the horseshoe map also implies this persistence (compare Figure 4.10), for instance, if this occurs in Hamiltonian systems.
- The structural stability also holds for the extended horeseshoe map on $\mathbb{S}^2$.
- Higher-dimensional analogues of the horseshoe map exist, with a similar symbolic dynamics, structural stability, and, hence, persistence.

4.4.3 Horseshoelike sets in basin boundaries

We now consider a modified form of the horseshoe map. Again a square $D = [0, 1] \times [0, 1]$ by a diffeomorphism f is 'mapped over itself' where now $D \cap f(D)$ consists of three vertical strips; compare Figure 4.11. Moreover we assume that the

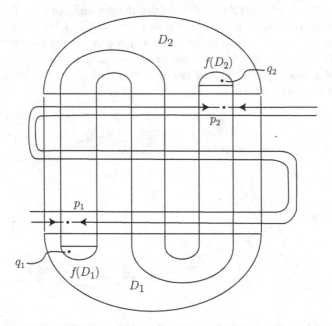

Fig. 4.11 Action of the horseshoe map f on D, D_1, and D_2 with hyperbolic point attractors q_1 and q_2 and the saddle fixed points p_1 and p_2.

half-circular discs D_1 and D_2 are mapped within themselves. This can be arranged such that within $\tilde{D} = D \cup D_1 \cup D_2$, there are exactly two hyperbolic point attractors q_1 and q_2, situated in D_1 and D_2, respectively. We now investigate the boundary of the corresponding basins.

As before, also two saddle fixed points p_1 and p_2 exist and it turns out that this basin boundary is formed by the closure

$$\overline{W^s(p_1)} = \overline{W^s(p_2)} \tag{4.8}$$

of the stable manifolds of each of the two saddle points p_1 and p_2; for a sketch see Figure 4.11. Also this modified horseshoe map can be extended as a diffeomorphism of $\mathbb{S}^2$, in which case both $W^s(p_1)$ and $W^s(p_2)$ 'connect' to the source at ∞.

We now describe the situation within $\tilde{D}$. The set $\bigcap_j f^j(D)$ and the dynamics inside, as for the 'ordinary' horseshoe map, can be described in terms of infinite sequences

$$\ldots, \beta_{-1}, \beta_0, \beta_1, \ldots,$$

where now the elements β_j take on the values $0, 1$, or 2. Almost all remarks made on the horseshoe map also apply here. The only exception is that here not all transversal homoclinic intersections give rise to this modified horseshoe dynamics. Moreover, the boundary between the basins of q_1 and q_2, as far as contained in $\tilde{D}$, exactly is formed by $\bigcap_{\ell \geq 0} f^{-\ell}(D)$, which would confirm the claim (4.8).

To prove this, we introduce some notation. Recall that the point attractor contained in D_j is called $q_j, j = 1, 2$. The saddle fixed points in the left and right vertical strips of $D \cap f(D)$ we call p_1, respectively, p_2. These saddle points correspond to symbol sequences that exclusively consist of 0s, respectively, 2s. It now should be clear that one branch of the unstable separatrix $W^u(p_1)$ is entirely contained in the basin of q_1, and similarly for p_2 and q_2. This implies that the stable separatrix $W^s(p_1)$ must lie in the boundary of the basin of q_2, because the points in $W^s(p_1)$, although they do not tend to q_1, are arbitrarily close to points that under iteration by f do tend to q_1. An analogous statement holds for p_2 and q_2. As for the standard horseshoe map,

$$W^s(p_j) \cap D \quad \text{is dense in} \quad \bigcap_{\ell \geq 0} f^{-\ell}(D),$$

for $j = 1, 2$. This means that $\bigcap_{\ell \geq 0} f^{-\ell}(D)$ must be part of the basin boundary of q_1 and of q_2.

Moreover if $r \in D \setminus \bigcap_{\ell \geq 0} f^{-\ell}(D)$, then for some n we have $f^n(r) \notin D$, so either $f^n(r) \in D_1$ or $f^n(r) \in D_2$. Hence r is in the interior of the basin of either q_1 or q_2. This finishes the proof.

Conclusive remarks. In the above example we have basins of q_1 and q_2 that have a joint boundary containing a set that is 'diffeomorphic' to the product of a Cantor set and an interval. That such a set really is 'fatter' than a (smooth) curve can be formalised as follows, using concepts (regarding fractal dimensions) that are treated more thoroughly in Chapter 6. We consider a bounded piece $\mathcal{L}$ of a smooth curve, and on the other hand we consider the product $\mathcal{C}$ of an interval and a Cantor set as this occurs in the above example. Both sets we regard as compact subsets of the plane $\mathbb{R}^2$. An essential difference between these two sets can be seen when considering the areas of their ε-neighbourhoods. Let $\mathcal{L}_\varepsilon$ be the ε-neighbourhood of $\mathcal{L}$ and, similarly, $\mathcal{C}_\varepsilon$ the ε-neighbourhood of $\mathcal{C}$. Then, for small values of $\varepsilon > 0$, the area of $\mathcal{C}_\varepsilon$ is much larger than that of $\mathcal{L}_\varepsilon$. In fact, it can be shown that

$$\lim_{\varepsilon \to 0} \frac{\ln(\text{area of } \mathcal{L}_\varepsilon)}{\ln(\varepsilon)} = 1$$

$$\lim_{\varepsilon \to 0} \frac{\ln(\text{area of } \mathcal{C}_\varepsilon)}{\ln(\varepsilon)} < 1;$$

compare [12]. This means that, if we can determine the initial states of evolutions only up to precision ε, the area where it is unknown whether the corresponding evolution converges to q_1 or q_2, is extremely large for small ε. This phenomenon, where it is very hard to determine an initial state that leads to a desired final state, is familiar from playing dice or tossing a coin. It therefore seems true that also for the dynamics of dice or coin, the basins belonging to the various final states (like the numbers $1, 2, \ldots, 6$ or the states 'heads' and 'tails'), have more or less similar basin boundaries.

4.5 Exercises

Generally we are dealing with a dynamical system (M, T, Φ) such that M is a topological space, with continuous Φ, and the evolutions are positively compact.

Exercise 4.1 (ω-limit set). Show that for any $x \in M$ the set $\omega(x)$ is nonempty, compact, and invariant under the evolution operator. The latter means that for any $y \in \omega(x)$ also $\Phi(y \times T) \subset \omega(x)$.

Exercise 4.2 (Property of attractor). Let $A \subset M$ be a compact invariant set for a dynamical system with evolution operator Φ. Assume that A has arbitrarily small neighbourhoods $U \subset M$ such that $\Phi^t(U) \subset U$ for all $0 < t \in T$ and such that for all $y \in U$ also $\omega(y) \subset A$. Then show that $\bigcap_{0 \le t \in T} \Phi^t(U) = A$ for any such U.

Exercise 4.3 (On attractors). Give an example of an attractor according to Definition 4.3, that contains no dense evolution, and hence does not satisfy the alternative definition.

Exercise 4.4 (Conservative systems have no attractors). Consider a dynamical system (M, T, Φ), where on M a probability measure μ is defined, such that any open subset $U \subset M$ has positive μ-measure. We call such a system measure-preserving (or conservative) if for any $t \in T$ the map Φ^t is a μ-preserving transformation. Show that for such systems no attractors are possible, except possibly the whole state space M.

Exercise 4.5 (Hyperbolicity of the Thom map). Let $\varphi : \mathbb{T}^2 \to \mathbb{T}^2$ be the Thom map; see §2.3.5. Show that $\mathbb{T}^2$ is a hyperbolic attractor for φ.

Exercise 4.6 (The solenoid revisited). In this exercise we use the coding of the discs $K_j \cap D_\varphi$ by binary numbers as introduced in § 4.2.1.

1. Show that the discs in $K_j \cap D_\varphi$, with binary expression $\alpha_{i-1}, \ldots, \alpha_0$, by f under the solenoid map (4.2) are put on the disc in $K_{j+1} \cap D_{2\varphi}$, with binary expression $\alpha_{i-1}, \ldots, \alpha_0, 0$.
2. Show that the disc in $K_j \cap D_\varphi$ with binary expression $\alpha_{i-1}, \ldots, \alpha_0$, as a disc in $K_j \cap D_{\varphi+2\pi}$, has the binary expression $\alpha'_{i-1}, \ldots, \alpha'_0$ whenever

$$\sum_{\ell=0}^{j-1} \alpha'_\ell 2^\ell = \left(\sum_{\ell=0}^{j-1} \alpha_\ell 2^\ell \right) - 1 \bmod 2^j.$$

Exercise 4.7 (The 2-adic metric). We define the 2-adic distance $\rho_2(i, j)$ between elements $i, j \in \mathbb{Z}_+$ as

$$\rho_2(i, j) = 2^{-n},$$

if n is the number of factors 2 in the absolute difference $|i - j|$. The corresponding norm on $\mathbb{Z}_+$ is denoted by $|z|_2 = \rho_2(z, 0)$.

1. Show that the triangle inequality

$$|x + y|_2 \le |x|_2 + |y|_2$$

holds true. (In fact, it is even true that $|x + y|_2 = \max\{|x|_2, |y|_2\}$.)
2. Show that

$$|x|_2 |y|_2 = |xy|_2.$$

Let $\mathbb{N}_2$ be the metric completion of $\mathbb{Z}_+$ with respect to the metric ρ_2. Then show that:

1. The elements of $\mathbb{N}_2$ can be represented in a unique way by infinite binary expressions

$$(\ldots, \alpha_2, \alpha_1, \alpha_0),$$

corresponding to the limit

$$\lim_{n \to \infty} \sum_{j=0}^{n} \alpha_j 2^j.$$

2. $\mathbb{N}_2$ is an additive group (where $-1 = (\ldots, 1, 1, 1)$).

3. The solenoid is homeomorphic to the quotient

$$K = \mathbb{N}_2 \times \mathbb{R}/\langle(-1, 2\pi)\rangle,$$

where $\langle(-1, 2\pi)\rangle$ is the subgroup generated by $(-1, 2\pi)$. The dynamics of the solenoid as an attractor, by this homeomorphism is conjugated to multiplication by 2 on K.

(Pro memori: This algebraic version of the solenoid is the format in which it was first introduced by D. van Dantzig in his study on topological groups [110].)

Exercise 4.8 (Attractors of gradient systems). Consider the gradient system $x' = -\mathrm{grad}\,V(x)$ with $x \in \mathbb{R}^2$, where the critical points of the smooth function V are nondegenerate.[7]

1. Show that this system cannot have any (nonconstant) periodic evolutions.
2. Show that for any $x \in \mathbb{R}^2$, for which the positive evolution starting at x is bounded, the set $\omega(x)$ consists of one point, which is a critical point of V.
3. ($*$) To what extent are the above items still true when we drop the requirement that the critical points of V are nondegenerate?

Exercise 4.9 (Periodic points of horseshoe map). Compute the number of points in D that have prime period 12 for the horseshoe map.

Exercise 4.10 (The unstable distribution of the solenoid). We revisit the unstable distribution for the solenoid, using the notation of §4.2. In 3-space we use cylinder coordinates (r, φ, z). For points in 3-space, for which $r \neq 0$, a basis of tangent vectors is given by $\partial_r, \partial_\varphi$, and ∂_z. We consider a 1-dimensional distribution $\mathcal{E}$ on solenoid A such that, at each point $p \in A$, the projection of $\mathcal{E}(p)$ on the φ-axis is surjective. Such a distribution can be uniquely represented as a vector field of the form

$$R\partial_r + \partial_\varphi + Z\partial_z,$$

where R and Z are continuous functions on A. For two such distributions $\mathcal{E}$ and $\mathcal{E}'$ we define the distance by

$$\rho(\mathcal{E}, \mathcal{E}') = \max_{p \in A} \sqrt{(R(p) - R'(p))^2 + (Z(p) - Z'(p))^2}.$$

Now show that:

1. By the transformation T_f the 1-dimensional distributions of the above type, are taken over into 1-dimensional distributions of this same type.
2. $\rho(T_f\mathcal{E}, T_f\mathcal{E}') = \frac{1}{2}\lambda\rho(\mathcal{E}, \mathcal{E}')$.

---

[7] Recall that we assume all evolutions to be positively compact.

3. The limit $E^u = \lim_{j \to \infty} (T_f)^j \mathcal{E}$ exists, is continuous, and is invariant under the derivative df and moreover has the property that vectors $v \in E^u(p)$ increase exponentially under repeated application of df.

Exercise 4.11 ((*) **Symbolic dynamics more general**). Consider the space $\mathcal{C}$ of maps $f : [0, 1] \setminus \{\frac{1}{2}\} \to [0, 1]$, such that

- $\lim_{x \uparrow 1/2} f(x) = 1$ and $\lim_{x \downarrow 1/2} f(x) = 0$.
- f is of class C^1 on the intervals $[0, \frac{1}{2})$ and $(\frac{1}{2}, 1]$, and for all such x one has $f'(x) > 1$.
- $f(0) > 0$ and $f(1) < 1$.

To each $x \in [0, 1]$ we assign a sequence of symbols 0 and 1 (in a not completely unique way) where the sequence

$$s_0(x), s_1(x), s_2(x), \ldots$$

corresponds to

$$s_j(x) = \begin{cases} 0 & \text{whenever } f^j(x) \in [0, \frac{1}{2}) \\ 1 & \text{whenever } f^j(x) \in (\frac{1}{2}, 1] \end{cases}$$

This definition fails when for a certain j_0 one has $f^{j_0}(x) = \frac{1}{2}$, in which case we continue the sequence as if the next iteration by f would yield the value 0 or 1 (which introduces a choice). Next to each symbol sequence $s_0(x), s_1(x), s_2(x), \ldots$ we assign the number

$$S(x) = \sum_j s_j(x) 2^j,$$

where, whenever the symbol sequence is nonunique, we define $S_-(x)$, respectively $S_+(x)$, as the minimum or the maximum that can occur for the different choices. The numbers $S(x)$, $S_-(x)$, and $S_+(x)$, thus defined, depend on x, but also on the map f. If we wish to express the latter dependence as well, we write $S^f(x)$, and so on.

Show that for $f_1, f_2 \in \mathcal{C}$ the following holds true. There exists a homeomorphism $h : [0, 1] \to [0, 1]$ with $h \circ f_1 = f_2 \circ h$, if and only if $S_-^{f_1}(0) = S_-^{f_2}(0)$ and $S_+^{f_1}(1) = S_+^{f_2}(1)$.

Exercise 4.12 ((*) **Symbolic dynamics of the Thom map**). (Compare [8].) Consider the Thom map $\varphi : \mathbb{T}^2 \to \mathbb{T}^2$ as introduced in §2.3.5, generated by the matrix

$$A = \begin{pmatrix} 2 & 1 \\ 1 & 1 \end{pmatrix};$$

Fig. 4.12 Markov partition
for the Thom map.

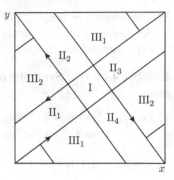

see (2.20). With help of the eigenspaces of A a φ-invariant partition of $\mathbb{T}^2$ can be given, as sketched in Figure 4.12.[8] The boundaries of the elements of the partition are obtained by taking intervals

$$I_s \subset W^s([0]) \quad \text{and} \quad I_u \subset W^u([0])$$

such that

1. $\varphi(I_s) \subset I_s$ and $\varphi(I_u) \supset I_u$.
2. $\partial I_s \subset I_u$ and $\partial I_u \subset I_s$.

Develop a symbolic dynamics for φ based on this partition. Hint: See §1.3.9.

Exercise 4.13 (($*$) Horseshoe and basin boundaries). Consider the horseshoe map, with its saddle fixed points p_1 and p_2; see §4.3.2.1. Show that there exists a diffeomorphism, with an attractor A such that the closure of the stable manifold $\overline{W^s(p_1)} = \overline{W^s(p_2)}$ corresponds to the basin boundary of A.

Exercise 4.14 (($*$) Like the Lorenz return maps). (Compare Sparrow [214].) We consider maps $f_\alpha : [-1, +1] \setminus \{0\} \to [-1, +1]$, where $\alpha \in (1, 2]$, given by

$$f_\alpha(x) = \alpha x - \text{sign}(x),$$

where

$$\text{sign}(x) = \frac{x}{|x|}.$$

Whenever desired we define $f_\alpha(0) = \pm 1$.

1. Show that, for $\sqrt{2} < \alpha < 2$, the map f_α has a dense evolution.
2. Show that for $1 < \alpha < \sqrt{2}$ the interval $I_\alpha = [1 - \alpha, \alpha - 1]$ and the map f_α have the following properties.

---

[8] Often called a Markov partition. Such a partition indicates that at a coarse level, the deterministic system resembles a discrete-time Markov process.

(a) $f_\alpha(I_\alpha)$ consists of two intervals that are both disjoint with I_α.

(b) $f_\alpha^2(I_\alpha) \subset I_\alpha$.

(c) The restriction $f_\alpha^2|_{I_\alpha}$, up to a linear rescaling, is equal to f_{α^2}.

(d) I_α is an attractor of f_{α^2}.

3. Investigate by induction which 'bifurcations' occur when $\alpha = 2^{1/2}, 2^{1/4}$, $2^{1/8}, \ldots$, and so on.

   (Nota bene: By a bifurcation occurring at $\alpha = \alpha_0$ we understand that the dynamics changes discontinuously when α passes through α_0.)

Exercise 4.15. ($*$) Show that any hyperbolic attractor of a dynamical system with a 1-dimensional state space either consists of one periodic or stationary evolution, or is the entire state space.

Chapter 5
On KAM theory

As a motivation for this chapter we refer to the historical remarks of §2.2.5 and the references given there. The main question is to what extent a quasi-periodic subsystem, as defined in §2.2.3, is persistent under a small perturbation of the whole system, or say, under variation of parameters; see §3.2.

This question belongs to Kolmogorov–Arnold–Moser (KAM) theory (see §2.2.5), which classically was mainly developed for conservative systems. Presently we restrict our attention to a case concerning certain circle maps, thereby motivating the *linear small-divisor problem*, that is discussed to some extent. In Appendix B we deal with the 'classical' KAM theorem regarding Lagrangian tori in Hamiltonian systems, as well as its 'dissipative' counterpart regarding a family of quasi-periodic attractors. Also we give elements of 'the' KAM proof, using the linear theory introduced in §5.5 below.

5.1 Introduction, setting of the problem

Our present setup of KAM theory uses two leading examples, one on periodically driven oscillators and the other on coupled oscillators. The former of these briefly revisits Example 2.2.3.2 on the driven Van der Pol equation.

Example 5.1 (Periodically driven oscillator I). The starting point of the present considerations is the periodically driven Van der Pol oscillator (2.7, see §2.2.3), in a slightly more general C^∞-format

$$y'' + cy' + ay + f(y, y') = \varepsilon g(y, y', \omega t; \varepsilon), \qquad (5.1)$$

where the function g is 2π-periodic in its third argument. It is important that f is such that the corresponding free oscillator has a hyperbolic attractor; for instance,

H.W. Broer and F. Takens, *Dynamical Systems and Chaos*,   173
Applied Mathematical Sciences 172, DOI 10.1007/978-1-4419-6870-8_5,
© Springer Science+Business Media, LLC 2011

Fig. 5.1 Phase portrait of the
free Van der Pol oscillator.

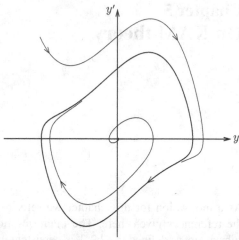

take $f(y, y') = by^2 y'$, for an appropriate value of b. For the phase portrait in the
(y, y')-plane see Figure 5.1. Passing to the system form

$$
\begin{aligned}
x' &= \omega \\
y' &= z \\
z' &= -ay - cz - f(y, z) + \varepsilon g(y, z, x; \varepsilon),
\end{aligned}
\qquad (5.2)
$$

we get a 3-dimensional state space coordinated by (x, y, z), with $x \in \mathbb{T}^1 = \mathbb{R}/(2\pi\mathbb{Z})$. Let us denote the corresponding vector field by X_ε. See Chapters 1 and 2 for this construction. As explained in §2.2, the unperturbed system X_0 has an attracting multiperiodic subsystem given by the restriction of X_0 to a 2-torus T_0, where the restriction $X_0|_{T_0}$ is smoothly conjugated to the translation system (2.10) here presented as

$$
\begin{aligned}
x_1' &= \omega_1 \\
x_2' &= \omega_2
\end{aligned}
\qquad (5.3)
$$

on the standard 2-torus $\mathbb{T}^2 = \mathbb{T}^2/(2\pi\mathbb{Z}^n)$. Note that $x_1 = x$ and $\omega_1 = \omega$; the value of ω_2 is not so easily identified.

A general question is what happens to this 2-torus attractor for $\varepsilon \neq 0$. By hyperbolicity of the periodic orbit, the unperturbed torus T_0 is normally hyperbolic. Thus, according to the normally hyperbolic invariant manifold theorem ([143], Theorem 4.1), the 2-torus T_0 is persistent as an invariant manifold. See Example 2.2.3.2 for a heuristic explanation of this argument. This means that for $|\varepsilon| \ll 1$, the vector field X_ε has a smooth invariant 2-torus T_ε, where the dependence on ε is also smooth. Here 'smooth' means finitely differentiable, where the degree of differentiability tends to ∞ as $\varepsilon \to 0$.

The remaining question then concerns the dynamics of $X_\varepsilon|_{T_\varepsilon}$. To what extent is the multiperiodicity of $X_0|_{T_0}$ also persistent for $\varepsilon \neq 0$? Before going deeper into this problem we first give the second leading example.

Example 5.2 (Coupled oscillators I). As a very similar example consider two non-linear oscillators with a weak coupling

$$y_1'' + c_1 y_1' + a_1 y_1 + f_1(y_1, y_1') = \varepsilon g_1(y_1, y_2, y_1', y_2')$$
$$y_2'' + c_2 y_2' + a_2 y_2 + f_2(y_2, y_2) = \varepsilon g_2(y_1, y_2, y_1, y_2),$$

$y_1, y_2 \in \mathbb{R}$. This yields the following vector field X_ε on the 4-dimensional phase space $\mathbb{R}^2 \times \mathbb{R}^2 = \{(y_1, z_1), (y_2, z_2)\}$:

$$y_j' = z_j$$
$$z_j' = -a_j y_j - c_j z_j - f_j(y_j, z_j) + \varepsilon g_j(y_1, y_2, z_1, z_2),$$

$j = 1, 2$. Note that for $\varepsilon = 0$ the system decouples into a system of two independent oscillators which has an attractor in the form of a two-dimensional torus T_0. This torus arises as the product of two (topological) circles, along which each of the oscillators has its periodic solution. The circles lie in the two-dimensional planes given by $z_2 = y_2 = 0$ and $z_1 = y_1 = 0$, respectively. The vector field X_0 has a multi-periodic subsystem as before, in the sense that the restriction $X_0|_{T_0}$ is smoothly conjugated to a translation system of the form (C.12) on $\mathbb{T}^2$.[1] The considerations and problems regarding the perturbation X_ε, the invariant 2-torus T_ε, and the multi-periodic dynamics of $X_\varepsilon|_{T_\varepsilon}$, are as in the periodically driven Van der Pol oscillator of Example 5.1.

5.2 KAM theory of circle maps

Continuing the considerations of §5.1, we now discuss the persistence of the dynamics in the X_ε-invariant 2-tori T_ε as this comes up in the above examples. After a first reduction to the standard 2-torus $\mathbb{T}^2$, we further reduce to maps of the circle $\mathbb{T}^1$ by taking a Poincaré section; compare §1.2.2.

5.2.1 Preliminaries

As said before, the tori T_ε are highly differentiable manifolds due to normal hyperbolicity [143]. Indeed, for $|\varepsilon| \ll 1$, all T_ε are diffeomorphic to the standard 2-torus $\mathbb{T}^2$, where the degree of differentiability increases to ∞ as ε tends to 0. To study the dynamics of X_ε inside the torus T_ε, we use this diffeomorphism to get a perturbation problem on $\mathbb{T}^2$. For simplicity we also use the name X_ε here for the system.

---

[1] In this case neither ω_1 nor ω_2 is identified.

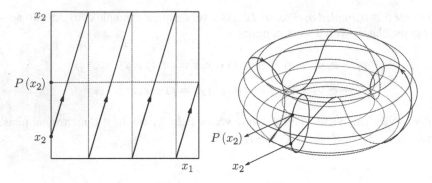

Fig. 5.2 Poincaré map associated with the section $x_1 = 0$ of $\mathbb{T}^2$.

The KAM theorem we are preparing allows for formulations in the world of C^k-systems for k sufficiently large, including $k = \infty$; compare [88, 194]. These versions are variations on the simpler setting where the systems are of class C^∞. For simplicity we therefore restrict to the case where the entire perturbation problem is C^∞.

The KAM setup at hand in principle looks for smooth conjugations between X_0 and X_ε, the existence of which would imply that X_ε also has a multiperiodic subsystem. However, from §2.2.2. we recall the fact that there exist arbitrary small perturbations that turn quasi-periodic into periodic and vice versa, and between such cases there cannot be a conjugation. Also see Exercise 2.13 on smooth conjugations between translation systems.

A rather standard way to deal with such problems systematically is to introduce parameters [2, 9]. Then, if trying to conjugate two different families of dynamical systems, parameter shifts may be used to avoid the difficulties just described. In the example (2.7) of §2.2.3, we may consider $\mu = (a, c)$ as a multiparameter and in the coupled oscillator case ending §5.1 even $\mu = (a_1, a_2, c_1, c_2)$. In principle, we could even take $\mu = (\omega_1, \omega_2)$ in the translation system format (C.12) as a multiparameter. To achieve this one usually requires that (ω_1, ω_2) depends in a submersive way on parameters as in the former two examples.

To simplify things a bit, we relax the KAM setup in the sense that we also allow for a smooth equivalence (see §3.4.2) between X_0 and X_ε, in which case only the frequency ratio $\omega_1 : \omega_2$ is an invariant; see the Exercises 5.1 and 5.2. For simplicity we consider $\beta = \omega_2/\omega_1$ itself as a parameter, so we study the family of vector fields $X_{\beta,\varepsilon}$ on $\mathbb{T}^2 = \{x_1, x_2\}$, both variables being counted modulo $2\pi\mathbb{Z}$.

To exploit this further we now turn to the corresponding family of Poincaré return maps $P_{\beta,\varepsilon}$ on the circle $x_1 = 0$, considered as a circle diffeomorphism $\mathbb{T}^1 \to \mathbb{T}^1$; again compare §2.2.3. Referring to a remark at the beginning of §2.2.1, we announce that throughout this chapter the convention $\mathbb{T}^1 = \mathbb{R}/(2\pi\mathbb{Z})$ is used. The map $P_{\beta,\varepsilon}$ has the form

$$P_{\beta,\varepsilon} : x_2 \mapsto x_2 + 2\pi\beta + \varepsilon a(x_2, \beta, \varepsilon). \tag{5.4}$$

Note that for $\varepsilon = 0$ the unperturbed map $P_{\beta,0}$ is just the discrete translation system given by the rigid rotation $R_{2\pi\beta} : x_2 \mapsto x_2 + 2\pi\beta$ of the circle $\mathbb{T}^1$. Compare §2.2.2.

For a proper formulation of our problem it is convenient to regard the family of circle maps as a 'fibre-preserving'[2] map of the cylinder by considering $P_\varepsilon : \mathbb{T}^1 \times [0,1] \to \mathbb{T}^1 \times [0,1]$ defined as

$$P_\varepsilon : (x_2, \beta) \mapsto (x_2 + 2\pi\beta + \varepsilon a(x_2, \beta, \varepsilon), \beta). \qquad (5.5)$$

Adopting the same fibre-preserving convention of writing X_ε at the level of vector fields, we observe that a conjugation between P_0 and P_ε gives rise to an equivalence between X_0 and X_ε.

To solve the conjugation problem for Poincaré maps, we start looking naively for a 'skew'[3] diffeomorphism $\Phi_\varepsilon : \mathbb{T}^1 \times [0,1] \to \mathbb{T}^1 \times [0,1]$ conjugating the unperturbed family P_0 to the perturbed family P_ε, this is, such that

$$P_\varepsilon \circ \Phi_\varepsilon = \Phi_\varepsilon \circ P_0. \qquad (5.6)$$

The conjugation equation (5.6) also is expressed by commutativity of the following diagram.

$$
\begin{array}{ccc}
\mathbb{T}^1 \times [0,1] & \xrightarrow{\ P_0\ } & \mathbb{T}^1 \times [0,1] \\
\Big\downarrow{\scriptstyle \Phi_\varepsilon} & & \Big\downarrow{\scriptstyle \Phi_\varepsilon} \\
\mathbb{T}^1 \times [0,1] & \xrightarrow[\ P_\varepsilon\ ]{} & \mathbb{T}^1 \times [0,1]
\end{array}
$$

When trying to solve the conjugation equation (5.6) we discover under what further conditions this can be done.

Remarks.

– At the level of circle maps (5.4), the conjugation Φ_ε maps a multiperiodic subsystem onto a multiperiodic subsystem.
  Conjugations between return maps directly translate to equivalences between the corresponding vector fields; see [2, 9]. In the case of the first example (5.1) these equivalences can even be made into conjugations.
– For orientation-preserving circle homeomorphisms, the rotation number is the average of rotation effected by the homeomorphism, which for the unperturbed map $P_{\beta,0} = R_{2\pi\beta}$ exactly coincides with the frequency ratio β. An important property of the rotation number is its invariance under (topological) conjugation [2, 4]. For a definition also see Exercise 5.3.

---

[2] Regarding $\mathbb{T}^1 \times [0,1]$ as a $\mathbb{T}^1$-bundle over $[0,1]$, the map P_ε is fibre-preserving [142].

[3] The map Φ_ε needs to send fibres to fibres and therefore is given this skew format.

– The Denjoy theorem [2, 4] asserts that for irrational β, whenever the rotation number of $P_{\beta',\varepsilon}$ equals β, a topological conjugation exists between $P_{\beta,0}$ and $P_{\beta',\varepsilon}$.

5.2.2   Formal considerations and small divisors

We now study equation (5.6) for the conjugation Φ_ε to some detail. To simplify the notation, first set $x = x_2$. Assuming that $\Phi_\varepsilon : \mathbb{T}^1 \times [0, 1] \to \mathbb{T}^1 \times [0, 1]$ has the general skew form

$$\Phi_\varepsilon(x, \beta) = (x + \varepsilon U(x, \beta, \varepsilon), \beta + \varepsilon \sigma(\beta, \varepsilon)), \tag{5.7}$$

we get the following nonlinear equation for the function U and the parameter shift σ,

$$U(x + 2\pi\beta, \beta, \varepsilon) - U(x, \beta, \varepsilon)$$
$$= 2\pi\sigma(\beta, \varepsilon) + a\left(x + \varepsilon U(x, \beta, \varepsilon), \beta + \varepsilon\sigma(\beta, \varepsilon), \varepsilon\right). \tag{5.8}$$

As is common in classical perturbation theory, we expand a, U, and σ as formal power series in ε and solve (5.8) by comparing coefficients. We only consider the coefficients of power zero in ε, not only because asymptotically these coefficients have the strongest effect, but also because the coefficients of higher ε-powers satisfy similar equations. So, writing

$$a(x, \beta, \varepsilon) = a_0(x, \beta) + O(\varepsilon), \quad U(x, \beta, \varepsilon) = U_0(x, \beta) + O(\varepsilon),$$

$$\sigma(\beta, \varepsilon) = \sigma_0(\beta) + O(\varepsilon),$$

substitution in equation (5.8) leads to the following linear, so-called homological, equation

$$U_0(x + 2\pi\beta, \beta) - U_0(x, \beta) = 2\pi\sigma_0(\beta) + a_0(x, \beta), \tag{5.9}$$

which has to be solved for U_0 and σ_0. Equation (5.9) is linear and therefore can be directly solved by Fourier series; also compare §1.3.5.3. Indeed, introducing

$$a_0(x, \beta) = \sum_{k \in \mathbb{Z}} a_{0k}(\beta)\, e^{ikx} \quad \text{and} \quad U_0(x, \beta) = \sum_{k \in \mathbb{Z}} U_{0k}(\beta)\, e^{ikx}$$

and comparing coefficients in (5.9), directly yields that

$$\sigma_0(\beta) = -\frac{1}{2\pi} a_{00}(\beta), \quad \text{and} \quad U_{0k}(\beta) = \frac{a_{0k}(\beta)}{e^{2\pi ik\beta} - 1}, \quad k \in \mathbb{Z} \setminus \{0\}, \tag{5.10}$$

and U_{00}, which corresponds to the position of the origin 0 on $\mathbb{T}^1$, remains arbitrary. This leads to the solution of (5.9) as a formal Fourier series:

$$U_0(x, \beta) = \sum_{k \in \mathbb{Z}} \frac{a_{0k}(\beta)}{e^{2\pi i k \beta} - 1} e^{ikx}. \tag{5.11}$$

In §5.5 we deal with linear small divisor problems problems in more detail.

We conclude from (5.10) that in general a formal solution exists if and only if β is irrational, which means that the translation system $P_{\beta,0} = R_{2\pi\beta}$ has quasi-periodic dynamics. Compare §2.2.

Even then one meets the problem of small divisors, caused by the accumulation of the denominators in (5.10) on 0, which makes the convergence of the Fourier series of U_0 problematic. This problem can be solved by a further restriction of β by so-called diophantine conditions, guaranteeing that β is sufficiently far away from all rationals p/q.

Definition 5.1 (Diophantine numbers). Let $\tau > 1$ and $\gamma > 0$ be given. We say that $\beta \in [0, 1]$ is $(\tau + 1, \gamma)$-diophantine, if for all $p, q \in \mathbb{Z}$ with $q > 0$ we have that

$$\left| \beta - \frac{p}{q} \right| \geq \frac{\gamma}{q^{\tau+1}}. \tag{5.12}$$

The set of β satisfying (5.12) is denoted by $[0, 1]_{\tau+1,\gamma} \subseteq [0, 1]$.[4] We soon show, that for sufficiently small $\gamma > 0$ this set is nonempty, but even a set of positive Lebesgue measure. First note that $[0, 1]_{\tau+1,\gamma} \subset \mathbb{R} \setminus \mathbb{Q}$, which implies that the translation system given by the rotation $P_{\beta,0}$ is quasi-periodic if $\beta \in [0, 1]_{\tau+1,\gamma}$.

Let us further examine the set $[0, 1]_{\tau+1,\gamma}$, for background in topology and measure theory referring to Appendix A. It is easily seen that $[0, 1]_{\tau+1,\gamma}$ is a closed set. From this, by the Cantor–Bendixson theorem [135] it follows that for small γ the nonempty set $[0, 1]_{\tau+1,\gamma}$ is the union of a perfect set and a countable set. The perfect set, for sufficiently small $\gamma > 0$, has to be a Cantor set, because it is compact and totally disconnected (or zero-dimensional). The latter means that every point of $[0, 1]_{\tau+1,\gamma}$ has arbitrarily small neighbourhoods with empty boundary, which directly follows from the fact that the dense set of rationals is in its complement. Note that, because $[0, 1]_{\tau+1,\gamma}$ is a closed subset of the real line, the property of being totally disconnected is equivalent to being nowhere dense. Anyhow, the set $[0, 1]_{\tau+1,\gamma}$ is small in the topological sense.[5] However, for small positive γ the Lebesgue measure of $[0, 1]_{\tau+1,\gamma}$ is not at all small, inasmuch as

$$\text{measure}\left([0, 1] \setminus [0, 1]_{\tau+1,\gamma}\right) \leq 2\gamma \sum_{q \geq 1} q^{-\tau} = O(\gamma), \tag{5.13}$$

---

[4] The fact that the exponent $\tau + 1$ is used in the definition of $(\tau + 1, \gamma)$-diophantine numbers has to do with the related notion of (τ, γ)-diophantine frequency vectors, introduced in §5.5; see Exercise 5.5.

[5] The complement $[0, 1] \setminus [0, 1]_{\tau+1,\gamma}$ is open and dense.

as $\gamma \downarrow 0$, by our assumption that $\tau > 1$ (e.g., compare [2, 87, 88], also for further reference). Note that the estimate (5.13) implies that the union

$$\bigcup_{\gamma > 0} [0, 1]_{\tau+1, \gamma},$$

for any fixed $\tau > 1$, is of full measure. For a thorough discussion of the discrepancy between the measure-theoretical and the topological notion of 'almost all' points in $\mathbb{R}$, see Oxtoby [184]. Also see Exercise 5.15.

Returning to (5.10), we now give a brief discussion on the convergence of Fourier series, for details referring to §5.5, in particular to the Paley–Wiener estimates on the decay of Fourier coefficients. For the C^∞-function a_0 the Fourier coefficients a_{0k} decay rapidly as $|k| \to \infty$. Also see Exercise 5.13. Second, a brief calculation reveals that for $\beta \in [0, 1]_{\tau+1, \gamma}$ it follows that for all $k \in \mathbb{Z} \setminus \{0\}$ we have

$$|e^{2\pi i k \beta} - 1| \geq \frac{2\gamma}{|k|^\tau};$$

see Exercise 5.5. It follows that the coefficients $U_{0,k}$ still have sufficient decay as $|k| \to \infty$, to ensure that the sum U_0 again is a C^∞-function. Of course this does not yet prove that the function $U = U(x, \beta, \varepsilon)$ exists in one way or another for $|\varepsilon| \ll 1$. Yet one has proved the following.

Theorem 5.2 (Circle map theorem). *For $\tau > 1$, for γ sufficiently small and for the perturbation εa in (5.4) sufficiently small in the C^∞-topology, there exists a C^∞ transformation $\Phi_\varepsilon : \mathbb{T}^1 \times [0, 1] \to \mathbb{T}^1 \times [0, 1]$, conjugating the restriction $P_0|_{[0,1]_{\tau+1, \gamma}}$ to a (diophantine) quasi-periodic subsystem of P_ε.*

Theorem 5.2 in the present structural stability formulation (compare Figure 5.3), is a special case of the results in [87, 88]; for details also see §5.5 and Appendix B. We speak here of *quasi-periodic stability*. For earlier versions see [57, 59].

Remarks.

– (Diophantine) rotation numbers are preserved by the map Φ_ε and these irrational rotation numbers correspond to quasi-periodicity. Theorem 5.2 thus ensures that typically quasi-periodicity occurs with positive measure in the parameter space. In terms of §§3.1 and 3.2 this means that quasi-periodicity for cylinder maps is weakly persistent under variation of parameters and persistent under variation of initial states (see Definition 3.3). Note that because Cantor sets are perfect, quasi-periodicity typically has a nonisolated occurrence.
– We like to mention that nonperturbative versions of Theorem 5.2 have been proven by Herman[6] and Yoccoz[7] [140, 141, 249].

---

[6] Michael Robert Herman 1942–2000.

[7] Jean-Christophe Yoccoz 1957–.

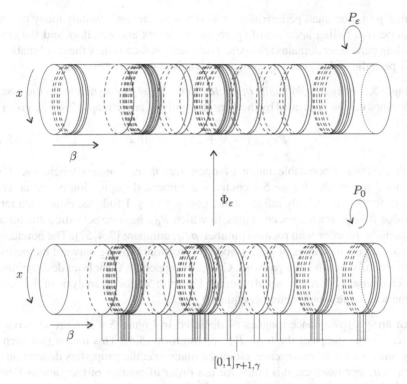

Fig. 5.3 Conjugation between the Poincaré maps P_0 and P_ε on $\mathbb{T}^1 \times [0,1]_{\tau+1,\gamma}$.

– For simplicity we formulated Theorem 5.2 under C^∞-regularity, noting that there
  exist many ways to generalise this. On the one hand there exist C^k-versions for
  finite k and there also exist subtleties of regularity in terms of real-analytic and
  Gevrey. For details we refer to [37, 87] and references therein. This same remark
  applies to other results in this section and in Appendix B on KAM theory.

5.2.3 Resonance tongues

We conclude this discussion with a few remarks on resonance, returning to general
circle maps of the form $P_{\beta,\varepsilon} : x \mapsto x + 2\pi\beta + \varepsilon a(x, \beta, \varepsilon)$; see equation (5.4).
The circle map theorem 5.2 deals with the circles carrying quasi-periodic motion
with Diophantine rotation number. Note that although the diffeomorphism Φ_ε only
is a conjugation on the diophantine Cantor set, outside this set it has no dynamical
meaning. Presently we are interested in the 'gap' dynamics, in particular, where
the rotation number is rational; compare the digression on the driven Van der Pol
equation in §2.2.3.

  We partly summarize this discussion. In the unperturbed setting $\varepsilon = 0$, for
$\beta = p/q$ all points of the circle are periodic with period q, even with rotation

number p/q. For small perturbations $\varepsilon \neq 0$ generically only finitely many periodic orbits persist, falling apart into hyperbolic attractors and repellors and the corresponding parameter domains are open. The question then is how the combination of quasi-periodic and resonant dynamics is organised in the (β, ε)–plane.

Example 5.3 (The Arnold family of circle maps). The 'gap' dynamics in the case of circle maps can be illustrated by the Arnold family of circle maps [2,4,57], given by

$$P_{\beta,\varepsilon}(x) = x + 2\pi\beta + \varepsilon \sin x \qquad (5.14)$$

which exhibits a countable union of open *resonance tongues* where the $P_{\beta,\varepsilon}$-dynamics is periodic. Figure 5.4 contains a numerical impression of the array of tongues for $0 \leq \beta \leq \frac{1}{2}$. By reflection the part $\frac{1}{2} \leq \beta \leq 1$ follows. From each rational value $\beta = p/q$ a tongue emanates, in which $P_{\beta,\varepsilon}$ has one periodic attractor and one periodic repellor with rotation number p/q; compare [2,4,57]. The boundaries of each tongue are curves of saddle-node bifurcation: moving outward the periodic orbits annihilate each other pairwise. Compare Appendix C for brief descriptions of such elementary bifurcations and compare Exercise 5.4 for an analysis of the 'main' tongue where the rotation number equals 0 mod $\mathbb{Z}$.

An array of resonance tongues as depicted in Figure 5.4 is representative for all 'reasonable' maps of the form $P_{\beta,\varepsilon}$ of the form (5.4). This means that such an array has a universal occurrence, although many specific properties depend on the function a. For instance, this holds for the order of contact of the various tongue boundaries at $\varepsilon = 0$. For details compare [91] and its references.

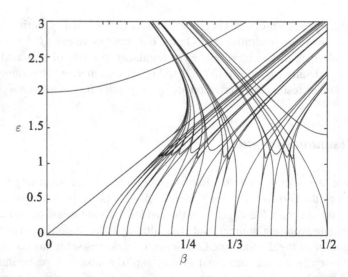

Fig. 5.4 Resonance tongues [91] in the Arnold family (5.14) of circle maps for $0 \leq \beta \leq \frac{1}{2}$. From each rational value $\beta = p/q$ a tongue emanates, in which the $P_{\beta,\varepsilon}$ is periodic with rotation number p/q; compare [2, 4, 57]. The boundaries of each tongue are curves of saddle-node bifurcations: moving outward the periodic points annihilate one another pairwise. Note that for $|\varepsilon| \geq 1$ the Arnold family (5.14) is only endomorphic.

In the above contexts of the Van der Pol system, either driven or coupled (see §5.1, the Examples 5.1 and 5.2), quasi-periodic circles of (5.14) correspond to quasi-periodic invariant 2-tori. In the resonant cases on the invariant 2-torus one finds periodic attractors and repellors, where the torus itself is the unstable manifold of the repellors. The only persistent motion takes place on a periodic attractor and, as in §2.2.3, here usually one speaks of *phase locking*.

Remarks.

– Note that for $|\varepsilon| \geq 1$ the maps $P_{\beta,\varepsilon}$ in the Arnold family (5.14) are only endomorphic and no longer diffeomorphic and hence beyond the context of this chapter. For $|\varepsilon| > 1$ curves of period doubling bifurcations can be detected inside the tongues. For a brief overview concerning elementary bifurcations see Appendix C.

– In the (β, ε)-plane, in between the resonance tongues there is a union of curves, emanating from parameter values with $\varepsilon = 0$ and β diophantine, on which the dynamics is quasi-periodic. By the circle map theorem 5.2 this union has positive Lebesgue measure, whereas the array of tongues is open and dense. For further background also see Appendices A and C and Oxtoby [184].

Example 5.4 (Huygens's clocks). We like to mention the famous observations made in 1665 by Christiaan Huygens[8] on the synchronizing behaviour of two clocks mounted next to each other on the same beam [147]. The two clocks do not differ too much regarding their essential characteristics. A partial explanation of this phenomenon can be found in the context of Example 5.2 where we take the two weakly coupled almost identical oscillators. In that case for the frequency ratio $\beta = \omega_2/\omega_1$ we find that $\beta \approx 1$. We consider the corresponding Poincaré map $P_{\beta,\varepsilon}$ of equation (5.4). In that case the fact that $\beta \approx 1$ and $|\varepsilon| \ll 1$ implies that the parameter point is situated in the 'main' tongue where the rotation number of $P_{\beta,\varepsilon}$ equals 1 exactly. In turn this means that the motion is in a $1 : 1$-resonance, in which the motions of the two clocks are phase locked.

It should be noted that Huygens's observation moreover includes the fact that the pendula of the clocks move in antiphase; exactly $180°$ out of phase. For a detailed discussion see [68] where interesting historical aspects are also taken into account.

5.3 KAM theory of area-preserving maps

We continue the previous section on the KAM theory of circle and cylinder maps, now turning to the setting of area-preserving maps of the annulus. In particular we discuss Moser's twist map theorem [173] (also see [37, 87, 88]), which is related to the conservative dynamics of frictionless mechanics.

---

[8] Christiaan Huygens 1629–1695.

Let $\Delta \subseteq \mathbb{R}^2 \setminus \{(0,0)\}$ be an annulus, with (polar) coordinates $(\varphi, I) \in \mathbb{T}^1 \times \mathbf{K}$, where $\mathbf{K}$ is an interval in $\mathbb{R}_+$. Moreover, let $\sigma = d\varphi \wedge dI$ be an area form on Δ. Compare Exercises 5.10 and 5.11.

We consider a σ-preserving smooth map $P_\varepsilon : \Delta \to \Delta$ of the form

$$P_\varepsilon(\varphi, I) = (\varphi + 2\pi\beta(I), I) + O(\varepsilon), \tag{5.15}$$

where we assume that the map $I \mapsto \beta(I)$ is a (local) diffeomorphism. This assumption is known as the *twist condition* and P_ε is called a *twist map*. For the unperturbed case $\varepsilon = 0$ we are dealing with a pure twist map and its dynamics is comparable to the unperturbed family of cylinder maps as met in §5.2. Indeed it is again a family of rigid rotations, parametrised by I and where $P_0(., I)$ has rotation number $\beta(I)$. In this case the question is what will be the fate of this family of invariant circles, as well as that of the corresponding rigid dynamics. In general the $O(\varepsilon)$-term in (5.15) will be φ-dependent.

Regarding the rotation number we again introduce diophantine conditions. Indeed, for $\tau > 1$ and $\gamma > 0$ the subset $[0,1]_{\tau+1,\gamma}$ is defined as in (5.12); that is, it contains all $\beta \in [0,1]$, such that for all rationals p/q

$$\left| \beta - \frac{p}{q} \right| \geq \frac{\gamma}{q^{\tau+1}}.$$

Pulling back $[0,1]_{\tau+1,\gamma}$ along the map β we obtain a subset $\Delta_{\tau+1,\gamma} \subseteq \Delta$.

Theorem 5.3 (Twist map theorem [173]). *For γ sufficiently small, $\tau > 1$ and for the perturbation $O(\varepsilon)$ sufficiently small in C^∞-topology, there exists a C^∞ transformation $\Phi_\varepsilon : \Delta \to \Delta$, conjugating the restriction $P_0|_{\Delta_{\tau,\gamma}}$ to a subsystem of P_ε.*

As in the case of Theorem 5.2 again we chose the formulation of [87,88]. Largely the remarks following Theorem 5.2 also apply here.

Remarks.

– Compare the format of the Theorems 5.2 and 5.3 and observe that in the latter case the role of the parameter β has been taken by the 'action' variable I. Theorem 5.3 implies that typically quasi-periodicity occurs with positive measure in phase space. Or, to rephrase this assertion in terms of §§3.1, 3.2, the property that quasi-periodicity is weakly persistent under variation of initial state, is persistent under variation of parameters (i.e., typical).

– In the resonance gaps of the diophantine Cantor set typically we have coexistence of periodicity, quasi-periodicity, and chaos [13,17,21,59,83,176,177]. The latter follows from transversality of homo- and heteroclinic connections which give rise to positive topological entropy.

Example 5.5 (Coupled oscillators II). Similar to the applications of Theorem 5.2 to driven and coupled nonlinear oscillators in §§5.1, 5.2, here direct applications are possible in the conservative setting. Indeed, consider a system of weakly coupled pendula

$$y_1'' + \alpha_1^2 \sin y_1 = -\varepsilon \frac{\partial U}{\partial y_1}(y_1, y_2)$$

$$y_2'' + \alpha_2^2 \sin y_2 = -\varepsilon \frac{\partial U}{\partial y_2}(y_1, y_2).$$

Writing $y_j' = z_j$, $j = 1, 2$ as before, we again get a vector field in the 4-dimensional phase space $\mathbb{R}^2 \times \mathbb{R}^2 = \{(y_1, y_2), (z_1, z_2)\}$ of the form

$$y_1' = z_1 = \frac{\partial H}{\partial z_1}(y_1, y_2, z_1, z_2)$$

$$z_1' = -\alpha_1^2 \sin y_1 - \varepsilon \frac{\partial U}{\partial y_1}(y_1, y_2) = -\frac{\partial H}{\partial y_1}(y_1, y_2, z_1, z_2)$$

$$y_2' = z_2 = \frac{\partial H}{\partial z_2}(y_1, y_2, z_1, z_2)$$

$$z_2' = -\alpha_2^2 \sin y_1 - \varepsilon \frac{\partial U}{\partial y_2}(y_1, y_2) = -\frac{\partial H}{\partial y_2}(y_1, y_2, z_1, z_2)$$

In this case the Hamiltonian (energy) H, which is given by

$$H_\varepsilon(y_1, y_2, z_1, z_2) = \tfrac{1}{2}z_1^2 + \tfrac{1}{2}z_2^2 - \alpha_1^2 \cos y_1 - \alpha_2^2 \cos y_2 + \varepsilon U(y_1, y_2)$$

is a constant of motion. We now restrict to a 3-dimensional energy hypersurface $H_\varepsilon^{-1} = $ const., where (locally) a 2-dimensional transversal section Δ is taken on which an *isoenergetic* Poincaré map P_ε is defined. One can show that $P_\varepsilon : \Delta \to \Delta$ is an area-preserving twist map. Application of Theorem 5.3 then yields the persistent occurrence of quasi periodicity with positive measure. For the system this means persistent occurrence of quasi-periodicity (on invariant 2-tori) with positive measure in the energy hypersurface of H_ε.

Remarks.

- The coefficients α_1 and α_2 are the imaginary parts of the eigenvalues of the linearisation at the equilibrium $(y_1, z_1, y_2, z_2) = (0, 0, 0, 0)$. The ratio $\alpha_1 : \alpha_2$, in particular whether it is rational or diophantine, becomes important only when considering a small neighborhood of this equilibrium.
- For a more general and systematic entrance to the literature on Hamiltonian dynamics we refer to [20–24]; also see [44]. A brief guide to these works is contained in Appendix E.

5.4 KAM **theory of holomorphic maps**

The sections §§5.2 and 5.3 both deal with smooth circle maps that are conjugated to rigid rotations. Presently the concern is with planar holomorphic (or complex analytic) maps that are conjugated to a rigid rotation on an open subset of the plane. Historically this was the first time that a small divisor problem was solved [2, 169, 211, 251, 252].

5.4.1 Complex linearisation

Given is a holomorphic germ $F : (\mathbb{C}, 0) \to (\mathbb{C}, 0)$ of the form $F(z) = \lambda z + f(z)$, with $f(0) = f'(0) = 0$. The problem is to find a biholomorphic germ $\Phi : (\mathbb{C}, 0) \to (\mathbb{C}, 0)$ such that

$$\Phi \circ F = \lambda \cdot \Phi.$$

Such a diffeomorphism Φ is called a *linearisation* of F near 0.

For completeness we include a definition of germs, a notion that is needed in a full mathematical treatment of the subject. Two holomorphic maps $F, G : (\mathbb{C}, 0) \to (\mathbb{C}, 0)$ are called *germ equivalent* if there is a neighbourhood U of 0 in $\mathbb{C}$, such that $F|_U = G|_U$. It directly follows that germ equivalence is an equivalence relation and a germ is just an equivalence class of this.

Remarks. Colloquially speaking a germ is a map defined locally near 0, where we don't want to specify the precise neighbourhood. Germs can be similarly defined for smooth or continuous maps, vector fields, and the like.

We first approach the linearisation problem formally. Given the series $f(z) = \sum_{j \geq 2} f_j z^j$, we look for $\Phi(z) = z + \sum_{j \geq 2} \phi_j z^j$. It turns out that a solution always exists whenever $\lambda \neq 0$ is not a root of unity. Indeed, direct computation reveals the following set of equations which can be solved recursively.

For $j = 2$:   get the equation $\lambda(1 - \lambda)\phi_2 = f_2$.
For $j = 3$:   get the equation $\lambda(1 - \lambda^2)\phi_3 = f_3 + 2\lambda f_2\phi_2$.
For $j = n$:   get the equation $\lambda(1 - \lambda^{n-1})\phi_n = f_n + \text{known}$.

The question now reduces to whether this formal solution has positive radius of convergence.

The hyperbolic case $0 < |\lambda| \neq 1$ was already solved by Poincaré; for a description see [2]. The elliptic case $|\lambda| = 1$ again has small divisors and was solved by Siegel[9] [211] when for some $\gamma > 0$ and $\tau > 1$ we have the diophantine nonresonance condition (for all $p \in \mathbb{Z}, q \in \mathbb{N}$)

$$|\lambda - e^{2\pi i p/q}| \geq \gamma |q|^{-(\tau+1)}.$$

The corresponding set of λ constitutes a set of full measure in the circle $\mathbb{T}^1$.

Remark. Yoccoz [251] completely solved the elliptic case using the Bruno-condition. If we write $\lambda = e^{2\pi i \alpha}$ and let p_n/q_n be the nth convergent in the continued fraction expansion of α, then the Bruno-condition reads

$$\sum_n \frac{\log(q_{n+1})}{q_n} < \infty.$$

---

[9] Carl Ludwig Siegel 1896–1981.

The Bruno condition, which is implied by the diophantine condition, turns out to be necessary and sufficient for Φ having positive radius of convergence; see Yoccoz [251,252]. Compare Exercise 5.16.

5.4.2 Cremer's example in Herman's version

As a simple example of a nonlinear holomorphic map $F : (\mathbb{C}, 0) \to (\mathbb{C}, 0)$ consider

$$F(z) = \lambda z + z^2,$$

where $\lambda \subset \mathbb{T}^1$ is not a root of unity. Let us consider the corresponding linearisation problem.

To this end observe that a point $z \in \mathbb{C}$ is a periodic point of F with period q if and only if $F^q(z) = z$, where obviously

$$F^q(z) = \lambda^q z + \cdots + z^{2^q}.$$

Writing

$$F^q(z) - z = z(\lambda^q - 1 + \cdots + z^{2^q - 1}),$$

the period q periodic points are precisely the roots of the right-hand side polynomial. Abbreviating $N = 2^q - 1$, it directly follows that, if $z_1, z_2, \ldots, z_N$ are the nontrivial roots, then for their product we have

$$z_1 \cdot z_2 \cdot \cdots \cdot z_N = \lambda^q - 1.$$

It follows that there exists a nontrivial root within radius

$$|\lambda^q - 1|^{1/N}$$

of $z = 0$.

Now consider the set of $\Lambda \subset \mathbb{T}^1$ defined as follows. $\lambda \in \Lambda$ whenever

$$\liminf_{q \to \infty} |\lambda^q - 1|^{1/N} = 0.$$

It can be directly shown that Λ is residual; see Exercise 5.15 and for background see Appendix A. It also follows that for $\lambda \in \Lambda$ linearisation is impossible. Indeed, because the rotation is irrational, the existence of periodic points in any neighbourhood of $z = 0$ implies zero radius of convergence.

Remarks.

- Notice that the residual set Λ is in the complement of the full measure set of all diophantine numbers; see Appendix A and [184]. Also compare Exercise 5.15.

– Considering $\lambda \in \mathbb{T}^1$ as a parameter, we see a certain analogy with Theorems 5.2 and 5.3. Indeed, in this case for a full measure set of λs on a λ-dependent neighbourhood of $z = 0$ the map $F = F_\lambda$ is conjugated to a rigid irrational rotation. Such a domain in the z-plane often is referred to as a Siegel disc. For a more general discussion of these and of Herman rings, see Milnor [169].

5.5 The linear small divisor problem

The linear small divisor problem turns out to be key problem useful for general KAM theory. This holds for Theorems 5.2 and 5.3 as discussed both in this chapter and elsewhere in the literature, including the case of flows; see Appendices B and C. The kind of proof we have in mind is a Newtonian iteration process for solving conjugation equations, the linearisations of which lead to linear small divisor problems. Compare the Newtonian iteration as described in Chapter 1 for finding roots of polynomials; the difference is that the present version takes place in an infinite-dimensional function space. This version also is somewhat adapted; compare the remark concluding §3.3.1 (for a thorough explanation see [177, 208]). Also compare [87, 88, 159, 194, 195, 250].

To be more precise we start briefly summarising §5.5.2, also referring to Appendix B, after which we deal with the linear small divisor problem for its own sake. Here we largely follow parts of [27, 87, 95].

5.5.1 Motivation

In this subsection we treat two motivating examples of linear small divisor problems, one for circle maps, as already met in §5.2, and the other for torus flows.

The linear small divisor problem: case of circle maps. We briefly recall the conjugation problem of §5.2 for families of circle maps. Given the fibre-preserving cylinder map (5.5) on $\mathbb{T}^1 \times [0, 1]$

$$P_\varepsilon : (x, \beta) \mapsto (x + 2\pi\beta + \varepsilon a(x, \beta, \varepsilon), \beta).$$

the search is for a skew cylinder diffeomorphism (5.7)

$$\Phi_\varepsilon(x, \beta) = (x + \varepsilon U(x, \beta, \varepsilon), \beta + \varepsilon\sigma(\beta, \varepsilon)),$$

satisfying the conjugation equation (5.6) $P_\varepsilon \circ \Phi_\varepsilon = \Phi_\varepsilon \circ P_0$, or equivalently the nonlinear equation (5.8), as far as possible. For $\beta \notin \mathbb{Q}$ we solved this equation, to first order in ε, by the formal Fourier series (5.11)

$$U(x, \beta, \varepsilon) = \sum_{k \in \mathbb{Z}} \frac{a_{0k}(\beta)}{e^{2\pi i k \beta} - 1} e^{ikx} + O(\varepsilon),$$

and by $\sigma(\beta, \varepsilon) = -(1/2\pi)a_{00}(\beta) + O(\varepsilon)$, where $a(x, \beta, \varepsilon) = \sum_{k \in \mathbb{Z}} a_{0k}(\beta)e^{ikx} + O(\varepsilon)$.

The linear small divisor problem: Case of torus flows. We now develop an analogous example for the case of flows, which relates to the flow analogue of Theorem 5.2 as this is formulated in [87], Theorem 6.1 (p. 142 ff.). For further details and a sketch of a proof, see Appendix B. This part of KAM theory deals with families of vector fields on the n-torus of the form

$$
\begin{aligned}
x_1' &= \omega_1 + \varepsilon f_1(x, \omega, \varepsilon) \\
x_2' &= \omega_2 + \varepsilon f_2(x, \omega, \varepsilon) \\
&\cdots \\
x_n' &= \omega_n + \varepsilon f_n(x, \omega, \varepsilon),
\end{aligned}
\tag{5.16}
$$

which for $\varepsilon = 0$ is a multiperiodic system, that is quasi-periodic in the case where the frequencies $\omega_1, \omega_2, \ldots, \omega_n$ are independent over $\mathbb{Q}$, that is, where no resonances occur; see §2.2. As in §5.2 the question is to what extent this 'unperturbed' dynamics is persistent for small $|\varepsilon|$. And just as in §5.2, this question is explored by looking for a smooth conjugation with the case $\varepsilon = 0$. This leads to another linear small divisor problem.

To be more precise, we rewrite the system (5.16) as a fibre-preserving vector field on $\mathbb{T}^n \times \mathbb{R}^n$ given by

$$X_\varepsilon(x, \omega) = [\omega + \varepsilon f(x, \omega, \varepsilon)]\partial_x, \tag{5.17}$$

where $\omega = (\omega_1, \omega_2, \ldots, \omega_n)$ is an n-dimensional parameter. As before, we naively try to conjugate (5.24) to the case $\varepsilon = 0$ by a skew diffeomorphism

$$\Phi_\varepsilon : \mathbb{T}^n \times \mathbb{R}^n \to \mathbb{T}^n \times \mathbb{R}^n; \ (x, \omega) \mapsto (x + \varepsilon U(x, \omega, \varepsilon), \omega + \varepsilon \Lambda(\omega, \varepsilon)), \tag{5.18}$$

meaning that we require that

$$(\Phi_\varepsilon)_* X_0 = X_\varepsilon, {}^{10} \tag{5.19}$$

as far as possible.

Writing $\xi = x + \varepsilon U(x, \omega, \varepsilon)$ and $\sigma = \omega + \Lambda(\omega, \varepsilon)$, by the chain rule,

---

[10] This is tensor notation for $D_{(x,\omega)}\Phi_\varepsilon \cdot X_0(x, \omega) \equiv X_\varepsilon(\Phi_\varepsilon(x, \omega))$; see Chapter 4, in particular § 3.5, equation (3.8) in Exercise 3.10.

$$\xi' = x' + \varepsilon \frac{\partial U}{\partial x}(x, \omega, \varepsilon) \quad x' = \left(\mathrm{Id} + \varepsilon \frac{\partial U}{\partial x}(x, \omega, \varepsilon)\right) \omega,$$

which turns (5.19) into the nonlinear equation

$$\frac{\partial U}{\partial x}(x, \omega, \varepsilon)\, \omega = \Lambda(\omega, \varepsilon) + f(x + \varepsilon U(x, \omega, \varepsilon), \omega + \varepsilon \Lambda(\omega, \varepsilon), \varepsilon). \qquad (5.20)$$

Next, expanding in powers of ε

$$f(x, \omega, \varepsilon) = f_0(x, \omega) + O(\varepsilon),$$
$$U(x, \omega, \varepsilon) = U_0(x, \omega) + O(\varepsilon),$$
$$\Lambda(\omega, \varepsilon) = \Lambda_0(\omega) + O(\varepsilon),$$

just as before, we obtain a linear (homological) equation

$$\frac{\partial U_0}{\partial x}(x, \omega)\, \omega = \Lambda_0(\omega) + f_0(x, \omega), \qquad (5.21)$$

which we can solve for U_0 and Λ_0 by using Fourier series

$$f_0(x, \omega) = \sum_{k \in \mathbb{Z}^n} f_{0k}(\omega) e^{i \langle k, x \rangle} \quad \text{and} \quad U_0(x, \omega) = \sum_{k \in \mathbb{Z}^n} U_{0k}(\omega) e^{i \langle k, x \rangle}.$$

Indeed, completely analogously to §5.2.2 we now find

$$\Lambda_0(\omega) = -f_{00}(\omega) \quad \text{and} \quad U_{0k}(\omega) = \frac{f_{0k}}{i \langle k, \omega \rangle}, \quad k \in \mathbb{Z}^n \setminus \{0\}, \qquad (5.22)$$

assuming that $\langle k, \omega \rangle \neq 0$ for all $k \in \mathbb{Z}^n \setminus \{0\}$, which means that $\omega_1, \omega_2, \ldots, \omega_n$ are independent over $\mathbb{Q}$. This leads to the formal series solution

$$U_0(x, \omega) = U_{00}(\omega) + \sum_{k \in \mathbb{Z}^n \setminus \{0\}} \frac{f_{0k}}{i \langle k, \omega \rangle} e^{i \langle k, x \rangle}, \qquad (5.23)$$

where $U_{00}(\omega)$ remains arbitrary.

5.5.2 Setting of the problem and formal solution

We now introduce two linear small divisor problems as follows, thereby slightly rephrasing the homological equations (5.9) and (5.21).

$$u(x + 2\pi\beta, \beta) - u(x, \beta) = f(x, \beta) \text{ (circle map case) and}$$
$$\frac{\partial u(x, \omega)}{\partial x} \omega = f(x, \omega) \text{ (torus flow case).} \tag{5.24}$$

In both cases u has to be solved in terms of the given function f, where in the map case $x \in \mathbb{T}^1$ and $u, f : \mathbb{T}^1 \to \mathbb{R}$ and in the flow case $x \in \mathbb{T}^n$ and $u, f : \mathbb{T}^n \to \mathbb{R}^n$. Moreover, $\beta \in [0, 1]$ and $\omega \in \mathbb{R}^n$ serve as parameters of the problems. By taking averages of both sides we obtain a necessary condition for the existence of a solution of both equations in (5.24), namely that

$$f_0(., \beta) \equiv 0 \quad \text{and} \quad f_0(., \omega) \equiv 0,$$

which we assume from now on.

Under the nonresonance conditions that β is irrational and that $\omega_1, \omega_2, \ldots, \omega_n$ are rationally independent, we find formal solutions (5.11) and (5.23) as summarised and described in the above motivation. In both cases the convergence of the Fourier series is problematic because of the corresponding small divisors.

Remarks.

- This setting has many variations; see below as well as the Exercises 5.6, 5.9, and 5.7 for other examples. One remaining case of interest is the analogue of (5.24) for general torus maps, that is, for translation systems on the n-torus with time set $\mathbb{Z}$; see §2.2.2. See Exercise 5.14.
- In [87] the linear problem (5.24) was coined as the 1-*bite small divisor problem*, because it can be solved at once by substitution of Fourier series. Compare this with the more complicated solution of the corresponding nonlinear equations (5.8) and (5.20) as described elsewhere in the literature. As said earlier we deal with the latter case in Appendix B sketching a Newtonian iteration procedure based on the linear theory developed in the present setting.

In the following example we meet a slightly different linear small divisor problem [27, 87].

Example 5.6 (Quasi-periodic responses in the linear case I). Consider a linear quasi-periodically forced oscillator

$$y'' + cy' + ay = g(t, a, c) \tag{5.25}$$

with y, $t \in \mathbb{R}$, where a and c with $a > 0, c > 0$, and $c^2 < 4a$ serve as parameters. The driving term g is assumed to be quasi-periodic in t, and real-analytic in all arguments. This means that $g(t, a, c) = G(t\omega_1, t\omega_2, \ldots, t\omega_n, a, c)$ for some real-analytic function $G : \mathbb{T}^n \times \mathbb{R}^2 \to \mathbb{R}$, $n \geq 2$, and a fixed nonresonant frequency vector $\omega \in \mathbb{R}^n$, that is, with $\omega_1, \omega_2, \ldots, \omega_n$ independent over $\mathbb{Q}$. The problem is to find a response solution $y = y(t, a, c)$ of (5.25) that is quasi-periodic in t with the same frequencies $\omega_1, \omega_2, \ldots, \omega_n$. This leads to a linear small divisor problem as follows.

The corresponding vector field on $\mathbb{T}^n \times \mathbb{R}^2$ in this case determines the system of differential equations

$$
\begin{aligned}
x' &= \omega \\
y' &= z \\
z' &= -ay - cz + G(x, a, c),
\end{aligned}
$$

where $x = (x_1, x_2, \ldots, x_n)$ modulo 2π. The response problem now translates to the search for an invariant torus which is a graph $(y, z) = (y(x, a, c), z(x, a, c))$ over $\mathbb{T}^n$ with dynamics $x' = \omega$.

For technical convenience we complexify the problem as follows.

$$
\zeta = z - \bar{\lambda} y \quad \text{and} \quad \lambda = -\frac{c}{2} + i \sqrt{a - \frac{c^2}{4}}. \tag{5.26}
$$

Notice that $\lambda^2 + c\lambda + a = 0$, $\Im \lambda > 0$, and that

$$
y'' + cy' + ay = \left[\frac{d}{dt} - \lambda \right]\left[\frac{d}{dt} - \bar{\lambda} \right] y.
$$

Now our vector field gets the form

$$
\begin{aligned}
x' &= \omega \\
\zeta' &= \lambda \zeta + G(x, \lambda),
\end{aligned}
$$

where we identify $\mathbb{R}^2 \cong \mathbb{C}$ and use complex multiplication. It is easily seen that the desired torus-graph $\zeta = \zeta(x, \lambda)$ with dynamics $x' = \omega$ now has to be obtained from the linear equation

$$
\left\langle \omega, \frac{\partial \zeta}{\partial x} \right\rangle = \lambda \zeta + G, \tag{5.27}
$$

which has a format that only slightly differs from the second equation of (5.24). Nevertheless it also provides a linear small divisor problem as we show now.

Again we expand G and ζ in the Fourier series

$$
G(x, \lambda) = \sum_{k \in \mathbb{Z}^n} G_k(\lambda) e^{i\langle x, k \rangle}, \quad \zeta(x, \lambda) = \sum_{k \in \mathbb{Z}^n} \zeta_k(\lambda) e^{i\langle x, k \rangle}
$$

and compare coefficients. It turns out that the following formal solution is obtained,

$$
\zeta(\lambda) = \sum_{k \in \mathbb{Z}^n} \frac{G_k(\lambda)}{i\langle \omega, k \rangle - \lambda} e^{i\langle x, k \rangle}, \tag{5.28}
$$

granted that $\lambda \neq i\langle \omega, k \rangle$, for all $k \in \mathbb{Z}^n$. Indeed, the denominators vanish at the resonance points $\lambda = i\langle \omega, k \rangle$, $k \in \mathbb{Z}^n$, and thus in a dense subset of the imaginary

λ-axis. In general a formal solution exists if and only if λ is not in this set, and, as in §5.5.2, the convergence is problematic, again because of the corresponding small divisors.

5.5.3 Convergence

We next deal with convergence of the formal solutions (5.10) and (5.23). First we discuss the regularity of the problem. Recall that Theorems 5.2 and 5.3 both are formulated in the world of C^{∞}-systems; for a discussion see the remark following Theorem 5.2. The proofs at hand, based on Newtonian iterations, all use the approximation properties of C^k-functions by real analytic ones [87, 88, 194, 244, 245, 253, 254]. It turns out that the linear small divisor problems mostly are solved in the world of real analytic systems. For that reason, and for simplicity, we here restrict to the case where in (5.24) the right-hand side function f is real-analytic in all its arguments. For variations on this to the cases of C^k-functions, $k \in \mathbb{N} \cup \{\infty\}$, see Exercise 5.13.

Recall that in §5.5.2 we required the rotation number $\beta \in [0, 1]$ to be $(\tau + 1, \gamma)$-diophantine (see (5.12)) for $\tau > 1$ and $\gamma > 0$, denoting the corresponding set by $[0, 1]_{\tau+1,\gamma} \subseteq [0, 1]$; we recall that this forms a Cantor set of positive measure. We give a similar definition for the frequency vector $\omega \in \mathbb{R}^n$, $n \geq 2$.

Definition 5.4 (Diophantine vectors). Let $\tau > n - 1$ and $\gamma > 0$. We say that the vector $\omega \in \mathbb{R}^n$ is (τ, γ)-diophantine if for all $k \in \mathbb{Z}^n \setminus \{0\}$ we have that

$$|\langle \omega, k \rangle| \geq \gamma |k|^{-\tau}, \tag{5.29}$$

where $|k| = \sum_{j=1}^n |k_j|$. The subset of all (τ, γ)-diophantine frequency vectors in $\mathbb{R}^n$ is denoted by $\mathbb{R}^n_{\tau,\gamma}$.

We give a few properties of the set of diophantine vectors.

Lemma 5.5 (Properties of diophantine vectors). *The closed subset $\mathbb{R}^n_{\tau,\gamma} \subset \mathbb{R}^n$ has the following properties.*

1. *Whenever $\omega \in \mathbb{R}^n_{\tau,\gamma}$ and $s \geq 1$, then also $s\omega \in \mathbb{R}^n_{\tau,\gamma}$, we say that $\mathbb{R}^n_{\tau,\gamma}$ has the closed halfline property.*
2. *Let $\mathbb{S}^{n-1} \subset \mathbb{R}^n$ be the unit sphere, then $\mathbb{S}^{n-1} \cap \mathbb{R}^n_{\tau,\gamma}$ is the union of a Cantor set and a countable set.*
3. *The complement $\mathbb{S}^{n-1} \setminus \mathbb{R}^n_{\tau,\gamma}$ has Lebesgue measure of order γ as $\gamma \downarrow 0$.*

For a proof see [87]. This proof is rather direct when using the Cantor–Bendixson theorem [135]; compare the details on $[0, 1]_{\tau,\gamma}$ in §5.2.2. In Figure 5.5 we sketch $\mathbb{R}^2_{\tau,\gamma} \subset \mathbb{R}^2$ and its position with respect to the unit circle $\mathbb{S}^1$. From Lemma 5.5 it follows that $\mathbb{R}^2_{\tau,\gamma}$ is a nowhere dense, uncountable union of closed half lines of positive measure; this and similar sets in the sequel are called *Cantor bundles*. It turns out

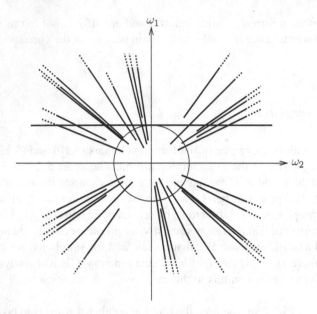

Fig. 5.5 Sketch of the set $\mathbb{R}^2_{\tau,\gamma}$ (and the unit circle and the horizontal line $\omega_1 = 1$).

that the sets $[0, 1]_{\tau+1,\gamma}$ and $\mathbb{R}^2_{\tau,\gamma}$ are closely related (see Exercise 5.5) where the horizontal line $\omega_1 = 1$ can be helpful.

Another ingredient for proving the convergence of the formal solutions (5.10) and (5.23) is the decay of the Fourier coefficients as $|k| \to \infty$. For brevity we now restrict to the second case of (5.24), leaving the first one to the reader.

Let $\Gamma \subset \mathbb{R}^n$ be a compact domain and put $\Gamma_\gamma = \Gamma \cap \mathbb{R}^n_\gamma$. By analyticity f has a holomorphic extension to a neighborhood of $\mathbb{T}^n \times \Gamma$ in $(\mathbb{C}/2\pi\mathbb{Z})^n \times \mathbb{C}^n$. For some constants κ, $\rho \in (0, 1)$, this neighborhood contains the set $\mathrm{cl}\,((\mathbb{T}^n + \kappa) \times (\Gamma + \rho))$ where 'cl' stands for the closure and

$$\Gamma + \rho := \bigcup_{\omega \in \Gamma} \{\omega' \in \mathbb{C}^n \mid |\omega' - \omega| < \rho\}, \tag{5.30}$$

$$\mathbb{T}^n + \kappa := \bigcup_{x \in \mathbb{T}^n} \{x' \in (\mathbb{C}/2\pi\mathbb{Z})^n \mid |x' - x| < \kappa\}.$$

Let M be the supremum of f over the set $\mathrm{cl}\,((\mathbb{T}^n + \kappa) \times (\Gamma + \rho))$, then one has the following.

Lemma 5.6 (Paley–Wiener, e.g., [2]). *Let $f = f(x, \omega)$ be real analytic as above, with Fourier series*

$$f(x, \omega) = \sum_{k \in \mathbb{Z}^n} f_k(\omega)e^{i\langle x, k \rangle}. \tag{5.31}$$

Then, for all $k \in \mathbb{Z}^n$ and all $\omega \in \mathrm{cl}\,(\Gamma + \rho)$,

$$|f_k(\omega)| \le M\,e^{-\kappa|k|}. \tag{5.32}$$

Remark. The converse of Lemma 5.6 also holds true (see Exercise 5.12): a function on $\mathbb{T}^n$ the Fourier coefficients of which decay exponentially is real analytic and can be extended holomorphically to the complex domain $(\mathbb{C}/2\pi\mathbb{Z})^n$ by a distance determined by the decay rate (i.e., κ in our notations). Also compare [2].

Proof. By definition

$$f_k(\omega) = (2\pi)^{-n} \oint_{\mathbb{T}^n} f(x, \omega) e^{-i\langle x, k\rangle} dx.$$

Firstly, assume that for $k = (k_1, \ldots, k_n)$ all the entries $k_j \neq 0$. Then take $z = (z_1, \ldots, z_n)$ with

$$z_j = x_j - i\kappa \, \mathrm{sign}(k_j), \quad 1 \le j \le n,$$

and use the Cauchy theorem to obtain

$$f_k(\omega) = (2\pi)^{-n} \oint f(z, \omega) \, e^{-i\langle z, k\rangle} dz = (2\pi)^{-n} \oint_{\mathbb{T}^n} f(z, \omega) \, e^{-i\langle x, k\rangle - \kappa|k|} dx,$$

whence $|f_k(\omega)| \le M e^{-\kappa|k|}$, as desired.

Secondly, if for some j $(1 \le j \le n)$ we have $k_j = 0$, we just do not shift the integration torus in that direction. □

As a direct consequence of the Paley–Wiener lemma we have the next theorem.

Theorem 5.7 (The linear small divisor problem in the real-analytic case). *Consider the linear small divisor problems (5.24) with a real analytic right-hand side f with average 0. Then, for any $\beta \in [0, 1]_{\tau+1,\gamma}$, $(\tau > 1)$ c.q., for $\omega \in \mathbb{R}^n_{\tau,\gamma}$, $(\tau > n-1)$ the formal solutions (5.10) and (5.23) converge to a solution that is real-analytic in x, varying either over $\mathbb{T}^1$ or $\mathbb{T}^n$.*

Proof. Indeed, now considering the torus flow case (5.23), by Lemma 5.6 the coefficients f_k decay exponentially with $|k|$ and by (5.29) the divisors $i\langle\omega, k\rangle$ in size only grow polynomially. Hence, the quotient in (5.23) still has exponentially decaying Fourier coefficients (with a somewhat smaller κ) and therefore by the remark following Lemma 5.6 this series converges to a real-analytic solution.

For a proof of the circle map case of (5.24), see Exercise 5.12. □

Example 5.7 (Quasi-periodic responses in the linear case II). We revisit Example 5.6, treating quasi-periodic responses in the linear case; we slightly adapt the above approach. Indeed, for a fixed nonresonant ω we study the formal solution (5.28)

$$\zeta(x, \lambda) = \sum_{k \in \mathbb{Z}^n} \frac{G_k(\lambda)}{i\langle\omega, k\rangle - \lambda} e^{i\langle x, k\rangle}.$$

The diophantine condition here is that for given $\tau > n - 1$ and $\gamma > 0$ and for all $k \in \mathbb{Z}^n \setminus \{0\}$,

$$|\lambda - i \langle \omega, k \rangle| \geq \gamma |k|^{-\tau}.$$

This condition excludes a countable union of open discs from the upper-half λ-plane, indexed by k; compare (5.26). This 'bead string' has a measure of the order of γ^2 as $\gamma \downarrow 0$. The complement of the 'bead string' is connected and its intersection with the imaginary λ-axis is a Cantor set.

It is of interest to study this problem for its own sake. A question is, what happens as $\gamma \downarrow 0$? The intersection of the 'bead strings' over all $\gamma > 0$ in the imaginary λ-axis leaves out a residual set of measure zero containing the resonance points $i \langle \omega, k \rangle, k \in \mathbb{Z}^n \setminus \{0\}$. Compare Appendix A and [184]. In the interior of the complement of this intersection our solution (5.28) is analytic, both in x and in λ. For $\Re \lambda \neq 0$ this also follows by writing it as $\zeta = T(G)$, where

$$(T(G))(x, \lambda) = \int_0^\infty e^{\lambda s} G(x - s \omega, \lambda) \, ds,$$

where, for simplicity, we restricted to the case $\Re \lambda < 0$. Notice that in the operator (supremum) norm we have $\|T\| = |\Re \lambda|^{-1}$, which explodes on the imaginary axis. For more details see [70].

Remark.

- Another matter is how regular the solutions (5.24), as mentioned in Theorem 5.7, are in the parameter directions: indeed, by the diophantine conditions, neither the β nor the ω directions contain any interior points. A similar question holds in the complement of the residual set of the response solution of Example 5.7. It turns out that here the notion of Whitney-differentiability applies as introduced in Appendix A; also see [244, 245]. In our real-analytic setting this one can show that all solutions are Whitney-C^∞ in the parameter directions. For further details and references see Appendix B.
- We already mentioned the fact that the KAM theory discussed here rests on analytic approximation techniques. Nevertheless it can be instructive to consider the linear small divisor problem in the finitely differentiable case. In Exercise 5.13 the details of this are explored, including the finitely differentiable analogue of the Paley–Wiener lemma 5.6 and its converse, where once more we restrict to the circle map case (5.23). In this case the decay of the Fourier coefficients $|f_k|$ is polynomial as $|k| \to \infty$, implying a finite loss of differentiability for the solution u of the equation (5.24), the circle map case

$$u(x + 2\pi \beta, \beta) - u(x, \beta) = f(x, \beta),$$

provided that β is a diophantine number and that f has average zero.

5.6 Exercises

Exercise 5.1 (Parallel vector fields on $\mathbb{T}^2$**).** On $\mathbb{T}^2$, with coordinates (x_1, x_2), both counted mod 2π, consider the constant vector field X, given by

$$x_1' = \omega_1$$
$$x_2' = \omega_2.$$

1. Suppose that ω_1 and ω_2 are rationally independent, then show that any integral curve of X is dense in $\mathbb{T}^2$.
2. Suppose that $\omega_1/\omega_2 = p/q$ with gcd $(p, q) = 1$. Show that $\mathbb{T}^2$ is foliated by periodic solutions, all of the same period.

Hint: Use the Poincaré (return) map of the circle $x_1 = 0$ and use Lemma 2.2.

Exercise 5.2 (Integer affine structure on the n**-torus (compare Exercise 2.13)).** Consider the n-torus $\mathbb{T}^n = \mathbb{R}^n/(2\pi\mathbb{Z}^n)$, with angles $x = (x_1, x_2, \ldots, x_n)$ modulo 2π.[11] Consider the translation system given by the constant vector field

$$\mathbb{X}_\omega(x) = \omega \, \partial_x \tag{5.33}$$

(or equivalently, the n-dimensional ordinary differential equation $x' = \omega$). Also consider torus-automorphisms Φ of the affine form

$$\Phi : x \mapsto a + Sx, \tag{5.34}$$

where $a \in \mathbb{R}^n$ and $S \in \mathrm{GL}(n, \mathbb{Z})$, an invertible $n \times n$-matrix with integer entries and det $S = \pm 1$.

1. Determine the transformed vector field $\Phi_*(\mathbb{X}_\omega)$.
2. Show that an individual vector field of the form (5.33) can never be structurally stable.

Remark. This motivates the definition of an integer affine structure on the n-torus, which is given by an atlas with transition maps of the form (5.34). Compare Appendix B, §B.1.

Exercise 5.3 (Rotation number). Let $\pi : \mathbb{R} \to \mathbb{T}^1$ be given by $\pi(x) = e^{2\pi i x}$. For an orientation-preserving homeomorphism $f : \mathbb{T}^1 \to \mathbb{T}^1$ of the circle let $\tilde{f} : \mathbb{R} \to \mathbb{R}$ that is a continuous lift of f, that is, such that $\pi \circ \tilde{f} = f \circ \pi$; see [4] and Appendix A.

1. Show that $\tilde{f}(x + 1) = \tilde{f}(x) + 1$.
2. Show that if $\tilde{f}_1$ and $\tilde{f}_2$ are two lifts of f, that $\tilde{f}_1 - \tilde{f}_2$ is an integer.

---

[11] This approach remains the same in the case where $\mathbb{T}^n = \mathbb{R}^n/\mathbb{Z}^n$.

3. Define the *rotation number*

$$\varrho_0(\tilde{f}) = \lim_{j \to \infty} \frac{1}{j} \left(\tilde{f} \right)^j (x)$$

and show that this limit, if it exists, is independent of x.

Next define the rotation number $\varrho(f)$ of f as the fractional part of $\varrho_0(\tilde{f})$, that is, as the unique number in $[0, 1)$ such that $\varrho_0(\tilde{f}) - \varrho(f)$ is an integer. Show that this definition is independent of the choice of lift $\tilde{f}$.

4. Show that the rotation number $\varrho(f)$ is invariant under topological conjugation: if f and g are both orientation-preserving circle homeomorphisms, then $\varrho(g^{-1} f g) = \varrho(f)$.

5. (∗) Show that the above limit exists, thereby showing that the rotation number is well defined.

Exercise 5.4 (Main tongue of the Arnold family). For the Arnold family (5.14)

$$P_{\beta,\varepsilon}(x) = x + 2\pi\beta + \varepsilon \sin x$$

of circle maps, consider the region in the (β, ε)-plane where the family has a fixed point. Compute its boundaries and describe (also sketch) the dynamics on both sides of the boundary and on a boundary curve. What kind of bifurcation occurs here?

Exercise 5.5 (On diophantine conditions). For $\tau > 1$ and $\gamma > 0$ we compare the sets $[0, 1]_{\tau+1,\gamma}$ and $\mathbb{R}^2_{\tau,\gamma}$; see Figure 5.5. These sets are determined by the diophantine conditions (5.12) and (5.29):

$$\left| \beta - \frac{p}{q} \right| \geq \frac{\gamma}{q^{\tau+1}},$$

for all rationals p/q with $\gcd(p, q) = 1$, and by

$$|\langle \omega, k \rangle| \geq \frac{\gamma}{|k|^{\tau}},$$

for all $k \in \mathbb{Z}^2 \setminus \{0\}$.

1. Show that the following holds. Whenever $\omega = (\omega_1, \omega_2)$, with $\omega_2 \neq 0$, is a (τ, γ)-diophantine vector, the ratio $\beta = \omega_1/\omega_2$ is a $(\tau + 1, \gamma)$-diophantine number. Similarly for $\beta' = \omega_2/\omega_1$ with $\omega_1 \neq 0$.

2. For $(\tau + 1, \gamma)$-diophantine number $\beta \in \mathbb{R}$ show that

$$\left| e^{2\pi i k \beta} - 1 \right| \geq \frac{4\gamma}{|k|^{\tau}}.$$

(Hint: Draw the complex unit circle and compare an appropriate arc with the corresponding chord.)

Exercise 5.6 (A problem of Sternberg [217], Appendix I). On $\mathbb{T}^2$, with coordinates (x_1, x_2) (mod 2π) a vector field X is given, with the following property. If C_1 denotes the circle $C_1 = \{x_1 = 0\}$, then the Poincaré return map $P : C_1 \to C_1$ with respect to X is a rigid rotation $x_2 \mapsto P(x_2) = x_2 + 2\pi\beta$, mod $2\pi\mathbb{Z}$. From now on we abbreviate $x = x_2$. Let $f(x)$ be the return time of the integral curve connecting the points x and $P(x)$ in C_1. A priori, f does not have to be constant. The problem now is to construct a (another) circle C_2, that does have a constant return time.

To be more precise:

1. Let Φ^t denote the flow of X and express P in terms of Φ^t and f. Then look for a circle C_2 of the form

$$C_2 = \{\Phi^{u(x)}(0, x) \mid x \in C_1\}.$$

In particular, the search is for a (periodic) function u and a constant c, such that

$$\Phi^c(C_2) = C_2.$$

Rewrite this equation as a linear small divisor problem in u and c.
2. Solve this equation formally in terms of Fourier series. What condition on β in general will be needed? Give conditions on β, such that for a real-analytic function f a real-analytic solution u exists.

Exercise 5.7 (Floquet problem on a codimension 1 torus). Consider a smooth system

$$x' = f(x, y)$$
$$y' = g(x, y),$$

with $(x, y) \in \mathbb{T}^n \times \mathbb{R}$. Assume that $f(x, y) = \omega + O(|y|)$, which expresses that $y = 0$ is a invariant n-torus, with a constant vector field on it with frequency-vector ω. Hence the torus $y = 0$ is multiperiodic. Put $g(x, y) = \Omega(x)y + O(|y|^2)$, for a map $\Omega : \mathbb{T}^n \to gl(1, \mathbb{R}) \cong \mathbb{R}$. The present problem is to find a transformation $\mathbb{T}^n \times \mathbb{R} \to \mathbb{T}^n \times \mathbb{R}$, of the form $(x, y) \mapsto (x, z) = (x, A(x)y)$, for some map $A : \mathbb{T}^n \to GL(1, \mathbb{R}) \cong \mathbb{R} \setminus \{0\}$, with the following property. The transformed system

$$x' = \omega + O(|z|)$$
$$z' = \Lambda z + O(|z|^2),$$

is on Floquet form, meaning that the matrix Λ is x-independent. By a computation show that

$$\Lambda = \Omega + \sum_{j=1}^{n} \omega_j \frac{\partial \log A}{\partial x_j}.$$

From this derive a linear small divisor problem in A, by requiring that Λ is constant in x. Formally solve this equation in A, given Ω. Give conditions on ω ensuring a formal solution. Also explain how to obtain a real-analytic solution A, assuming real-analyticity of Ω.

Exercise 5.8 ((∗) **A quasi-periodic normal form I (compare [87], Lemma 5.9)).**
Let

$$X_\lambda(x) = [\omega_0 + f(x, \lambda)]\partial_x,$$

$x \in \mathbb{T}^n, \lambda \in \mathbb{R}^s$, be a real analytic family of vector fields on the n-torus with fixed $\omega_0 \in \mathbb{R}^n_{\tau,\gamma}$, with constants $\tau > n - 1$ and $\gamma > 0$. Assume that $f(x, 0) \equiv 0$. The aim is to show the following. Given $N \in \mathbb{Z}_+$, there exists a family of transformations $\Phi_\lambda : \mathbb{T}^n \to \mathbb{T}^n$ of the form $\xi = x + \varphi(x, \lambda)$ such that the transformed vector field $\overline{X}_\lambda = (\Phi_\lambda)_* X_\lambda$ has the form

$$\overline{X}_\lambda(\xi) = [\omega_0 + \overline{f}(\lambda) + O(\lambda^{N+1})]\partial_\xi,$$

with $\overline{f}(0) = 0$, as $\lambda \to 0$ uniformly in ξ. Note that all ξ-dependence of $\overline{X}$ has disappeared in the $O(\lambda^{N+1})$-term.

The proof runs by induction on $N \in \mathbb{Z}_+$. First check the case $N = 0$.

Next assume that the statement holds for $N - 1$; then $X = X_\lambda(x)$ has the form

$$X_\lambda(x) = [\omega_0 + \overline{f}(\lambda) + \tilde{f}(x, \lambda)]\partial_x,$$

where $\tilde{f}(x, \lambda) = O(\lambda^N)$ uniformly in x with $\overline{f}(0) = 0$. Then, concerning the Nth step, find a transformation $\xi = x + \varphi(x, \lambda)$ with

$$\varphi(x, \lambda) = \lambda^N u(x, \lambda),$$

using multi-index notation, which makes the Nth-order part of the transformed vector field ξ-independent. To this end:

1. Show that the transformed vector field obtains the form

$$\xi' = \omega_0 + \overline{f}(\lambda) + \tilde{f}(x, \lambda) + \frac{\partial\varphi(x, \lambda)}{\partial x}\omega_0 + O(\lambda^{N+1}).$$

   (Hint: Use the chain rule.)
2. Considering the homogeneous Nth-order part, obtain an equation for φ of the form

$$\frac{\partial\varphi(x, \lambda)}{\partial x}\omega_0 + \tilde{f}(x, \lambda) \equiv \lambda^N c \pmod{O(\lambda^{N+1})},$$

   for an appropriate constant $c \in \mathbb{R}^n$, as $\lambda \to 0$.
3. Setting $\tilde{f} = \lambda^N g(x, \lambda)$ obtain a linear small divisor problem for u in terms of g. Give the appropriate value of c.

Exercise 5.9 ((∗) A quasi-periodic normal form II (compare Exercise 5.8)).
Given is a 1-parameter family of circle maps $P_\lambda : \mathbb{T}^1 \to \mathbb{T}^1$ of the form

$$P_\lambda : x \mapsto x + 2\pi\beta + f(x, \lambda),$$

where x is counted mod 2π and where $f(x, 0) \equiv 0$. One has to show that by
successive transformations of the form

$$H_\lambda : x \mapsto x + h(x, \lambda)$$

the x-dependence of P can be pushed away to higher and higher order in λ. For this
appropriate conditions on β will be needed. Carry out the corresponding inductive
process. What do you think the first step is? Then, concerning the Nth step, consider

$$P_{N,\lambda}(x) = x + 2\pi\beta + g(\lambda) + f_N(x, \lambda) + O(|\lambda|^{N+1}),$$

with $f_N(x, \lambda) = \tilde{f}(x)\lambda^N$ and look for a transformation $H = \mathrm{Id} + h$, with
$h(x, \lambda) = \tilde{h}(x, \lambda)\lambda^N$, such that in $H^{-1} \circ P_N \circ H$ the Nth order part in λ is x-
independent. Formulate sufficient conditions on β, ensuring that the corresponding
equation can be formally solved, in terms of Fourier series. Finally give conditions
on β, such that in the real-analytic case we obtain real-analytic solutions h. Explain
your arguments.

(Hint: Use the mean value theorem to obtain an expression of the form

$$H_\lambda^{-1} \circ P_{N,\lambda} \circ H_\lambda (x)$$
$$= x + 2\pi\beta + g(\lambda) + h(x, \lambda) - h(x + 2\pi\beta, \lambda) + f_N(x, \lambda) + O(|\lambda|^{N+1}),$$

where we require for the Nth-order part $h(x, \lambda) - h(x + 2\pi\beta, \lambda) + f_N(x, \lambda) = c\lambda^N$,
for an appropriate constant c. Show that this is a linear small divisor problem where
h can be solved given f_N, provided that $c = f_{N,0}$, the $\mathbb{T}^1$-average of f_N.)

Exercise 5.10 ((∗) Action-angle variables for the autonomous pendulum). Consider the equation of motion of the pendulum

$$y'' = -\alpha^2 \sin y.$$

Writing $z = y'$ consider the energy $H(y, z) = \frac{1}{2}z^2 - \alpha^2 \cos y$. In the (y, z)-plane
consider the set $\Delta = \{(y, z) \mid -\alpha^2 < H(y, z) < \alpha^2\}$. For $-\alpha^2 < E < \alpha^2$, in the
energy level $H^{-1}(E)$ the autonomous pendulum carries out a periodic motion of
period $T = T(E)$.

1. Define

$$I(E) = \frac{1}{2\pi} \oint_{H^{-1}(E)} z \, dy$$

and show that
$$T(E) = 2\pi \frac{\mathrm{d}I}{\mathrm{d}E}(E).$$

2. Let t be the time the motion in $H^{-1}(E)$ takes to get from the line $z = 0$ to the point (y, z). Defining
$$\varphi(y, z) = -\frac{2\pi}{T(E)}t,$$

show that $\mathrm{d}I \wedge \mathrm{d}\varphi = \mathrm{d}y \wedge \mathrm{d}z$

Remark. We call (I, φ) a pair of action-angle variables for oscillations of the pendulum.

Exercise 5.11 (... and perturbations). The equation of motion of a periodically driven pendulum is
$$y'' + \alpha^2 \sin y = \varepsilon \cos(\omega t).$$
Write this as a vector field in $\mathbb{R}^2 \times \mathbb{T}^1 = \{(y, z), x\}$, with $z = y'$ and $x = \omega t$. Then define an appropriate Poincaré map P_ε, which has the form
$$P_\varepsilon(\varphi, I) = (\varphi + 2\pi\beta(I), I) + O(\varepsilon)$$

and derive an integral expression for $\beta(I)$. See Exercise 5.10. Show that P_ε is a perturbed twist map and describe the occurrence of quasi-periodic invariant circles. (Hint: Compare Exercise 3.5.)

Exercise 5.12 (Real-analytic linear small divisor problem (5.24: Paley–Wiener and its converse)). Consider the circle map case of the linear problem (5.24), which is the linear difference equation
$$u(x + 2\pi\beta, \beta) - u(x, \beta) = f(x, \beta)$$

where $x \in \mathbb{T}^1 = \mathbb{R}/(2\pi\mathbb{Z})$, $\beta \in \mathbb{R}$, and where $u, f : \mathbb{T}^1 \to \mathbb{R}$. Let f have $\mathbb{T}^1$-average zero and let β satisfy the diophantine conditions (5.12)
$$\left| \beta - \frac{p}{q} \right| \geq \gamma |q|^{-(\tau+1)},$$

for all $p \in \mathbb{Z}$, $q \in \mathbb{Z} \setminus \{0\}$, with $\tau > 1$, $\gamma > 0$. By expanding f and u in Fourier series
$$f(x) = \sum_{k \in \mathbb{Z}} f_k e^{ikx}, \quad u(x) = \sum_{k \in \mathbb{Z}} u_k e^{ikx},$$

the linear equation transforms to an infinite system of equations in the coefficients u_k and f_k as described in §§5.2 and 5.5 leading to a formal solution.

First assume that f is real-analytic and bounded by M on the complex strip of width $\kappa > 0$ around $\mathbb{T}^1$. That is, for

$$x \in \mathbb{T}^1 + \kappa = \{x \in \mathbb{C}/(2\pi\mathbb{Z}) \mid |\operatorname{Im} x| < \kappa\},$$

assume that

$$\sup_{x \in \mathbb{T}^1 + \kappa} |f(x)| \le M.$$

The Paley–Wiener lemma 5.6 gives estimates of the kth Fourier coefficient f_k in terms of σ, M, and k.

1. Use these and the diophantine conditions to obtain estimates of u_k. (Hint: Also use Exercise 5.5.)
2. Show that for some $0 < \varrho < \sigma$, the formal Fourier series $\sum_k u_k e^{ikx}$ converges on $\mathbb{T}^1 + \varrho$ and defines an analytic function $u(x)$.
3. Derive a bound for u on $\mathbb{T}^1 + \varrho$ that depends explicitly on σ and ϱ.
4. Adapt this exercise for the linear small divisor problem 5.24, the case of torus flows.

Exercise 5.13 ((∗) C^ℓ **linear small divisor problem (5.24, circle map case): Paley–Wiener and its converse).** Consider the same setup as Exercise 5.12, but now assume that f is only of class C^ℓ. Recall that $f(x) = \sum_k f_k e^{ikx}$ where $f_k = (2\pi)^{-1} \oint_{\mathbb{T}^1} f(x) e^{-ikx} dx$.

1. Show that for a constant $C > 0$ that only depends on n, we have

$$|f_k| \le C|k|^{-\ell} \quad \text{for all } k \in \mathbb{Z} \setminus \{0\},$$

which is the C^ℓ-equivalent of Paley–Wiener lemma 5.6.
2. Conversely, show that for the formal series $\sum_k f_k e^{ikx}$ to converge uniformly, it is sufficient that $|f_k| \le C'|k|^{-2}$ for all $k \in \mathbb{Z} \setminus \{0\}$. Next show that sufficient for convergence to a C^ℓ-function f is that

$$|f_k| \le C|k|^{-(\ell+2)} \quad \text{for all } k \in \mathbb{Z} \setminus \{0\}.$$

3. Reconsidering the linear difference equation

$$u(x + 2\pi\beta, \beta) - u(x, \beta) = f(x, \beta),$$

for a C^ℓ-function f with zero average, using the first item and the diophantine conditions (5.12), estimate the rate of decay of the coefficients u_k as $|k| \to \infty$.
4. What is the least value ℓ_0 of ℓ, such that the formal Fourier series $\sum u_k e^{ikx}$ converges absolutely and so defines a continuous function?
5. Given that f is C^ℓ with $\ell > \ell_0$, what is the amount of differentiability of u?
6. Adapt this exercise for the linear small divisor problem 5.24, the case of torus flows.

Exercise 5.14 ((∗) **The linear small divisor problem for n-torus maps**). Develop the linear small divisor problem for families of general torus maps, that are perturbations of translation systems on the n-torus with time set $\mathbb{Z}$; see Chapter 2, §2.2.2.

Exercise 5.15 ((∗) **Residual sets of zero measure**). For definitions see Appendix A; also compare [184].

1. Use the sets $[0, 1]_{\tau+1,\gamma} \subset [0, 1]$ of diophantine numbers (5.12) to construct a subset of $[0, 1]$ that is both residual and of zero measure.
2. Show that the subset $\Lambda \subset \mathbb{T}^1$ as defined in Cremer's example of Chapter 2, in particular §2.4.2, is both residual and of zero measure.

Exercise 5.16 ((∗∗) **diophantine numbers are Bruno**). Show that any diophantine number also satisfies the Bruno condition; for background see [168].

Chapter 6
Reconstruction and time series analysis

Until now the emphasis has been on the analysis of the evolutions and their organisation in state space, in particular their geometric (topological) and measure-theoretic structure, given the system. In this chapter, the question is which properties of the system one can reconstruct, given a time series of an evolution, just assuming that it has a deterministic evolution law. We note that this is exactly the historical way that information was obtained on the solar system.

6.1 Introduction

Since it was realised that even simple and deterministic dynamical systems can produce trajectories which look like the result of a random process, there occurred some obvious questions: how can one distinguish 'deterministic chaos' from 'randomness' and, in the case of determinism, how can we extract from a time series relevant information about the dynamical system generating it? It turned out that these questions could not be answered in terms of conventional linear time series analysis, that is, in terms of autocovariances and power spectra. In this chapter we review the new methods of time series analysis which were introduced in order to answer these questions. We mainly restrict to the methods which are based on embedding theory and correlation integrals: these methods, and the relations of these methods with the linear methods, are now well understood and supported by mathematical theory. Though they are applicable to a large class of time series, it is often necessary to supplement them for special applications by ad hoc refinements. We do not deal in this chapter with such ad hoc refinements, but concentrate on the main concepts. In this way we hope to make the general theory as clear as possible. For the special methods which are required for the various applications, we refer to the monographs [115, 125, 150] as well as to the collections of research papers [11, 243].

We start in the next section with an illustrative experimental example, a dripping faucet, taken from [106] which gives a good heuristic idea of the method of reconstruction. Then, in §6.3, we describe the reconstruction theorem which gives, amongst others, a justification for this analysis of the dripping faucet. It also motivates a (still rather primitive) method for deriving numerical invariants which can be

H.W. Broer and F. Takens, *Dynamical Systems and Chaos*,
Applied Mathematical Sciences 172, DOI 10.1007/978-1-4419-6870-8_6,
© Springer Science+Business Media, LLC 2011

used to characterise aspects of the dynamics of the system generating a time series; this is discussed in §6.4. After that we introduce in §6.5 some notions related with stationarity, in particular the so-called reconstruction measures, which describe the statistics of the behaviour of a time series as far as subsegments (of small lengths) are concerned. After that we discuss the correlation integrals and the dimensions and entropies which are based on them in §6.6 and their estimation in §6.7. In §6.8 we relate the methods in terms of correlation integrals with linear methods. Finally, in §6.9, we give a short discussion of related topics which do not fall within the general framework of the analysis of time series in terms of correlation integrals.

6.2 An experimental example: The dripping faucet

Here we give an account of a heuristic way in which the idea of reconstruction was applied to a simple physical system, which was first published in *Scientific American* [106] and which was based on earlier work by the same authors in [185]. We also use this example to introduce the notion of a (deterministic) dynamical system as a generator of a time series.

We consider a sequence of successive measurements from a simple physical experiment: one measures the lengths of the time intervals between successive drops falling from a dripping faucet. They form a finite, but long, sequence of real numbers, a *time series* $\{y_0, y_1, \ldots\}$. The question is whether this sequence of real numbers contains an indication about whether it is generated by a deterministic system (the faucet was set in such a way that the dripping was not 'regular', that is, not periodic; for many faucets such a setting is hard to obtain). In order to answer this question, the authors of [106] considered the set of 3-vectors $\{(y_{i-1}, y_i, y_{i+1})\}_i$, each obtained by taking three successive values of the time series. These vectors form a 'cloud of points' in $\mathbb{R}^3$. It turned out that, in the experiment reported in [106], this cloud of points formed a (slightly thickened) curve C in 3-space. This gives an indication that the process is deterministic (with small noise corresponding to the thickening of the curve). The argument is based on the fact that the values y_{n-1} and y_n of the time series, together with the (thickened) curve C, enable in general a good prediction for the next value y_{n+1} of the time series.

This prediction of y_{n+1} is obtained in the following way. We consider the line ℓ_n in 3-space which is in the direction of the third coordinate axis and whose first two coordinates are y_{n-1} and y_n. This line ℓ_n has to intersect the cloud of 3-vectors, because (y_{n-1}, y_n, y_{n+1}) belongs to this cloud (even though the cloud is composed of vectors which were collected in the past, we assume the past to be so extensive that new vectors do not substantially extend the cloud). If we 'idealise' this cloud to a curve C, then generically ℓ_n and C only have this one point in common. Later we make all this mathematically rigorous, but the idea is that two (smooth) curves in 3-space have in general, or generically, no intersection (in dimension 3 there is enough space for two curves to become disjoint by a small perturbation); by construction, C and ℓ_n must have one point in common, namely (y_{n-1}, y_n, y_{n+1}), but there is no reason for other intersections. For the definition of 'generic' see Appendix A. This means that we expect that we can predict the value y_{n+1} from the previous values y_{n-1} and y_n as the third coordinate of the intersection $C \cap \ell_n$. Even if we do not idealise the cloud of points in $\mathbb{R}^3$ to the curve, we conclude that the next value y_{n+1} should be in the range given by the intersection of this cloud with ℓ_n; usually this range is small and hence the expected error of the prediction will be small. The fact that such predictions are possible means that we deal with a deterministic system or an almost deterministic system if we take into account the fact that the cloud of vectors is really a thickened curve.

So, in this example, we can consider the cloud of vectors in 3-space, or its idealisation to the curve C, as the *statistical information* concerning the time series which is extracted from a long sequence of observations. It describes the dynamical law generating the time series in the sense that it enables us to predict, from a short sequence of recent observations, the next observation.

A mathematical justification of the above method to detect a deterministic structure in an apparently random time series, is given by the reconstruction theorem; see §6.3. This was a main starting point of what is now called nonlinear time series analysis. The fact that many systems are not strictly deterministic but can be interpreted as deterministic with a small contamination of noise gives extra complications to which we return later.

Note that we only considered here the prediction of one step in the future. So even if this type of prediction works, the underlying dynamical system still may have a positive dispersion exponent; see §2.3.2.

6.3 The reconstruction theorem

In this section we formulate the reconstruction theorem and discuss some of its generalisations. In the next section we give a first example of how it is relevant for the questions raised in the introduction of this chapter.

We consider a dynamical system, generated by a diffeomorphism $\varphi : M \to M$ on a compact manifold M, together with a smooth 'read-out' function $f : M \to \mathbb{R}$. This setup is proposed as the general form of mathematical models for deterministic dynamical systems generating time series in the following sense. Possible *evolutions*, or *orbits*, of the dynamical system are sequences of the form $\{x_n = \varphi^n(x_0)\}$, where usually the index n runs through the nonnegative integers. We assume that what is observed, or measured, at each time n is not the whole state x_n but just one real number $y_n = f(x_n)$, where f is the read-out function which assigns to each state the value that is measured (or observed) when the system is in that state. So corresponding to each evolution $\{x_n = \varphi^n(x_0)\}$ there is a time series Y of successive measurements $\{y_n = f(\varphi^n(x_0)) = f(x_n)\}$.

In the above setting of a deterministic system with measurements, we made some restrictions which are not completely natural. The time could have been continuous: the dynamics could have been given by a differential equation (or vector field) instead of a diffeomorphism (in other words, the time set could have been the reals $\mathbb{R}$ instead of the natural numbers $\mathbb{N}$); instead of measuring only one value, one could measure several values (or a finite-dimensional vector); also the dynamics could have been given by an endomorphism instead of a diffeomorphism; finally we could allow M to be noncompact. We return to such generalisations after treating the situation as proposed, namely with the dynamics given by a diffeomorphism on a compact manifold and a 1-dimensional read-out function. In this situation we have the following.

Theorem 6.1 (Reconstruction theorem). *For a compact m-dimensional manifold M, $\ell \geq 1$, and $k > 2m$ there is an open and dense subset $\mathcal{U} \subset \mathrm{Diff}^\ell(M) \times C^\ell(M)$, the product of the space of C^ℓ-diffeomorphisms on M and the space of C^ℓ-functions on M, such that for $(\varphi, f) \in \mathcal{U}$ the following map is an embedding of M into $\mathbb{R}^k$:*

$$M \ni x \mapsto (f(x), f(\varphi(x)), \ldots, f(\varphi^{k-1}(x))) \in \mathbb{R}^k.$$

Observe that the conclusion of Theorem 6.1 holds for generic pairs (φ, f). A proof, which is an adaptation of the Whitney embedding theorem [246], first appeared in [221]; also see [55]. For a later and more didactical version, with some extensions, see [209]. For details we refer to these publications.

We introduce the following notation: the map from M to $\mathbb{R}^k$, given by

$$x \mapsto (f(x), f(\varphi(x)), \ldots, f(\varphi^{k-1}(x))),$$

is denoted by Rec_k and vectors of the form $(f(x), f(\varphi(x)), \ldots, f(\varphi^{k-1}(x)))$ are called k-dimensional *reconstruction vectors* of the system defined by (φ, f). The image of M under Rec_k is denoted by $\mathcal{X}_k$.

The meaning of the reconstruction theorem is the following. For a time series $Y = \{y_n\}$ we consider the sequence of its k-dimensional *reconstruction vectors* $\{(y_n, y_{n+1}, \ldots, y_{n+k-1})\}_n \subset \mathbb{R}^k$; this sequence is 'diffeomorphic' to the evolution of the deterministic system producing the time series Y provided that the following conditions are satisfied.

– The number k is sufficiently large, for example, larger than twice the dimension of the state space M.
– The pair (φ, f), consisting of the deterministic system and the read-out function, is generic in the sense that it belongs to the open and dense subset as in Theorem 6.1.

The sense in which the sequence of reconstruction vectors of the time series is diffeomorphic to the underlying evolution is as follows. The relation between the time series $Y = \{y_n\}$ and the underlying evolution $\{x_n = \varphi^n(x_0)\}$ is that $y_n = f(x_n)$. This means that the reconstruction vectors of the time series Y are just the images under the map Rec_k of the successive points of the evolution $\{x_n\}$. So Rec_k, which is, by the theorem, an embedding of M onto its image $\mathcal{X}_k \subset \mathbb{R}^k$, sends the orbit of $\{x_n\}$ to the sequence of reconstruction vectors $\{(y_n, y_{n+1}, \ldots, y_{n+k-1})\}$

of the time series Y. We note that a consequence of this fact, and the compactness of M, is that for any metric d on M (derived from some Riemannian metric), the evolution $\{x_n\}$ and the sequence of reconstruction vectors of Y are metrically equal up to *bounded distortion*. This means that the quotients

$$\frac{d(x_n, x_m)}{\|(y_n, \ldots, y_{n+k-1}) - (y_m, \ldots, y_{m+k-1})\|}$$

are uniformly bounded and bounded away from zero. So, all the recurrence properties of the evolution $\{x_n\}$ are still present in the corresponding sequence of reconstruction vectors of Y.

6.3.1 Generalisations

6.3.1.1 Continuous time

Suppose now that we have a dynamical system with continuous time, that is, a dynamical system given by a vector field (or a differential equation) X on a compact manifold M. For each $x \in M$ we denote the corresponding orbit by $t \mapsto \varphi^t(x)$ for $t \in \mathbb{R}$. In that case there are two alternatives for the definition of the reconstruction map Rec_k. One is based on a discretisation: we take a (usually small) time interval $h > 0$ and define the reconstruction map Rec_k in terms of the diffeomorphism φ^h, the time h map of the vector field X. Also then one can show that for $k > 2m$ and generic triples (X, f, h), the reconstruction map Rec_k is an embedding. Another possibility is to define the reconstruction map in terms of the derivatives of the function $t \mapsto f(\varphi^t(x))$ at $t = 0$, that is, by taking

$$\text{Rec}_k(x) = (f(\varphi^t(x)), \frac{\partial}{\partial t} f(\varphi^t(x)), \ldots, \frac{\partial^{k-1}}{\partial t^{k-1}} f(\varphi^t(x))) \mid_{t=0} .$$

Also for this definition of the reconstruction map the conclusion is the same: for $k > 2m$ and generic pairs (X, f) the reconstruction map is an embedding of M into $\mathbb{R}^k$. The proof for both these versions is in [221]; in fact in [55] it was just the case of continuous time that was treated through the discretisation mentioned above.

We note that the last form of reconstruction (in terms of derivatives) is not very useful for experimental time series inasmuch as the numerical evaluation of the derivatives introduces in general a considerable amount of noise.

6.3.1.2 Multidimensional measurements

In the case that at each time one measures a multidimensional vector, so that we have a read-out function with values in $\mathbb{R}^s$, the reconstruction map Rec_k has values in $\mathbb{R}^{sk}$. In this case the conclusion of the reconstruction theorem still holds for generic pairs (φ, f) whenever $ks > 2m$. As far as is known to us, there is no reference for this result, but the proof is the same as in the case with 1-dimensional measurements.

6.3.1.3 Endomorphisms

If we allow the dynamics to be given by an endomorphism instead of a diffeomorphism, then the obvious generalisation of the reconstruction theorem is not correct. A proper generalisation in this case, which can serve as a justification for the type of analysis which we described in §6.2 and also of the numerical estimation of dimensions and entropies in §6.7, was given in [224]. For $k > 2m$ one proves that there is, under the usual generic assumptions, a map $\pi_k : \mathcal{X}_k \to M$ such that $\pi_k \circ \mathrm{Rec}_k = \varphi^{k-1}$. This means that a sequence of k successive measurement determines the state of the system at the end of the sequence of measurements. However, there may be different reconstruction vectors corresponding to the same final state; in this case $\mathcal{X}_k \subset \mathbb{R}^k$ is in general not a submanifold, but still the map π_k is differentiable, in the sense that it can be extended to a differentiable map from a neighbourhood of $\mathcal{X}_k$ in $\mathbb{R}^k$ to M.

6.3.1.4 Compactness

If M is not compact, the reconstruction theorem remains true, but not the remark about 'bounded distortion'. The reason for this restriction was, however, the following. When talking about predictability, as in §6.2, based on the knowledge of a (long) segment of a time series, an implicit assumption is always: what will happen next already happened (exactly or approximately) in the past. So this idea of predictability is only applicable to evolutions $\{x_n\}_{n\geq 0}$ which have the property that for each $\varepsilon > 0$ there is an $N(\varepsilon)$ such that for each $n > N(\varepsilon)$, there is some $0 < n'(n) < N(\varepsilon)$ with $d(x_n, x_{n'(n)}) < \varepsilon$. This is a mathematical formulation of the condition that after a sufficiently long segment of the evolution (here length $N(\varepsilon)$) every new state, like x_n, is approximately equal (here equal up to a distance ε) to one of the past states, here $x_{n'(n)}$. It is not hard to see that this condition on $\{x_n\}$, assuming that the state space is a complete metrisable space, is equivalent with the condition that the positive evolution $\{x_n\}_{n\geq 0}$ has a compact closure; see §2.3.2. Such an assumption is basic for the main applications of the reconstruction theorem, therefore it is no great restriction of the generality to assume, as we did, that the state space M itself is a compact manifold because we only want to deal with a compact part of the state space anyway. In this way we also avoid the complications of defining the topology on spaces of functions (and diffeomorphisms) on noncompact manifolds.

There is another generalisation of the reconstruction theorem which is related to the above remark. For any evolution $\{x_n\}_{n\geq 0}$ of a (differentiable) dynamical system with compact state space, one defines its ω-limit $\omega(x_0)$ as

$$\omega(x_0) = \{x \in M \mid \exists n_i \to \infty, \text{ such that } x_{n_i} \to x\};$$

compare §4.1. (Recall that often such an ω-limit is an attractor, but that is of no immediate concern to us here.) Using the compactness of M, or of the closure of the evolution $\{x_n\}_{n \geq 0}$, one can prove that for each $\varepsilon > 0$ there is some $N'(\varepsilon)$ such that for each $n > N'(\varepsilon)$, $d(x_n, \omega(x_0)) < \varepsilon$. So the ω-limit describes the dynamics of the evolution starting in x_0 without the peculiarities (transients) which are only due to the specific choice of the initial state. For a number of applications, one does not need the reconstruction map Rec_k to be an embedding of the whole of M, but only of the ω-limit of the evolution under consideration. Often the dimension of such an ω-limit is much lower than the dimension of the state space. For this reason it was important that the reconstruction theorem was extended in [209] to this case, where the condition on k could be weakened: k only has to be larger than twice the dimension of the ω-limit. It should, however, be observed that these ω-limit sets are in general not very nice spaces (like manifolds) and that for that reason the notion of dimension is not so obvious (one has to take in this case the box-counting dimension; see §6.4.1); in the conclusion of the theorem, one gets an injective map of the ω-limit into $\mathbb{R}^k$, but the property of bounded distortion has not been proven for this case (and is probably false).

6.3.2 Historical note

This reconstruction theorem was obtained independently by Aeyels and Takens. In fact it was D. L. Elliot who pointed out, at the Warwick conference in 1980 where the reconstruction theorem was presented, that Aeyels had obtained the same type of results in his thesis (Washington University, 1978) which was published as [55]. His motivation came from systems theory, and in particular from the observability problem for generic nonlinear systems.

6.4 Reconstruction and detecting determinism

In this section we show how, from the distribution of reconstruction vectors of a given time series, one can obtain an indication whether the time series was generated by a deterministic or a random process. In some cases this is very simple; see §6.2. As an example we consider two time series: one is obtained from the Hénon system and the second is a randomised version of the first one. We recall (see also §1.2.3) that the Hénon system has a 2-dimensional state space $\mathbb{R}^2$ and is given by the map

$$(x_1, x_2) \mapsto (1 - ax_1^2 + bx_2, x_1);$$

we take for a and b the usual values $a = 1.4$ and $b = 0.3$. As a read-out function we take $f(x_1, x_2) = x_1$. For an evolution $\{(x_1(n), x_2(n))\}_n$, with $(x_1(0), x_2(0))$

close to $(0, 0)$, we consider the time series $y_n = f(x_1(n), x_2(n)) = x_1(n)$ with $0 \leq n \leq N$ for some large N. In order to get a time series of an evolution inside the Hénon attractor (within the limits of the resolution of the graphical representation) we omitted the first 100 values. This is our first time series. Our second time series is obtained by applying a random permutation to the first time series. So the second time series is essentially a random iid (identically and independently distributed) time series with the same histogram as the first time series. From the time series itself it is not obvious which of the two is deterministic; see Figure 6.1. However, if we plot the 'cloud of 2-dimensional reconstruction vectors,' the situation is completely different; compare Figure 6.2. In the case of the reconstruction vectors from the Hénon system we clearly distinguish the well-known picture of the Hénon attractor; see Figure 1.13 in §1.3. In the case of the reconstruction vectors of the randomised time series we just see a structureless cloud.

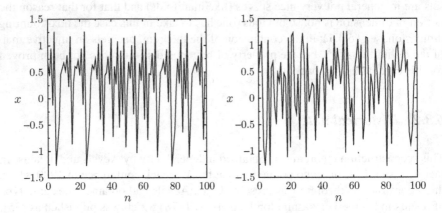

Fig. 6.1 Time series: (left) from the Hénon system with n running from 100 to 200 and (right) a randomised version of this time series.

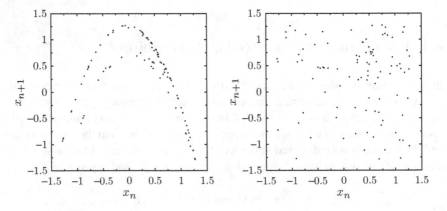

Fig. 6.2 The clouds of 2-dimensional reconstruction vectors for the two time series in Figure 6.1 The box-counting dimensions are approximately 1.2 (left) and ∞ (right).

Of course the above example is rather special, but, in terms of the reconstruction theorem, it may be clear that there is a more general method behind this observation. In the case that we use a time series generated by a deterministic process, the reconstruction vectors fill a 'dense' subset of the differentiable image (under the reconstruction map) of the closure of the corresponding evolution of the deterministic system (or of its ω-limit if we omit a sufficiently long segment from the beginning of the time series); of course, for these vectors to fill the closure of the evolution really densely, the time series should be infinite. If the reconstruction vectors are of sufficiently high dimension, one generically gets even a diffeomorphic image. In the case of the Hénon system the reconstruction vectors form a diffeomorphic image of the Hénon attractor which has, according to numerical evidence, a dimension strictly between 1 and 2 (for the notion of dimension see below). This means that the reconstruction vectors also densely fill a subset of dimension strictly smaller than 2; this explains that the 2-dimensional reconstruction vectors are concentrated on a 'thin' subset of the plane (this is also explained below). Such a concentration on a 'thin' subset is clearly different from the diffuse distribution of reconstruction vectors which one sees in the case of the randomised time series. A diffuse distribution, filling out some open region is typical for reconstruction vectors generated by a random process, for example, this is what one gets for a time series generated by any (nondeterministic) Gaussian process.

In the above discussion the, still not defined, notion of dimension played a key role. The notion of dimension which we have to use here is not the usual one from topology or linear algebra which only can have integer values: we use here so-called fractal dimensions. In this section we limit ourselves to the box-counting dimension. We first discuss its definition, or rather a number of equivalent definitions, and some of its basic properties. It then turns out that the numerical estimation of these dimensions provides a general method for discrimination between time series as discussed here, namely those generated by a low-dimensional attractor of a deterministic system, and those generated by a random process.

6.4.1 Box-counting dimension and its numerical estimation

For introductory information on the various dimensions used in dynamical systems we refer the reader to [12] and to references therein. For a more advanced treatment see [122, 123] and [188].

The box-counting dimension, which we discuss here, is certainly not the most convenient for numerical estimation, but it is conceptually the simplest one. The computationally more relevant dimensions are discussed in §6.6

Let K be a compact metric space. We say that a subset $A \subset K$ is ε-spanning if each point in K is contained in the ε-neighbourhood of one of the points of A. The smallest cardinality of an ε-spanning subset of K is denoted by $a_\varepsilon(K)$; note that this number is finite due to the compactness of K. For the same compact metric space

K a subset $B \subset K$ is called ε-separated if, for any two points $b_1 \neq b_2 \in B$, the distance between b_1 and b_2 is at least ε. The greatest cardinality of an ε-separated subset of K is denoted by $b_\varepsilon(K)$.

It is not hard to prove that

$$a_{\frac{\varepsilon}{2}}(K) \geq b_\varepsilon(K) \geq a_\varepsilon(K);$$

see the exercises. Also it may be clear that, as $\varepsilon \downarrow 0$, both a_ε and b_ε tend to infinity, except if K consists of a finite number of points only. The speed of this growth turns out to provide a definition of dimension.

Definition 6.2 (Box-counting dimension). The *box-counting dimension* of a compact metric space K is defined as

$$d(K) = \limsup_{\varepsilon \downarrow 0} \frac{\ln(a_\varepsilon(K))}{-\ln(\varepsilon)} = \limsup_{\varepsilon \downarrow 0} \frac{\ln(b_\varepsilon(K))}{-\ln(\varepsilon)}.$$

Note that the equality of the two expressions in the theorem follows from the above inequalities.

It is not hard to show that if K is the unit cube, or any compact subset of the Euclidean n-dimensional space with nonempty interior, then its box-counting dimension is $d(K) = n$; also, for any compact subset $K \subset \mathbb{R}^n$, we have $d(K) \leq n$ (also see the exercises). This is what we should expect from any quantity which is some sort of a dimension. There are, however, also examples where this dimension is not an integer. Main examples of this are the middle α Cantor sets. These sets can be constructed as follows. We start with the interval $K_0 = [0, 1] \subset \mathbb{R}$; each K_{i+1} is obtained from K_i by deleting from each of the intervals of K_i an open interval in its middle of length α times the length of the interval K_i. So K_i will have 2^i intervals of length $((1 - \alpha)/2)^i$. For the box-counting dimension of the middle α Cantor set, see Exercise 6.8.

Next we consider a compact subset $K \subset \mathbb{R}^k$. With the Euclidean distance function this is also a metric space. For each $\varepsilon > 0$ we consider the partition of $\mathbb{R}^k$ into ε-boxes of the form

$$\{n_1 \varepsilon \leq x_1 < (n_1 + 1)\varepsilon, \ldots, n_k \varepsilon \leq x_k < (n_k + 1)\varepsilon\},$$

with $(n_1, \ldots, n_k) \in \mathbb{Z}^k$. The number of boxes of this partition containing at least one point of K, is denoted by $c_\varepsilon(K)$. We then have the following inequalities,

$$a_{\varepsilon\sqrt{k}}(K) \leq c_\varepsilon(K) \leq 3^k a_\varepsilon(K).$$

This means that in the definition of box-counting dimension, we may replace the quantities a_n or b_n by c_n whenever K is a subset of a Euclidean space. This explains the name 'box-counting dimension'; other names are also used for this dimension; for a discussion, with some history, compare [123].

It follows easily from the definitions that if K' is the image of K under a map f with bounded expansion, or a Lipschitz map, that is, a map such that for some constant C we have for all pairs $k_1, k_2 \in K$ that

$$\rho'(f(k_1), f(k_2)) \leq C\rho(k_1, k_2),$$

where ρ, ρ' denote the metrics in K and K', respectively, then $d(K') \leq d(K)$. A differentiable map, restricted to a compact set, has bounded expansion. Thus we have proven the following.

Lemma 6.3 (Dimension of differentiable image). *Under a differentiable map the box-counting dimension of a compact set does not increase, and under a diffeomorphism it remains the same.*

Remark. One should not expect that this lemma remains true without the assumption of differentiability. This can be seen from the well-known Peano curve which is a continuous map from the (1-dimensional) unit interval onto the (2-dimensional) square. Compare [12].

6.4.2 Numerical estimation of the box-counting dimension

Suppose $K \subset \mathbb{R}^k$ is a compact subset and $k_1, k_2, \ldots \in K$ is a dense sequence in K. If such a dense sequence is numerically given, in the sense that we have (good approximations of) $k_1, \ldots, k_N$ for some large value N, this can be used to estimate the box-counting dimension of K in the following way. First we estimate the quantities $c_\varepsilon(K)$ for various values of ε by counting the number of ε-boxes containing at least one of the points $k_1, \ldots, k_N$. Then we use these quantities to estimate the limit (or limsup) of $\ln(c_\varepsilon(K))/-\ln(\varepsilon)$ (which is the box-counting dimension).

We have to use the notion of 'estimating' here in a loose sense, inasmuch as there is certainly no error bound. However, we can say somewhat more: first, we should not try to make ε smaller than the precision with which we know the successive points k_i; then there is another lower bound for ε which is much harder to assess, and which depends on N: if we take ε very small, then $c_\varepsilon(K)$ will usually become larger than N which implies that we will get too low an estimate for c_ε (because our estimate will at most be N). These limitations on ε imply that the estimation of the limit (or limsup) of $\ln(c_\varepsilon(K))/-\ln(\varepsilon)$ cannot be more than rather 'speculative'. The 'speculation' which one usually assumes is that the graph of $\ln(c_\varepsilon(K))$, as a function of $-\ln(\varepsilon)$, approaches a straight line for decreasing ε, that is, for increasing $-\ln(\varepsilon)$. Whenever this speculation is supported by numerical evidence, one takes the slope of the estimated straight line as estimate for the box-counting dimension. It turns out that this procedure often leads to reproducible results in situations which one encounters in the analysis of time series of (chaotic) dynamical systems (at least if the dimension of the ω-limit is sufficiently low).

Later, we show other definitions of dimension which lead to estimation procedures that are better than the present estimate of the box-counting dimension in the sense that we have a somewhat better idea of the variance of the estimates. Still, dimension estimations have been carried out along the line as described above; see [128]. In that paper, dimension estimates were obtained for a 1-parameter family of dimensions, the so-called D_q-dimensions introduced by Renyi; see [199]. The box-counting dimension as defined here coincides with the D_0-dimension. The estimated value of this dimension for the Hénon attractor is $D_0 = 1.272 \pm .006$, where the error bound of course is not rigourous: there is not even a rigourous proof for the fact that D_0 is greater than 0 in this case!

6.4.3 Box-counting dimension as an indication for 'thin' subsets

As remarked above, a compact subset of the plane which has a box-counting dimension smaller than 2 cannot have interior points. So the 'thin appearance' of the reconstructed Hénon attractor is in agreement with the fact that the box-counting dimension is estimated to be smaller than 2.

From the above arguments it may be clear that if we apply the dimension estimation to a sequence of reconstruction vectors from a time series which is generated by a deterministic system with a low-dimensional attractor, then we should find a dimension which does not exceed the dimension of this attractor. So an estimated dimension, based on a sequence of reconstruction vectors $\{(y_i, \dots, y_{i+k-1}) \in \mathbb{R}^k\}$, which is significantly lower than the dimension k in which the reconstruction takes place, is an indication in favour of a deterministic system generating the time series.

Here we have to come back to the remark at the end of §6.2 also in relation with the above remark warning against too small values of ε. Suppose we consider a sequence of reconstruction vectors generated by a system which we expect to be essentially deterministic, but contaminated with some small noise (as in the case of the dripping faucet of §6.2). If, in such cases, we consider values of ε which are smaller than the amplitude of the noise we should find that the 'cloud' of k-dimensional reconstruction vectors has dimension k. So such values of ε should be avoided.

Remark. So far we have given a partial answer how to distinguish between deterministic and random time series. In §6.8, and in particular §§6.8.2 and 6.8.3, we are able to discuss this question in greater depth.

6.4.4 Estimation of topological entropy

The topological entropy of dynamical systems is a measure for the sensitiveness of evolutions on initial states, or of their unpredictability (we note that this notion is related to, but different from, the notion of dispersion exponent introduced in

§2.3.2). This quantity can be estimated from time series with a procedure that is very much related to the above procedure for estimating the box-counting dimension. It can also be used as an indication whether we deal with a time series generated by a deterministic or by a random system; see [221]. However, we do not discuss this matter here. The reason is that, as we observed before, the actual estimation of these quantities is difficult and often unreliable. We introduce in §6.6 quantities which are easier to estimate and carry the same type of information. In the context of these quantities we also discuss several forms of entropy.

6.5 Stationarity and reconstruction measures

In this section we discuss some general notions concerning time series. These time series $\{y_i\}_{i \geq 0}$ are supposed to be of infinite length; the elements y_i can be in $\mathbb{R}$, in $\mathbb{R}^k$, or in some manifold. This means that also a (positive) evolution of a dynamical system can be considered as a time series from this point of view. On the other hand we do not assume that our time series are generated by deterministic processes. Also, we do not consider our time series as realisations of some stochastic process as is commonly done in texts in probability theory; that is, we do not consider y_i as a function of $\omega \in \Omega$, where Ω is some probability space. In §6.5.2, however, we indicate how our approach can be interpreted in terms of stochastic processes.

The main notion which we discuss here is the notion of *stationarity*. This notion is related to the notion of predictability as we discussed it before: it means that the 'statistical behaviour' of the time series stays constant in time, so that knowledge of the past can be used to predict the future, or to conclude that the future cannot be (accurately) predicted. Equivalently, stationarity means that all kinds of averages, quantifying the 'statistical behaviour' of the time series, are well defined.

Before giving a formal definition, we give some (counter) examples. Standard examples of nonstationary time series are economical time series such as prices as a function of time (say measured in years): due to inflation such a time series will, on average, increase, often at an exponential rate. This means that, for example, the average of such a time series (as an average over the infinite future) is not defined, at least not as a finite number. Quotients of prices in successive years have a much better chance of forming a stationary time series. These examples have to be taken with a grain of salt: such time series are not (yet) known for the whole infinite future. However, more sophisticated mathematical examples can be given. For example we consider the time series $\{y_i\}$ with:

- $y_i = 0$ if the integer part of $\ln(i)$ is even.
- $y_i = 1$ if the integer part of $\ln(i)$ is odd.

It is not hard to verify that $\lim_{n \to \infty} 1/n \sum_{i=0}^{n} y_i$ does not exist; see the exercises. This nonexistence of the average for the infinite future means that this time series is not stationary. Though this example may look pathological, we show that examples of this type occur as evolutions of dynamical systems.

6.5.1 Probability measures defined by relative frequencies

We consider a topological space M with a countable sequence of points $\{p_i\}_{i \in \mathbb{Z}_+} \subset M$. We want to use this sequence to define a Borel probability measure μ on M which assigns to each (reasonable) subset $A \subset M$ the average fraction of points of the sequence $\{p_i\}$ which are in A, so

$$\mu(A) = \lim_{n \to \infty} \frac{1}{n} \#\{i \mid 0 \le i < n \text{ and } p_i \in A\},$$

where as before, # denotes the cardinality. If this measure is well defined we call it the *measure of relative frequencies* of the sequence $\{p_i\}$. One problem is that these limits may not be defined; another is that the specification of the 'reasonable' sets is not so obvious. This is the reason that, from a mathematical point of view, another definition of this measure of relative frequencies is preferred. For this approach we need our space M to be metrisable and to have a countable basis of open sets (these properties hold for all the spaces one encounters as state spaces of dynamical systems). Next we assume that for each continuous function $\psi : M \to \mathbb{R}$ with compact support, the limit

$$\hat{\mu}(\psi) = \lim_{n \to \infty} \frac{1}{n} \sum_{i=0}^{n-1} \psi(p_i)$$

exists; if this limit does not exist for some continuous function with compact support, then the measure of relative frequencies is not defined. If all these limits exist, then $\hat{\mu}$ is a continuous positive linear functional on the space of continuous functions with compact support. According to the Riesz representation theorem (e.g. see [202]), there is then a unique (nonnegative) Borel measure μ on M such that for each continuous function with compact support ψ we have

$$\hat{\mu}(\psi) = \int \psi \mathrm{d}\mu.$$

It is not hard to see that for this measure we always have $\mu(M) \le 1$. If $\mu(M) = 1$, we call μ *the probability measure of relative frequencies defined by* $\{p_i\}$. Note that it may happen that $\mu(M)$ is strictly smaller than 1; for example, if we have $M = \mathbb{R}$ and $p_i \to \infty$, as $i \to \infty$, think of prices as a function of time in the above example. In that case $\hat{\mu}(\psi) = 0$ for each function ψ with compact support, so that $\mu(M) = 0$. If $\mu(M) < 1$, the probability measure of relative frequencies is not defined.

6.5.2 Definition of stationarity and reconstruction measures

We first consider a time series $\{y_i\}$ with values in $\mathbb{R}$. For each k we have a corresponding sequence of k-dimensional reconstruction vectors

$$\{Y_i^k = (y_i, \ldots, y_{i+k-1}) \in \mathbb{R}^k\}_{i \ge 0}.$$

We say that the time series $\{y_i\}$ is stationary if for each k the probability measure of relative frequencies of $\{Y_i^k\}_i$ is well defined. So here we need the time series to be of infinite length. In the case that the time series has values in a manifold M (or in a vector space E), the definition is completely analogous. The only adaptation one has to make is that the elements $Y_i^k = (y_i, \ldots, y_{i+k-1})$ are not k-vectors, but elements of M^k (or of E^k).

Assuming that our time series is stationary, the probability measures of relative frequencies of k-dimensional reconstruction vectors are called k-*dimensional reconstruction measures* and are denoted by μ^k. These measures satisfy some obvious compatibility relations. For time series with values in $\mathbb{R}$ these read as follows. For any subset $A \subset \mathbb{R}^k$:

$$\mu^{k+1}(\mathbb{R} \times A) = \mu^k(A) = \mu^{k+1}(A \times \mathbb{R});$$

for any sequence of measures μ^k on $\mathbb{R}^k$, satisfying these compatibility relations (and an ergodicity property, which is usually satisfied for reconstruction measures), there is a well-defined time series in the sense of stochastic processes. This is explained in [71], Chapter 2.11.

6.5.3 Examples of nonexistence of reconstruction measures

We have seen that the existence of probability measures of relative frequencies, and hence the validity of stationarity, can be violated in two ways: one is that the elements of the sequence, or of the time series, under consideration move off to infinity; the other is that limits of the form

$$\lim_{n \to \infty} \frac{1}{n} \sum_{i=0}^{n-1} \psi(p_i)$$

do not exist. If we restrict to time series generated by a dynamical system with compact state space, then nothing can move off to infinity, so then there is only the problem of existence of the limits of these averages. There is a general belief that for evolutions of generic dynamical systems with generic initial state, the probability measures of relative frequency are well defined. A mathematical justification of this belief has not yet been given. This is one of the challenging problems in the ergodic theory of smooth dynamical systems; see also [206, 226].

There are however nongeneric counterexamples. One is the map $\varphi(x) = 3x$ modulo 1, defined on the circle $\mathbb{R}$ modulo $\mathbb{Z}$. If we take an appropriate initial point, then we get an evolution which is much like the 0-, 1-sequence which we gave as an example of a nonstationary time series. The construction is the following. As initial state we take the point x_0 which has in the ternary system the form $0.\alpha_1\alpha_2\ldots$ with

- $\alpha_i = 0$ if the integer part of $\ln(i)$ is even.
- $\alpha_i = 1$ if the integer part of $\ln(i)$ is odd.

If we consider the partition of $\mathbb{R}$ modulo $\mathbb{Z}$, given by

$$I_i = \left[\frac{i}{3}, \frac{i+1}{3}\right), \quad i = 0, 1, 2,$$

then $\varphi^i(x_0) \in I_{\alpha_i}$; compare this with the doubling map and symbolic dynamics as described in §1.3.8. Both I_0 and I_1 have a unique expanding fixed point of φ: 0 and 0.5, respectively; long sequences of 0s or 1s correspond to long sojourns very near 0, respectively 0.5. This implies that for any continuous function ψ on the circle, which has different values in the two fixed points 0 and 0.5, the partial averages

$$\frac{1}{n} \sum_{i=0}^{n-1} \psi(\varphi^i(x_0))$$

do not converge for $n \to \infty$.

We note that for any $\tilde{\varphi}$, C^1-close to φ, there is a similar initial point leading to a nonstationary orbit. For other examples see [206, 226] and references therein.

6.6 Correlation dimensions and entropies

As observed before, the box-counting dimension of an attractor (or ω-limit) can be estimated from a time series, but the estimation procedure is not very reliable. In this section we describe the correlation dimension and entropy, which admit better estimation procedures. Correlation dimension and entropy are special members of 1-parameter families of dimensions and entropies, which we define in §6.6.1.2. The present notion of dimension (and also of entropy) is only defined for a metric space *with a Borel probability measure*. Roughly speaking, the definition is based on the probabilities for two randomly and independently chosen points, for a given probability distribution, to be at a distance of at most ε for $\varepsilon \to 0$. In an n-dimensional space with probability measure having a positive and continuous density, one expects this probability to be of the order ε^n: once the first of the two points is chosen, the probability that the second point is within distance ε is equal to the measure of the ε-neighbourhood of the first point, and this is of the order ε^n.

6.6.1 Definitions

This is formalised in the following definitions, in which K is a (compact) metric space, with metric ρ and Borel probability measure μ.

Definition 6.4 (Correlation Integral). The ε-*correlation integral* $C(\varepsilon)$ of (K, ρ, μ) is the $\mu \times \mu$ measure of

$$\Delta_\varepsilon = \{(p, q) \in K \times K \mid \rho(p, q) < \varepsilon\} \subset K \times K.$$

We note that the above definition is equivalent to

$$C(\varepsilon) = \int \mu(D(x,\varepsilon))\mathrm{d}\mu(x),$$

where $D(x,\varepsilon)$ is the ε-neighbourhood of x: it is the μ-average of the μ-measure $\mu(D(x,\varepsilon))$ of ε-neighbourhoods.

Definition 6.5 (Correlation Dimension). The *correlation dimension*, or D_2-dimension, of (K,ρ,μ) is

$$D_2(K) = \limsup_{\varepsilon \to 0} \frac{\ln(C_\varepsilon)}{\ln \varepsilon}.$$

As in the case of the box-counting dimension, one can prove that here also the unit cube in $\mathbb{R}^n$, with the Lebesgue measure, has D_2-dimension n; see the exercises. Also, any Borel probability measure on $\mathbb{R}^n$ has its D_2-dimension at most equal to n. An indication of how to prove this is given at the end of this section where we discuss the 1-parameter family of Renyi dimensions of which our D_2-dimension is a member. For more information on this and other dimensions, in particular in relation to dynamical systems, we refer to [188].

Next we come to the notion of entropy. This notion quantifies the sensitive dependence of evolutions on initial states. So it is related with the notion of 'dispersion exponent' as discussed in §§2.3.2 and 2.3.3. For the definition we need a compact metric space (K,ρ) with Borel probability measure μ and a (continuous) map $\varphi : K \to K$ defining the dynamics. Though not necessary for the definitions which we consider here, we note that usually one assumes that the probability measure μ is invariant under the map φ; that is, one assumes that for each open set $U \subset K$ we have $\mu(U) = \mu(\varphi^{-1}(U))$. We use the map φ to define a family of metrics ρ_n on K in the following way,

$$\rho_n(p,q) = \max_{i=0,\ldots,n-1} \rho(\varphi^i(p), \varphi^i(q)).$$

So $\rho_1 = \rho$ and if φ is continuous, then each one of the metrics ρ_n defines the same topology on K. If moreover φ is a map with bounded expansion, in the sense that there is a constant C such that $\rho(\varphi(p), \varphi(q)) \le C\rho(p,q)$ for all $p,q \in K$, then $\rho(p,q) \le \rho_n(p,q) \le C^{n-1}\rho(p,q)$. Note that any C^1-map, defined on a compact set, always has bounded expansion.

In this situation we consider the ε, n-*correlation integral* $C^{(n)}(\varepsilon)$, which is just the ordinary correlation integral, with the metric ρ replaced by ρ_n. Then we define the H_2-*entropy* by

$$H_2(K,\varphi) = \limsup_{\varepsilon \to 0} \limsup_{n \to \infty} \frac{-\ln(C^{(n)}(\varepsilon))}{n}.$$

Here recall that $C^{(n)}(\varepsilon)$ is the μ-average of $\mu(D^{(n)}(x, \varepsilon))$ and that $D^{(n)}(x, \varepsilon)$ is the ε-neighbourhood of x with respect to the metric ρ_n.

As we explain, this entropy is also a member of a 1-parameter family of entropies. For this and related definitions, see [227]. In order to see the relation between the entropy as defined here and the entropy as originally defined by Kolmogorov, we recall the Brin–Katok theorem.

Theorem 6.6 (Brin–Katok [72]). *In the above situation with μ a nonatomic Borel probability measure which is invariant and ergodic with respect to φ, for μ almost every point x the following limit exists and equals the Kolmogorov entropy.*

$$\lim_{\varepsilon \to 0} \lim_{n \to \infty} \frac{-\ln(\mu(D^{(n)}(x, \varepsilon)))}{n},$$

Theorem 6.6 means that the Kolmogorov entropy is roughly equal to the exponential rate at which the measure of an ε-neighbourhood with respect to ρ_n decreases as a function of n. This is indeed a measure for the sensitive dependence of evolutions on initial states: it measures the decay, with increasing n, of the fraction of the initial states whose evolutions stay within distance ε after n iterations. In our definition it is, however, the exponential decay of the *average* measure of ε-neighbourhoods with respect to the ρ_n metric which is taken as entropy. In general the Kolmogorov entropy is greater than or equal to the H_2 entropy as we defined it.

6.6.2 Miscellaneous remarks

6.6.2.1 Compatibility of the definitions of dimension and entropy with reconstruction

In the next section we consider the problem of estimating dimension and entropy, as defined here, based on the information of one time series. One then deals not with the original state space, but with the set of reconstruction vectors. One has to assume then that the (generic) assumptions in the reconstruction theorem are satisfied so that the reconstruction map Rec_k, for k sufficiently large, is an embedding, and hence a map of bounded distortion. Then the dimension and entropy of the ω-limit of an orbit (with probability measure defined by relative frequencies) are equal to the dimension and entropy of that same limit set *after reconstruction*. This is based on the fact, which is easy to verify, that if h is a homeomorphism $h : K \to K'$ between metric spaces (K, ρ) and (K', ρ') such that the quotients

$$\frac{\rho(p, q)}{\rho'(h(p), h(q))},$$

with $p \neq q$, are uniformly bounded and bounded away from zero, and if μ, μ' are Borel probability measures on K, K', respectively, such that $\mu(U) = \mu'(h(U))$ for each open $U \subset K$, then the D_2-dimensions of K and K' are equal. If moreover $\varphi : K \to K$ defines a dynamical system on K and if $\varphi' = h\varphi h^{-1} : K' \to K'$, then also the H_2-entropies of φ and φ' are the same.

In the case where the dynamics is given by an endomorphism instead of a diffeomorphism, similar results hold; see [224]. If the assumptions in the reconstruction theorem are not satisfied, this may lead to underestimation of both dimension and entropy.

6.6.2.2 Generalised correlation integrals, dimensions, and entropies

As mentioned before, both dimension and entropy, as we defined them, are members of a 1-parameter family of dimensions, respectively, entropies. These notions are defined in a way which is very similar to the definition of the Rényi information of order q; hence they are referred to as Rényi dimensions and entropies. These 1-parameter families can be defined in terms of generalised correlation integrals. The (generalised) correlation integrals of order q of (K, ρ, μ), for $q > 1$, are defined as:

$$C_q(\varepsilon) = \left(\int (\mu(D(x,\varepsilon)))^{q-1} d\mu(x)\right)^{1/q-1},$$

where $D(x, \varepsilon)$ is the ε-neighbourhood of x. We see that $C_2(\varepsilon) = C(\varepsilon)$. This definition can even be used for $q < 1$ but that is not very important for our purposes. For $q = 1$ a different definition is needed, which can be justified by a continuity argument:

$$C_1(\varepsilon) = \exp\left(\int \ln(\mu(D(x,\varepsilon))) d\mu(x)\right).$$

Note that for $q < q'$ we have $C_q \leq C_{q'}$.

The definitions of the generalised dimensions and entropies of order q, denoted by D_q and H_q, are obtained by replacing in the definitions of dimension and entropy the ordinary correlation integrals by the order q correlation integrals. This implies that for $q < q'$ we have $D_q \geq D_{q'}$. We note that it can be shown that the box-counting dimension equals the D_0-dimension; this implies that for any Borel probability measure on $\mathbb{R}^n$ (with compact support) the D_2 dimension is at most n.

For a discussion of the literature on these generalised quantities and their estimation, see at the end of §6.7.

6.7 Numerical estimation of correlation integrals, dimensions, entropies

We are now in the position to describe how the above theory can be applied to analyse a time series which is generated by a deterministic system. As described before, in §6.3, the general form of a model generating such time series is a dynamical system, given by a compact manifold M with a diffeomorphism (or endomorphism) $\varphi : M \to M$, and a read-out function $f : M \to \mathbb{R}$. We assume the generic conditions in the reconstruction theorem to be satisfied. A time series generated by such a system is then determined by the initial state $x_0 \in M$. The corresponding evolution is $\{x_n = \varphi^n(x_0)\}$ with time series $\{y_n = f(x_n)\}$. It is clear that from such a time series we only can get information about this particular orbit, and, in particular, about its ω-limit set $\omega(x_0)$ and the dynamics on that limit set given by $\varphi \mid \omega(x_0)$. As discussed in the previous section, for the definition of the D_2-dimension and the H_2-entropy we need some (φ-invariant) probability measure concentrated on the set to be investigated. Preferably this measure should be related to the dynamics. The only natural choice seems to be the measure of relative frequencies, as discussed in §6.5.2, of the evolution $\{x_n\}$, *if that measure exists*. We now assume that this measure of relative frequencies does exist for the evolution under consideration. We note that this measure of relative frequencies is φ-invariant and that it is concentrated on the ω-limit set $\omega(x_0)$.

Given the time series $\{y_n\}$, we consider the corresponding time series of reconstruction vectors $\{Y_n^{k_0} = (y_n, \ldots, y_{n+k_0-1})\}$ for k_0 sufficiently large, that is, so that the sequence of reconstruction vectors and the evolution $\{x_n\}$ are metrically equal up to bounded distortion; see §6.3. Then we also have the measure of relative frequencies for this sequence of reconstruction vectors. This measure is denoted by μ^{k_0}; its support is the limit set

$$\Omega = \{Y \mid \text{for some } n_i \to \infty, \lim_{i \to \infty} Y_{n_i}^{k_0} = Y\}.$$

Clearly, $\Omega = \text{Rec}_{k_0}(\omega(x_0))$ and the map Φ on Ω, defined by $\text{Rec}_{k_0} \circ \varphi \circ (\text{Rec}_{k_0} \mid \omega(x_0))^{-1}$ is the unique continuous map which sends each $Y_n^{k_0}$ to $Y_{n+1}^{k_0}$.

Both dimension and entropy have been defined in terms of correlation integrals. As observed before, we may substitute the correlation integrals of $\omega(x_0)$ by those of Ω, as long as the transformation Rec_{k_0}, restricted to $\omega(x_0)$, has bounded distortion. An estimate of a correlation integral of Ω, based on the first N reconstruction vectors is given by:

$$\hat{C}(\varepsilon, N) = \frac{1}{\frac{1}{2}N(N-1)} \#\{(i,j) \mid i < j \text{ and } \|Y_i^{k_0} - Y_j^{k_0}\| < \varepsilon\}.$$

It was proved in [112] that this estimate converges to the true value as $N \to \infty$, in the statistical sense. In that paper information about the variance of this estimator was also obtained. In the definition of this estimate, one may take for $\| \cdot \|$ the Euclidean norm, but in view of the usage of these estimates for the estimation of the entropy, it is better to take the *maximum norm*: $\|(z_1, \ldots, z_{k_0})\|_{\max} = \max_{i=1\ldots k_0} |z_i|$. The reason is the following. If the distance function on Ω is denoted by d, then, as in the previous section, when discussing the definition of entropy, we use the metrics

$$d_n(Y, Y') = \max_{i=0,\ldots,n-1} d(\Phi^i(Y), \Phi^i(Y')).$$

If we now take for the distance function d on Ω the maximum norm (for k_0-dimensional reconstruction vectors), then d_n is the corresponding maximum norm for $(k_0 + n - 1)$-dimensional reconstruction vectors. Note also that the transition from the Euclidean norm to the maximum norm is a 'transformation with bounded distortion,' so that it should not influence the dimension estimates. When necessary, we express the dimension of the reconstruction vectors used in the estimation $\hat{C}(\varepsilon, N)$ by writing $\hat{C}^{(k_0)}(\varepsilon, N)$; as in §6.6, the correlation integrals based on k_0-dimensional reconstruction vectors are denoted by $C^{(k_0)}(\varepsilon)$.

In order to obtain estimates for the D_2-dimension and the H_2-entropy of the dynamics on the ω-limit set, one needs to have estimates of $C^{(k)}(\varepsilon)$ for many values of both k and ε. A common form to display such information graphically is a diagram showing the estimates of $\ln(C^{(k)}(\varepsilon))$, or rather the logarithms of the estimates of $C^{(k)}(\varepsilon)$, for the various values of k, as a function of $\ln(\varepsilon)$. In Figure 6.3 we show such estimates based on a time series of length 4000 which was generated with the Hénon system (with the usual parameter values $a = 1.4$ and $b = .3$). We discuss various aspects of this figure and then give references to the literature for the algorithmic aspects for extracting numerical information from the estimates as displayed.

First we observe that we normalised the time series in such a way that the difference between the maximal and minimal value equals 1. For ε we took the values $1, 0.9, (0.9)^2, \ldots, (0.9)^{42} \sim 0.011$. It is clear that for $\varepsilon = 1$ the correlation integrals have to be 1, independently of the reconstruction dimension k. Because $C^{(k)}(\varepsilon)$

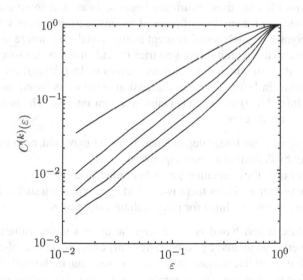

Fig. 6.3 Estimates for the correlation integrals of the Hénon attractor. For $k = 1, \ldots, 5$ the estimates of $C^{(k)}(\varepsilon)$ are given as function of ε, with logarithmic scale along both axes. (This time series was normalised so that the difference between the maximal and the minimal values is 1.) Because it follows from the definitions that $C^{(k)}(\varepsilon)$ decreases for increasing values of k the highest curve corresponds to $k = 1$ and the lower curves to $k = 2, \ldots, 5$ respectively.

is decreasing as a function of k, the various graphs of (the estimated values of) $\ln(C^{(k)}(\varepsilon))$ as function of $\ln(\varepsilon)$, are ordered monotonically in k: the higher graphs belong to the lower values of k. For $k = 1$ we see that $\ln(C^{(1)}(\varepsilon))$, as a function of $\ln(\varepsilon)$, is very close to a straight line, at least for values of ε smaller than say .2; the slope of this straight line is approximately equal to 1 (N.B.: To see this, one has to rescale the figure in such a way that a factor of 10 along the horizontal and vertical axes corresponds to the same length!). This is an indication that the set of 1-dimensional reconstruction vectors has a D_2-dimension equal to 1. This is to be expected: inasmuch as this set is contained in a 1-dimensional space, the dimension is at most 1, and the dimension of the Hénon attractor is, according to numerical evidence, somewhat larger than 1. Now we consider the curves for $k = 2, \ldots, 5$. These curves also approach straight lines for small values of ε (with fluctuations due, at least in part, to estimation errors). The slope of these lines is somewhat larger than in the case of $k = 1$ indicating that the reconstruction vectors, in the dimensions $k > 1$, form a set with D_2-dimension also somewhat larger than 1. The slope indicates a dimension around 1.2. Also we see that the slopes of these curves do not seem to depend significantly on k for $k \geq 2$; this is an indication that this slope corresponds to the true dimension of the Hénon attractor itself. This agrees with the fact that the reconstruction maps Rec_k, $k > 1$, for the Hénon system are embeddings (if we use one of the coordinates as read-out function). We point out again that the estimation of dimensions, as carried out on the basis of estimated correlation integrals, can never be rigourously justified: the definition of the dimension is in terms of the behaviour of $\ln(C^{(k)}(\varepsilon))$ for $\varepsilon \to 0$ and if we have a time series of a given (finite) length, these estimates become more and more unreliable as ε tends to zero because the number of pairs of reconstruction vectors which are very close will become extremely small (except in the trivial case where the time series is periodic). Due to these limitations, one tries to find an interval, a so-called *scaling interval*, of ε values in which the estimated values of $\ln(C^{(k)}(\varepsilon))$ are linear, up to small fluctuations, in $\ln(\varepsilon)$. Then the dimension estimate is based on the slope of the estimated $\ln(C^{(k)}(\varepsilon))$ versus In (ε) curves in that interval. The justifications for such restrictions are that for

- Large values of ε no linear dependence is to be expected because correlation integrals are by definition at most equal to 1
- Small values of ε, the estimation errors become too large
- Many systems that we know there is in good approximation such a linear dependence of $\ln(C^{(k)}(\varepsilon))$ on $\ln(\varepsilon)$ for intermediate values of ε.

So the procedure is just based on an extrapolation to smaller values of ε. In the case of time series generated by some (physical) experiment one also has to take into account that small fluctuations are always present due to thermal fluctuations or other noise; this provides another argument for disregarding values of ε which are too small.

An estimate for the H_2-entropy also can be obtained from the estimated correlation integrals. The information in Figure 6.3 also suggests here how to proceed. We observe that the parallel lines formed by the estimated $\ln(C^{(k)}(\varepsilon))$ versus $\ln(\varepsilon)$

curves, for $k \geq 2$, are not only parallel, but also approximately at equal distances. This is an indication that the differences

$$\ln(C^{(k)}(\varepsilon)) - \ln(C^{(k+1)}(\varepsilon))$$

are approximately independent of ε and k, at least for small values of ε, here smaller than say .2, and large values of k, here larger than 1. Assuming that this constancy holds in good approximation, this difference should be equal to the H_2-entropy as defined before.

We note that this behaviour of the estimated $\ln(C^{(k)}(\varepsilon))$ versus $\ln(\varepsilon)$ curves, namely that they approach equidistant parallel lines for small values of ε and large values of k, turns out to occur quite generally for time series generated by deterministic systems. It may, however, happen that, in order to reach the 'good' ε and k values with reliable estimates of the correlation integrals, one has to take extremely long time series (with extremely small noise). For a quantification of these limitations, see [205].

This way of estimating the dimension and entropy was proposed in [222] and applied by [126, 127]. There is an extensive literature on the question of how to improve these estimations. We refer to the (survey) publications and proceedings [108, 109, 120, 230, 233], and to the textbook [150]. As we mentioned above, the D_2-dimension and H_2-entropy are both members of families of dimensions and entropies, respectively. The theory of these generalised dimensions and entropies and their estimation was considered in a number of publications; see [61, 62, 132, 138, 228, 230] and [235] as well as the references in these works.

6.8 Classical time series analysis, correlation integrals, and predictability

What we here consider as classical time series analysis is the analysis of time series in terms of autocovariances and power spectra. This is a rather restricted view, but still it makes sense to compare that type of analysis with the analysis in terms of correlation integrals which we discussed so far. In §6.8.1 we summarise this classical time series analysis insofar as we need it; for more information see, for example, [196]. Then, in §6.8.2 we show that the correlation integrals provide information which cannot be extracted from the classical analysis; in particular, the autocovariances cannot be used to discriminate between time series generated by deterministic systems and by stochastic systems. Finally in §6.8.3 we discuss predictability in terms of correlation integrals.

6.8.1 Classical time series analysis

We consider a stationary time series $\{y_i\}$. Without loss of generality we may assume that the average of this time series is 0. The kth autocovariance is defined as the average R_k = average$_i(y_i \cdot y_{i+k})$. Note that R_0 is called the variance of the time series. From the definition it follows that for each k we have $R_k = R_{-k}$.

We do not make use of the power spectrum but, because it is used very much as a means to give graphical information about a signal (or time series) we just give a brief description of its meaning and its relation with the autocovariances. The power spectrum gives information about how strong each (angular) frequency ω is represented in the signal: $P(\omega)$ is the energy, or squared amplitude, of the (angular) frequency ω. Note that for a discrete time series, which we are discussing here, these frequencies only have values in $[-\pi, \pi]$ and that $P(\omega) = P(-\omega)$. The formal definition is somewhat complicated due to possible difficulties with convergence: in general the power spectrum only exists as a generalised function, in fact as a measure. It can be shown that the collection of autocovariances is the Fourier transform of the power spectrum:

$$R_k = \int_{-\pi}^{\pi} e^{i\omega k} P(\omega) d\omega.$$

This means that not only the autocovariances are determined by the power spectrum, but that also, by the inverse Fourier transform, the power spectrum is determined by the autocovariances. So the information, which this classical theory can provide, is exactly the information contained in the autocovariances.

We note that under very general assumptions the autocovariances R_k tend to zero as k tends to infinity. (The most important exceptions are (quasi)-periodic time series.) This is what we assume from now on. If the autocovariances converge sufficiently fast to zero, then the power spectrum will be continuous, smooth, or even analytic. For time series generated by a deterministic system, the autocovariances converge to zero if the system is mixing; see [59].

6.8.1.1 Optimal linear predictors

For a time series as discussed above, one can construct the optimal linear predictor of order k. This is given by coefficients $\alpha_1^{(k)}, \ldots, \alpha_k^{(k)}$; it makes a prediction of the nth element as a linear combination of the k preceding elements:

$$\hat{y}_n^{(k)} = \alpha_1^{(k)} y_{n-1} + \cdots + \alpha_k^{(k)} y_{n-k}.$$

The values $\alpha_1^{(k)}, \ldots, \alpha_k^{(k)}$ are chosen in such a way that the average value σ_k^2 of the squares of the prediction errors $(\hat{y}_n^{(k)} - y_n)^2$ is minimal. This last condition explains why we call this an *optimal* linear predictor. It can be shown that the autocovariances $R_0, \ldots, R_k$ determine, and are determined by, $\alpha_1^{(k)}, \ldots, \alpha_k^{(k)}$ and σ_k^2. Also if we generate a time series $\{z_i\}$ by using the autoregressive model (or system)

$$z_n = \alpha_1^{(k)} z_{n-1} + \cdots + \alpha_k^{(k)} z_{n-k} + \varepsilon_n,$$

where the values of ε_n are independently chosen from a probability distribution, usually a normal distribution, with zero mean and variance σ_k^2, then the autocovariances

$\tilde{R}_i$ of this new time series satisfy, with probability 1, $\tilde{R}_i = R_i$ for $i = 0, \ldots, k$. This new time series, with ε_n taken from a normal distribution, can be considered as the 'most unpredictable' time series with the given first $k + 1$ autocovariances. By performing the above construction with increasing values of k one obtains time series which are, as far as power spectrum or autocovariances are concerned, converging to the original time series $\{y_i\}$.

Remark. If $\{y_n\}$ is a periodic time series with period N, then $\hat{y}_n = y_{n-N}$ is a perfect predictor which is linear. So predictors being linear here is not a serious restriction. Also if $\{y_n\}$ is quasi-periodic (i.e., obtained from a quasi-periodic evolution through a read-out function) then, for any $\varepsilon > 0$, and integer $T(\varepsilon)$ exists such that for all n we have $|y_{n+T(\varepsilon)} - y_n| < \varepsilon$. Hence the prediction $\hat{y}_n = y_{n-T(\varepsilon)}$, which is linear, is correct up to an error of at most ε. For time series from chaotic systems the performance of linear predictors in general is much worse; see the doubling map and the tent maps.

6.8.1.2   Gaussian time series

We note that the analysis in terms of power spectra and autocovariances is in particular useful for *Gaussian time series*. These time series are characterised by the fact that their reconstruction measures are normal; one can show that then:

$$\mu_k = |B_k|^{-1/2}(2\pi)^{-k/2}e^{-\langle x, B_k^{-1}x \rangle/2}dx,$$

where B_k is a symmetric and invertible matrix which can be expressed in terms of $R_0, \ldots, R_{k-1}$:

$$B_k = \begin{pmatrix} R_0 & R_1 & \cdots & R_{k-1} \\ R_1 & R_0 & \cdots & R_{k-2} \\ \vdots & \vdots & \cdots & \vdots \\ R_{k-1} & R_{k-2} & \cdots & R_0 \end{pmatrix}.$$

(For an exceptional situation where these formulae are not valid see the footnote in §6.8.2.) So these times series are, as far as reconstruction measures are concerned, completely determined by their autocovariances. Such time series are generated, for example, by the above type of autoregressive systems whenever the ε_ns are normally distributed.

6.8.2   Determinism and autocovariances

We have discussed the analysis of time series in terms of correlation integrals in the context of time series which are generated by deterministic systems, that is, systems with the property that the state at a given time determines all future states and for which the space of all possible states is compact and has a low, or at least finite, dimension. We call such time series 'deterministic'. The Gaussian time series belong to a completely different class. These time series are prototypes of what we call 'random' time series. One can use the analysis in terms of correlation integrals to distinguish between deterministic and Gaussian time series in the following way.

If one tries to estimate numerically the dimension of an underlying attractor (or ω-limit) as in §6.7, one makes estimates of the dimension of the set of k-dimensional reconstruction vectors for several values of k. If one applies this algorithm to a Gaussian time series, then one should find that the set of k-dimensional reconstruction vectors has dimension k for all k. This contrasts with the case of deterministic time series where these dimensions are bounded by a finite number, namely the dimension of the relevant attractor (or ω-limit). Also if one tries to estimate the entropy of a Gaussian time series by estimating the limits

$$H_2(\varepsilon) = \lim_{k \to \infty} (\ln(C^{(k)}(\varepsilon)) - \ln(C^{(k+1)}(\varepsilon))),$$

one finds that this quantity tends to infinity as ε tends to zero, even for a fixed value of k. This contrasts with the deterministic case, where this quantity has a finite limit, at least if the map defining the time evolution has bounded expansion.

It should be clear that the above arguments are valid for a much larger class of nondeterministic time series than just the Gaussian ones.

Next we want to argue that the autocovariances cannot be used to make the distinction between random and deterministic systems. Suppose that we are given a time series (of which we still assume that the average is 0) with autocovariances $R_0, \ldots, R_k$. Then such a time series (i.e., a time series with these same first $k + 1$ autocovariances) can be generated by an autoregressive model, which is of course random.[1] On the other hand, one can also generate a time series with the same autocovariances by a deterministic system. This can be done by taking in the autoregressive model for the values ε_n not random (and independent) values from some probability distribution, but values generated by a suitable deterministic system. Though deterministic, $\{\varepsilon_n\}$ should still have (i) average equal to zero, (ii) variance equal to the variance of the prediction errors of the optimal linear predictor of order k, and (iii) all the other autocovariances equal to zero; that is, $R_i = 0$ for $i > 0$. These conditions namely imply that the linear theory cannot 'see' the difference between such a deterministic time series and a truly random time series of independent and identically distributed noise. The first two of the above conditions are easy to satisfy: one just has to apply an appropriate affine transformation to all elements of the time series.

---

[1] There is a degenerate exception where the variance of the optimal linear predictor of order k is zero. In that case the time series is multiperiodic and has the form

$$y_n = \sum_{i \leq s} (a_i \sin(\omega_i n) + b_i \cos(\omega_i n)),$$

with $2s \leq k + 1$. Then also the matrices B_k (see Section 8.1) are not invertible.

In order to make an explicit example where all these conditions are satisfied, we start with the *tent map* (e.g., see [4]). This is a dynamical system given by the map φ on $[-\frac{1}{2}, \frac{1}{2}]$:

$$\varphi(x) = \begin{cases} 2x + \frac{1}{2} & \text{for } x \in [-\frac{1}{2}, 0) \\ -2x + \frac{1}{2} & \text{for } x \in [0, \frac{1}{2}). \end{cases}$$

It is well known that for almost any initial point $x \in [-\frac{1}{2}, \frac{1}{2}]$, in the sense of Lebesgue, the measure defined by the relative frequencies of $x_n = \varphi^n(x)$, is just the Lebesgue measure on $[-\frac{1}{2}, +\frac{1}{2}]$. We show that such a time series satisfies (up to scalar multiplication) the above conditions. For this we have to evaluate

$$R_i = \int_{-(1/2)}^{+(1/2)} x\varphi^i(x)\,dx.$$

Because for $i > 0$ we have that $\varphi^i(x)$ is an even function of x, this integral is 0. The variance of this time series is 1/12. So in order to make the variance equal to σ^2 one has to transform the state space (i.e., the domain of φ) to $[-\sqrt{3\sigma^2}, +\sqrt{3\sigma^2}]$. We denote this rescaled evolution map by φ_{σ^2}. Now we can write down explicitly the (deterministic) system which generates time series with the same autocovariances as the time series generated by the autoregressive model:

$$y_n = \alpha_1^{(k)} y_{n-1} + \cdots + \alpha_k^{(k)} y_{n-k} + \varepsilon_n,$$

where the ε_n are independent samples of a distribution with mean 0 and variance σ^2. The dynamics is given by:

$$u_1 \mapsto \sum_{i=1}^{k} \alpha_i^{(k)} u_i + x$$
$$u_2 \mapsto u_1$$
$$u_3 \mapsto u_2$$
$$\cdots \quad \cdots \quad \cdots$$
$$u_k \mapsto u_{k-1}$$
$$x \mapsto \varphi_{\sigma^2}(x).$$

The time series consists of the successive values of u_1; that is, the read-out function is given by $f(u_1, \ldots, u_k, x) = u_1$.

We note that, instead of the tent map, we could have used the logistic map $x \mapsto 1 - 2x^2$ for $x \in [-1, 1]$.

We produced this argument in detail because it proves that the correlation integrals, which can be used to detect the difference between time series generated by deterministic systems and Gaussian time series, contain information which is not contained in the autocovariances, and hence also not in the power spectrum. We should expect that deterministic time series are much more predictable than random (or Gaussian) time series with the same autocovariances. In the next subsection we give an indication how the correlation integrals give an indication about the predictability of a time series.

6.8.3 Predictability and correlation integrals

The prediction problem for chaotic time series has attracted a lot of attention. The following proceedings volumes were devoted, at least in part, to this, [99, 232, 233, 243].

The prediction methods proposed in these papers all substantially use the stationarity of the time series: the generating model is supposed to be unknown, so the only way to make predictions about the future is to use the behaviour in the past, assuming (by stationarity) that the future behaviour will be like the behaviour in the past. Roughly speaking there are two ways to proceed.

1. One can try to estimate a (nonlinear) model for the system generating the time series and then use this model to generate predictions. The disadvantage of this method is that estimating such a model is problematic. This is, amongst other things, due to the fact that the orbit, corresponding to the time series which we know, explores only a (small) part of the state space of the system so that only a (small) part of the evolution equation can be known.
2. For each prediction, one can compare the most recent values of the time series with values in the past: if one finds in the past a segment which is very close to the segment of the most recent values, one predicts that the continuation in the future will be like the continuation in the past (we give a more detailed discussion below). In §2.3, this principle was coined as *l'histoire se répète*. The disadvantage is that for each prediction one has to inspect the whole past, which can be very time consuming.

6.8.3.1 L'histoire se répète

Although the second method also has its disadvantages it is less ad hoc than the first one and hence more suitable for a theoretical investigation. So we analyse the second method and show how the correlation integrals, and in particular the estimates of dimension and entropy, give indications about the predictability.

Note that in this latter procedure there are two ways in which we use the past: we have the 'completely known past' to obtain information about the statistical properties, that is, the reconstruction measures, of the time series; using these statistical properties, we base our prediction of the next value on a short sequence of the most recent values. Recall also our discussion of the dripping faucet in §6.2.

In order to discuss the predictability in terms of this second procedure in more detail, we assume that we know a finite, but long, segment $\{y_0, \ldots, y_N\}$ of a stationary time series. The question is how to predict the next value y_{N+1} of the time series. For this procedure, in its simplest form, we have to choose an integer k and a positive real number ε. We come back to the problem of how to choose these values. (Roughly speaking, the k most recent values $y_{N-k+1}, \ldots, y_N$ should contain the relevant information as far as predicting the future is concerned, given the statistical information about the time series which can be extracted from the whole past; ε should be of the order of the prediction error which we are willing to accept.) We now inspect the past, searching for a segment $y_m, \ldots, y_{m+k-1}$ which is ε-close to the last k values, in the sense that $|y_{m+i} - y_{N-k+1+i}| < \varepsilon$ for $i = 0, \ldots, k-1$. If there is no such segment, we should try again with a smaller value of k or a larger value of ε. If there is such a segment, then we predict the next value y_{N+1} to be $\hat{y}_{N+1} = y_{m+k}$.

Here we refer back to our discussion on predictability in Chapter 2, in particular in §2.3. If our time series were obtained in a situation where the reconstruction theorem 6.1, say in dimension k, applies, then the reconstruction vectors $(y_0, \ldots, y_{k-1}), \ldots, (y_{N-k+1}, \ldots, y_N)$ should form a 'sufficiently dense' subset of the reconstructed orbit (or its ω-limit) in the sense of property A.

This is about the most primitive way to use the past to predict the future and later we discuss an improvement; see the discussion below of local linear predictors. For the moment we concentrate on this prediction procedure and try to find out how to assess the reliability of the prediction. We assume, for simplicity, that a prediction which is correct within an error of at most ε is considered to be correct; in the case of a larger error, the prediction is considered to be incorrect. Now the question is: what is the probability of the prediction being correct? This probability can be given in terms of correlation integrals. Namely the probability that two segments of length k, respectively, $k+1$, are equal up to an error of at most ε is $C^{(k)}(\varepsilon)$, respectively, $C^{(k+1)}(\varepsilon)$. So the probability that two segments of length $k+1$, given that the first k elements are equal up to an error of at most ε, have also their last elements equal up to an error of at most ε, is $C^{(k+1)}(\varepsilon)/C^{(k)}(\varepsilon)$. Note that this quotient decreases in general with increasing values of k. So in order to get good predictions we should take k so large that this quotient has essentially its limit value. This is the main restriction on how to take k, which we want otherwise to be as small as possible.

We have already met this quotient, or rather its logarithm, in the context of the estimation of entropy. In §6.6 we defined

$$H_2(\varepsilon) = \lim_{k \to \infty} (\ln(C^{(k)}(\varepsilon)) - \ln(C^{(k+1)}(\varepsilon))) = \lim_{k \to \infty} -\ln\left(\frac{C^{(k+1)}(\varepsilon)}{C^{(k)}(\varepsilon)}\right).$$

So, assuming that k is chosen as above, the probability of giving a correct prediction is $e^{-H_2(\varepsilon)}$. This means that $H_2(\varepsilon)$ and hence also $H_2 = \lim_{\varepsilon \to 0} H_2(\varepsilon)$ is a measure of the *unpredictability* of the time series.

Also the estimate for the dimension gives an indication about the (un)predictability in the following way. Suppose we have fixed the value of k in the prediction procedure so that $C^{(k+1)}(\varepsilon)/C^{(k)}(\varepsilon)$ is essentially at its limiting value (assuming for simplicity that this value of k is independent of ε). Then, in order to have a reasonable probability of finding in the past a sequence of length k which agrees sufficiently with the sequence of the last k values, we should have that $1/N$, the inverse length of the time series, is small compared with $C^{(k)}(\varepsilon)$. This is the main limitation when choosing ε, which we otherwise want to be as small as possible. The dependence of $C^{(k)}(\varepsilon)$ on ε is, in the limiting situation (i.e. for small ε and sufficiently large k), proportional to ε^{D_2}, so in the case of a high dimension, we cannot take ε very small and hence we cannot expect predictions to be very accurate. To be more precise, if we want to improve the accuracy of our predictions by a factor 2, then we expect to need a segment of past values which is 2^{D_2} times longer.

6.8.3.2 Local linear predictors

As we mentioned before, the above principle *l'histoire se répète* to predict the future is rather primitive. We here briefly discuss a refinement which is based on a combination of optimal linear prediction and the above principle. This type of prediction was the subject of [98] which appeared in [233]. In this case we start in the same way as above, but now we collect all the segments from the past which are ε-close to the last segment of k elements. Let $m_1, \ldots, m_s$ be the first indices of these segments. We then have s different values, namely $y_{m_1+k}, \ldots, y_{m_s+k}$, which we can use as prediction for y_{N+1}. This collection of possible predictions already gives a better idea of the variance of the possible prediction errors.

We can, however, go further. Assuming that there should be some functional dependence $y_n = F(y_{n-1}, \ldots, y_{n-k})$ with differentiable F (and such an assumption is justified by the reconstruction theorem if we have a time series which is generated by a smooth and deterministic dynamical system and if k is sufficiently large), then F admits locally a good linear, or rather affine, approximation (given by its derivative). In the case that s, the number of nearby segments of length k, is sufficiently large, one can estimate such a linear approximation. This means estimating the constants $\alpha_0, \ldots, \alpha_k$ such that the variance of $\{(\hat{y}_{m_i+k} - y_{m_i+k})\}$ is minimal, where $\hat{y}_{m_i+k} = \alpha_0 + \alpha_1 y_{m_i+k-1} + \cdots + \alpha_k y_{m_i}$. The determination of the constants $\alpha_0, \ldots, \alpha_k$ is done by linear regression. The estimation of the next value is then $\hat{y}_{N+1} = \alpha_0 + \alpha_1 y_N + \cdots + \alpha_k y_{N-k+1}$. This means that we use essentially a linear predictor, which is, however, only based on 'nearby segments' (the fact that we have here a term α_0, which was absent in the discussion of optimal linear predictors, comes from the fact that here there is no analogue of the assumption that the average is zero).

Finally we note that if nothing is known about how a (stationary) time series is generated, it is not a priori clear that this method of local linear estimation will give better results. Also a proper choice of k and ε is less obvious in that case; see [98].

6.9   Miscellaneous subjects

In this concluding section we consider two additional methods for the analysis of
time series which are of a somewhat different nature. Both methods can be motivated by the fact that the methods which we discussed up to now cannot be used to
detect the difference between a time series and its time reversal. The *time reversal*
of a time series $\{y_i\}_{i \in \mathbb{Z}}$ is the time series $\{\tilde{y}_i = y_{-i}\}_{i \in \mathbb{Z}}$ (for finite segments of a
time series the definition is similar). It is indeed obvious from the definitions that
autocorrelations and correlation integrals do not change if we reverse time. Still we
can show with a simple example that a time series may become much more unpredictable by reversing the time; this also indicates that the predictability, as measured
by the entropy, is a somewhat defective notion.

Our example is a time series generated by the tent map (using a generic initial
point); for the definition see §6.8.2. If we know the initial state of this system, we
can predict the future, but not the past. This is a first indication that time reversal has
drastic consequences in this case. We make this more explicit. If we know the initial
state of the tent map up to an error ε, then we know the next state up to an error 2ε,
so that a prediction of the next state has a probability of 50% to be correct if we
allow for an error of at most ε. (This holds only for small ε and even then to first
approximation due to the effect of boundary points and the 'turning point' at $x = 0$.)
For predicting the previous state, which corresponds to ordinary prediction for the
time series with the time reversed, we also expect a probability of 50% of a correct
prediction, that is, with an error not more than ε, but in this latter case the prediction
error is in general much larger: the inverse image of an ε-interval consists in general
of two intervals of length $\frac{1}{2}\varepsilon$ at a distance which is on average approximately $\frac{1}{2}$.

In the next subsection we discuss the *Lyapunov exponents* and their estimation
from time series. They give another indication for the (un)predictability of a time series; these Lyapunov exponents are not invariant under time reversal. The estimation
of the Lyapunov exponents is discussed in the next subsection. In the final subsection we consider the *(smoothed) mixed correlation integrals* which were originally
introduced in order to detect the possible differences between a time series and its
time reversal. It can, however, be used much more generally as a means to test
whether two time series have the same statistical behaviour.

6.9.1   *Lyapunov exponents*

A good survey on Lypunov exponents is the third chapter of [120] to which we refer
for details. Here we give only a heuristic definition of these exponents and a sketch
of their numerical estimation.

Consider a dynamical system given by a differentiable map $f : M \to M$, where
M is a compact m-dimensional manifold. For an initial state $x \in M$ the Lyapunov
exponents describe how evolutions, starting infinitesimally close to x diverge from
the evolution of x. For these Lyapunov exponents to be defined, we have to assume

that there are subspaces of the tangent space at x: $T_x = E_1(x) \supset E_2(x) \supset \cdots \supset E_s(x)$ and real numbers $\lambda_1(x) > \lambda_2(x) > \cdots > \lambda_s(x)$ such that for a vector $v \in T_x \setminus \{0\}$ the following are equivalent,

$$\lim_{n \to \infty} \frac{1}{n} \ln(\|\mathrm{d}f^n(v)\|) = \lambda_i \iff v \in E_i(x) - E_{i+1}(x).$$

The norm $\| \cdot \|$ is taken with respect to some Riemannian metric on M. The existence of these subspaces (and hence the convergence of these limits) for 'generic' initial states is the subject of Oseledets's multiplicative ergodic theorem (see [183] and [120]). Here 'generic' does not mean belonging to an open and dense subset, but belonging to a subset which has full measure, with respect to some f-invariant measure. The numbers $\lambda_1, \ldots, \lambda_s$ are called the Lyapunov exponents at x. They are to a large extent independent of x: in general they depend mainly on the attractor to which the orbit through x converges. In the notation we drop the dependence of the Lyapunov exponents on the point x. We say that the Lyapunov exponent λ_i has multiplicity k if

$$\dim(E_i(x)) - \dim(E_{i+1}(x)) = k.$$

Often each Lyapunov exponent with multiplicity $k > 1$ is repeated k times, so that we have exactly m Lyapunov exponents with $\lambda_1 \geq \lambda_2 \geq \cdots \geq \lambda_m$.

When we know the map f and its derivative explicitly, we can estimate the Lyapunov exponents using the fact that for a generic basis $v_1, \ldots, v_m$ of T_x and for each $s = 1, \ldots, m$

$$\sum_{i=1}^{s} \lambda_i = \lim_{n \to \infty} \frac{1}{n} \ln(\|\mathrm{d}f^n(v_1, \ldots, v_s)\|_s),$$

where $\|\mathrm{d}f^n(v_1, \ldots, v_s)\|_s$ denotes the s-dimensional volume of the parallelepiped spanned by the s vectors $\mathrm{d}f^n(v_1), \ldots, \mathrm{d}f^n(v_s)$.

It should be clear that in the case of sensitive dependence on initial states one expects the first (and largest) Lyapunov exponent λ_1 to be positive. There is an interesting conjecture relating Lyapunov exponents to the dimension of an attractor, the Kaplan–Yorke conjecture [151], which can be formulated as follows. Consider the function L, defined on the interval $[0, m]$ such that $L(s) = \sum_{i=1}^{s} \lambda_i$ for integers s (and $L(0) = 0$) and which interpolates linearly on the intervals in between. Because Lyapunov exponents are by convention nonincreasing as a function of their index, this function is concave. We define the Kaplan–Yorke dimension D_{KY} as the largest value of $t \in [0, m]$ for which $L(t) \geq 0$. The conjecture claims that $D_{KY} = D_1$, where D_1 refers to the D_1-dimension, as defined in §6.6, of the attractor (or ω-limit) to which the evolution in question converges. For some 2-dimensional systems this conjecture was proven (see [157]); in higher dimensions there are counterexamples, but so far no persistent counterexamples have been found, so that it might still be true generically. Another important relation in terms of Lyapunov exponents is that, under 'weak' hypotheses (for details see [120] or [157]), the Kolmogorov (or H_1) entropy of an *attractor* is the sum of its positive Lyapunov exponents.

6.9.2 Estimation of Lyapunov exponents from a time series

The main method of estimating Lyapunov exponents, from time series of a deterministic system, is due to Eckmann et al. [119]; for references to later improvements, see [11].

This method makes use of the reconstructed orbit. The main problem here is to obtain estimates of the derivatives of the mapping (in the space of reconstruction vectors). This derivative is estimated in a way which is very similar to the construction of the local linear predictors; see §6.8.3. Working with a k-dimensional reconstruction, and assuming that Rec_k is an embedding from the state space M into $\mathbb{R}^k$, we try to estimate the derivative at each reconstruction vector Y_i^k by collecting a sufficiently large set of reconstruction vectors $\{Y_{j(i,s)}^k\}_{1 \leq s \leq s(i)}$ which are close to Y_i^k. Then we determine, using a least squares method, a linear map A_i which sends each $Y_{j(i,s)}^k - Y_i^k$ to $Y_{j(i,s)+1}^k - Y_{i+1}^k$ plus a small error $\varepsilon_{i,s}$ (of course so that the sum of the squares of these $\varepsilon_{i,s}$ is minimised). These maps A_i are used as estimates of the derivative of the dynamics at Y_i^k. The Lyapunov exponents are then estimated, based on an analysis of the growth of the various vectors under the maps $A_N A_{N-1} \cdots A_1$. A major problem with this method, and also with its later improvements, is that the linear maps which we are estimating are too high-dimensional: we should use the derivative of the map on $\mathcal{X}_k$ defined by $F = \text{Rec}_k \circ f \circ \text{Rec}_k^{-1}$, with $\mathcal{X}_k = \text{Rec}_k(M)$, while we are estimating the derivative of a (nondefined) map on the higher-dimensional $\mathbb{R}^k$. In fact the situation is even worse: the reconstruction vectors only give information on the map F restricted to the reconstruction of one orbit (and its limit set) which usually is of even lower dimension. Still, the method often gives reproducible results, at least as far as the largest Lyapunov exponent(s) is (are) concerned.

6.9.3 The Kantz–Diks test: Discriminating between time series and testing for reversibility

The main idea of the Kantz–Diks test can be described in a context which is more general than that of time series. We assume to have two samples of vectors in $\mathbb{R}^k$, namely $\{X_1, \ldots, X_s\}$ and $\{Y_1, \ldots, Y_t\}$. We assume that the X-vectors are chosen independently from a distribution P_X and the Y-vectors independently from a distribution P_Y. The question is whether these samples indicate in a statistically significant way that the distributions P_X and P_Y are different. The original idea of Kantz was to use the correlation integrals in combination with a 'mixed correlation integral' [149]. We denote the correlation integrals of the distributions P_X and P_Y, respectively, by $C_X(\varepsilon)$ and $C_Y(\varepsilon)$; they are as defined in §6.6. The mixed correlation integral $C_{XY}(\varepsilon)$ is defined as the probability that the distance between a P_X-random vector and a P_Y-random vector is componentwise smaller than ε. If the distributions P_X and P_Y are the same, then these three correlation integrals are equal. If not, one expects the mixed correlation integral to be smaller (this need not be the case, but this problem is removed by using the smoothed correlation integrals which we define below). So if an estimate of $C_X(\varepsilon) + C_Y(\varepsilon) - 2C_{XY}(\varepsilon)$ differs in a significant way from zero, this is an indication that P_X and P_Y are different.

A refinement of this method was given in Diks [116]. It is based on the notion of a *smoothed correlation integral*. In order to explain this notion, we first observe that the correlation integral $C_X(\varepsilon)$ can be defined as the expectation value, with respect to the measure $P_X \times P_X$, of the function $H_\varepsilon(v, w)$ on $\mathbb{R}^k \times \mathbb{R}^k$ which is 1 if the vectors v and w differ, componentwise, less than ε and 0 otherwise. In the definition of the smoothed integral $S_\varepsilon(X)$, the only difference is that the function H_ε is replaced by

$$G_\varepsilon(v, w) = e^{-(\|v-w\|^2)/\varepsilon^2}$$

where $\| \cdot \|$ denotes the Euclidean distance. The smoothed correlation integrals $S_Y(\varepsilon)$ and $S_{XY}(\varepsilon)$ are similarly defined. This modification has the following consequences.

- The quantity $S_X(\varepsilon) + S_Y(\varepsilon) - 2S_{XY}(\varepsilon)$ is always nonnegative.
- $S_X(\varepsilon) + S_Y(\varepsilon) - 2S_{XY}(\varepsilon) = 0$ if and only if the distributions P_X and P_Y are equal.

The estimation of $S_X(\varepsilon) + S_Y(\varepsilon) - 2S_{XY}(\varepsilon)$ as a test for the distribution P_X and P_Y to be equal or not can be used for various purposes. In the present context it is used to test whether two (finite segments of) time series show a significantly different behaviour by applying it to the distributions of their reconstruction vectors. One has to be aware, however, of the fact that these vectors cannot be considered as statistically independent; this is discussed in [115].

Originally this method was proposed as a test for 'reversibility', that is, as a test whether a time series and its time reversal are significantly different. Another application is a test for stationarity. This is used when monitoring a process in order to obtain a warning whenever the dynamics changes. Here one uses a segment during which the dynamics is as desired. Then one collects, say every 10 minutes, a segment and applies the test to see whether there is a significant difference between the last segment and the standard segment. See [182] for an application of this method in chemical engineering.

Finally we note that for Gaussian time series the k-dimensional reconstruction measure, and hence the correlation integrals up to order k, are completely determined by the autocovariances $\rho_0, \ldots, \rho_{k-1}$. There is, however, no explicit formula for correlation integrals in terms of these autocovariances. For the smoothed correlation integrals such formulas are given in [225] where also the dimensions and entropies in terms of smoothed correlation integrals are discussed.

6.10 Exercises

Exercise 6.1 (When Rec$_k$ is a diffeomorphism). We consider the Hénon system (see §1.3.2); as a read-out function we take one of the coordinate functions. Show that in this case Rec$_k$ is a diffeomorphism for $k \geq 2$.

Exercise 6.2 (Reconstructing a linear system). We consider a dynamical system with state space $\mathbb{R}^k$ given by a linear map $L : \mathbb{R}^k \to \mathbb{R}^k$; we take a linear read-out map $\chi : \mathbb{R}^k \to \mathbb{R}$. What are the necessary and sufficient conditions on L and χ for the conclusion of the reconstruction theorem to hold? Verify that these conditions determine an open and dense subset in the product space of the space of linear maps in $\mathbb{R}^k$ and the space of linear functions on $\mathbb{R}^k$.

Exercise 6.3 (Reconstruction for circle maps I). We consider a dynamical system with state space $\mathbb{S}^1 = \mathbb{R}/2\pi\mathbb{Z}$ given by the map $\varphi(x) = x + a \mod 2\pi\mathbb{Z}$.

1. Show that for $a = 0$ or $\pi \mod 2\pi$ and for any function f on $\mathbb{S}^1$ the conclusion of the reconstruction theorem does not hold for any k.
2. Show that for $a \neq 0$ or $\pi \mod 2\pi$ and $f(x) = \sin(x)$, the map Rec$_k$ is an embedding for $k \geq 2$.

Exercise 6.4 (Reconstruction for circle maps II). We consider a dynamical system, again with state space $\mathbb{S}^1$ but now with map $\varphi(x) = 2x \mod 2\pi$. Show that for any (continuous) $f : \mathbb{S}^1 \to \mathbb{R}$ the map Rec$_k$ is not an embedding for any k.

Exercise 6.5 (Reconstruction in practice). Here the purpose is to see how reconstruction works in practice. The idea is to use a program such as MATHEMATICA, Maple, or MATLAB to experiment with a time series from the Lorenz attractor.

1. Generate a (long) solution of the Lorenz equations. It should reasonably 'fill' the butterfly shape of the Lorenz attractor.
   We obtain a time series by taking only one coordinate, as a function of time, of this solution.
2. Make reconstructions, in dimension two (so that it can easily be represented graphically), using different delay times.

One should notice that very short and very long delay times give bad figures (why?). For intermediate delay times one can recognise a projection of the butterfly shape.

Exercise 6.6 (Equivalent box-counting dimensions). Prove the inequalities (concerning a_ε, b_ε, and c_ε) needed to show that the various definitions of the box-counting dimension are equivalent.

Exercise 6.7 (Box-counting dimension of compact subset $\mathbb{R}^n$). Prove that for any compact subset $K \subset \mathbb{R}^n$ the box-counting dimension satisfies $d(K) \leq n$.

Exercise 6.8 (Box-counting dimension of Cantor set). Prove that the box-counting dimension of the middle α Cantor set is

$$\frac{\ln 2}{\ln 2 - \ln(1 - \alpha)}.$$

Exercise 6.9 (Nonexistence of a limit). Prove that the limit of

$$\frac{1}{n+1} \sum_{i=0}^{n} y_i, \quad \text{for } n \to \infty$$

does not exist for $y_i = 0$, respectively, 1 if the integer part of $\ln(i)$ is even, respectively odd.

Exercise 6.10 (On averages). In $\mathbb{S}^1 = \mathbb{R}/\mathbb{Z}$ we consider the sequence $p_i = i\alpha$ with α irrational. Show that for each continuous function $f : \mathbb{S}^1 \to \mathbb{R}$ we have

$$\lim_{N \to \infty} \frac{1}{N} \sum_{i=0}^{N-1} f(p_i) = \int_{\mathbb{S}^1} f.$$

Hint: See [59].

Exercise 6.11 (Reconstruction measure for doubling map). We consider the doubling map φ on $\mathbb{S}^1 = \mathbb{R}/\mathbb{Z}$ defined by $\varphi(x) = 2x \mod \mathbb{Z}$. Show that for a positive orbit of this system we can have, depending on the initial point:

1. The orbit has a reconstruction measure which is concentrated on k points where k is any positive integer.
2. The reconstruction measure of the orbit is the Lebesgue measure on $\mathbb{S}^1$.
3. The reconstruction measure is not defined.

Are there other possibilities? Which of the three possibilities occurs with the 'greatest probability'?

Exercise 6.12 (On D_2-dimension I). Let $K \subset \mathbb{R}^n$ be the unit cube with Lebesgue measure and Euclidian metric. Show that the correlation dimension of this cube is n.

Exercise 6.13 (On D_2-dimension II). Let $K = [0,1] \subset \mathbb{R}$ with the usual metric and with the probability measure ρ defined by:

$$\rho(\{1\}) = \tfrac{1}{2};$$
$$\rho([a,b)) = \tfrac{1}{2}(b - a).$$

Show that $D_2(K) = 0$.

Exercise 6.14 (On D_2-dimension III). Let (K_i, ρ_i, μ_i), for $i = 1, 2$, be compact spaces with metric ρ_i, probability measure μ_i, and D_2-dimension d_i.

On the disjoint union of K_1 and K_2 we take a metric ρ with $\rho(x, y)$ equal to:

$$\rho(x, y) = \begin{cases} \rho_i(x, y) & \text{if both } x, y \in K_i; \\ 1 & \text{if } x, y \text{ belong to different } K_1 \text{ and } K_2. \end{cases}$$

The measure μ on $K_1 \cup K_2$ is such that for any subset $A \subset K_i$, $\mu(A) = \frac{1}{2}\mu_i(A)$. Show that the D_2-dimension of this disjoint union satisfies $d = \min\{d_1, d_2\}$. On the product $K_1 \times K_2$ we take the product metric, defined by

$$\rho_p((x_1, x_2), (y_1, y_2)) = \max\{\rho_1(x_1, x_2), \rho_2(y_1, y_2)\},$$

and the product measure μ_p defined by

$$\mu_p(A \times B) = \mu_1(A)\mu_2(B)$$

for $A \subset K_1$ and $B \subset K_2$. Show that the D_2-dimension of $K_1 \times K_2$ is at most equal to $d_1 + d_2$.

Exercise 6.15. We consider the middle third Cantor set K with the usual metric (induced from the metric in $[0, 1] \subset \mathbb{R}$). In order to define the measure $\mu_{\alpha,\beta}$, with $\alpha, \beta > 0$ and $\alpha + \beta = 1$ on this set we recall the construction of this Cantor set as intersection $K = \cap K_i$ with $K_0 = [0, 1]$, so that K_i contains 2^i intervals of length 3^{-i}. for each interval of K_i, the two intervals of K_{i+1} contained in it have measure α times, respectively, β times, the measure of the interval of K_i. So K_i has $i!/k!(i-k)!$ intervals of measure $\alpha^k \beta^{i-k}$.

1. Prove that $C_q(3^{-n}) = (\alpha^q + \beta^q)^{n/q-1}$.
2. Compute the D_q-dimensions of these Cantor sets.
3. The map $\varphi : K \to K$ is defined by $\varphi(x) = 3x \mod \mathbb{Z}$. Compute the H_q entropies of this map.

Exercise 6.16 (D_2-dimension and entropy in practice I). Write programs, in MATLAB, MATHMETICA, or Maple to estimate, from a given time series the D_2-dimension and -entropy; see also the next exercise.

Exercise 6.17 (D_2-dimension and entropy in practice II). The idea of this exercise is to speed up the computation of (estimates of) correlation integrals.

According to the definition for the calculation of $\hat{C}(\varepsilon, N)$ one has to check for all $\frac{1}{2}N(N-1)$ pairs $1 \le i < j \le N$ whether the distance between the corresponding reconstruction vectors is smaller than ε. Because the number of such pairs is usually very large, it might be attractive to use only a smaller number of such pairs; these can be selected using the random generator.

Generate a time series of length 4000 from the Hénon attractor. Estimate $C^{(2)}(0.2)$ using m random pairs $1 \le i < j \le N$ for various values of m and try to determine for which value of m we get an estimate which is as reliable as using all pairs.

The fact that in general one can take m as above much, smaller than $\frac{1}{2}N(N-1)$ is to be expected on the basis of the theory of U-statistics; see [111, 112].

Exercise 6.18 (Reconstruction measure of Gaussian process). Prove the statement in §6.8.2 that the set of k-dimensional reconstruction vectors of a time series generated by a (nondeterministic) Gaussian process defines a reconstruction measure which has D_2-dimension k.

Exercise 6.19 (ε Entropies of Gaussian process). Prove that for a time series generated by a (nondeterministic) Gaussian process the ε entropies $H_2(\varepsilon)$ diverge to ∞ for $\varepsilon \to 0$.

Appendix A
Differential topology and measure theory

In this appendix we discuss notions from general topology, differential topology, and measure theory which were used in the previous chapters. In general proofs are omitted. We refer the reader for proofs and a systematic treatment of these subjects to:

Topology:

> [152] J.L. Kelley, *General Topology*, Van Nostrand 1955; GTM **27** Springer-Verlag 1975
>
> [121] R. Engelking, *General Topology*, Second Edition, Heldermann Verlag 1989
>
> [134] K.P. Hart, J. Nagata, and J.E. Vaughan (eds.), *Encyclopædia of General Topology*, North-Holland 2004

Differential Topology:

> [142] M.W. Hirsch, *Differential Topology*, Springer-Verlag 1976

Measure Theory:

> [131] P.R. Halmos, *Measure Theory*, Van Nostrand 1950
>
> [184] J.C. Oxtoby, *Measure and Category*, Springer-Verlag 1971.

A.1 Topology

Topological space. A topological space consists of a set X together with a collection $\mathcal{O}$ of subsets of X such that

- $\emptyset$ and X are in $\mathcal{O}$.
- $\mathcal{O}$ is closed under the formation of arbitrary unions and finite intersections.

The elements in $\mathcal{O}$ are called *open sets*; the complement of an open set is called *closed*. A *basis* for such a topology is a subset $\mathcal{B} \subset \mathcal{O}$ such that all open sets can be obtained by taking arbitrary unions of elements of $\mathcal{B}$. A *subbase* is a set $\mathcal{S} \subset \mathcal{O}$ such that all open sets can be obtained by first taking finite intersections of elements of $\mathcal{S}$ and then taking arbitrary unions of these.

We call a subset A of a topological space X *dense* if for each open set $U \subset X$, the intersection $A \cap U$ is nonempty.

H.W. Broer and F. Takens, *Dynamical Systems and Chaos*,
Applied Mathematical Sciences 172, DOI 10.1007/978-1-4419-6870-8,
© Springer Science+Business Media, LLC 2011

Continuity. A map $f : X \to Y$ from a topological space X to a topological space Y is called *continuous* if for each open $U \subset Y$, $f^{-1}(U)$ is open in X. If moreover such a map is a bijection and if f^{-1} is also continuous, then f is called a *homeomorphism* and the spaces X and Y are called *homeomorphic*.

Metric space. A metric space is a set X together with a function ρ from $X \times X$ to the nonnegative real numbers such that:

- $\rho(p,q) = \rho(q,p)$ for all $p, q \in X$;
- $\rho(p,q) = 0$ if and only if $p = q$;
- $\rho(p,q) + \rho(q,r) \geq \rho(p,r)$ for all $p, q, r \in X$.

The function ρ as above is called a *metric* or a *distance function*.

Each metric space (X, ρ) has an associated topology which is generated by the base consisting of the ε-neighbourhoods $U_{p,\varepsilon} = \{x \in X | \rho(p,x) < \varepsilon\}$ for all $p \in X$ and $\varepsilon > 0$. As an example we consider $\mathbb{R}^n$ with the Euclidean metric $\rho((x_1, \ldots, x_n), (y_1, \ldots, y_n)) = \sqrt{\sum_i (x_i - y_i)^2}$. The corresponding topology is the 'usual topology'.

Two metrics ρ_1 and ρ_2 on the same set X are called *equivalent* if there is a constant $C \leq 1$ such that for each pair of points $p, q \in X$ we have

$$C\rho_1(p,q) \leq \rho_2(p,q) \leq C^{-1}\rho_1(p,q).$$

Clearly, if two metrics are equivalent, they define the same topology. However, inequivalent metrics may define the same topology; for example, if ρ is any metric then $\bar{\rho}(p,q) = \min\{\rho(p,q), 1\}$ is in general not equivalent with ρ, but defines the same topology. So when using a metric for defining a topology it is no restriction to assume that the diameter (i.e., the largest distance) is at most 1; this is useful when defining product metrics. A topology which can be obtained from a metric is called *metrisable*.

A metric space (X, ρ) is called *complete* if each sequence $\{x_i\}_{i \in \mathbb{N}}$ in X which is Cauchy (i.e., for each $\varepsilon > 0$ there is an N such that for al $i, j > N$ we have $\rho(x_i, x_j) < \varepsilon$) has a *limit* in X (i.e., a point $x \in X$ such that $\rho(x_i, x) \to 0$ for $i \to \infty$). A topological space which can be obtained from a complete metric is called *complete metrisable*. Complete metric spaces have important properties:

- *Contraction principle.* If $f : X \to X$ is a contracting map of a complete metric space to itself (i.e., a map such that for some $c < 1$ we have $\rho(f(p), f(q)) \leq c\rho(p,q)$ for all $p, q \in X$), then there is a unique point $x \in X$ such that $f(x) = x$; such a point x is called a *fixed point*. This property is used in the proof of some existence and uniqueness theorems where the existence of the object in question is shown to be equivalent with the existence of a fixed point of some contracting map, for example, the usual proof of existence and uniqueness of solutions of ordinary differential equations.
- *Baire property.* In a complete metric space (and hence in a complete metrisable space) the countable intersection of open and dense subsets is dense.

This needs some explanation. In a topological space the notion 'almost all points' can be formalised as 'the points, belonging to a set containing an open and dense subset'. Indeed let $U \subset G \subset X$ be an open and dense subset of X containing G. Then any point of X can be approximated (in any of its neighbourhoods) by a point of G and even a point of U. Such a point of U even has a neighbourhood which is completely contained in U and hence in G. So each point of X can be approximated by a point in U which not only belongs to G, but even remains in G (and even in U) after a sufficiently small perturbation. We use here 'a set containing an open and dense subset' instead of just 'an open and dense subset' because we want to formalise the notion 'almost all', so we cannot exclude the set to be larger than open and dense. (We note that also in measure theory there is a notion of 'almost all' which, however, in cases where there is both a natural topology and a natural measure, does not in general agree with the present notion; we come back to this in the end of this appendix). The Baire property means that if we have a countable number of properties P_i which may or may not hold for elements of a complete metric space X, defining subsets $G_i \subset X$, that is, $G_i = \{x \in X \mid x \text{ has property } P_i\}$, which each contains an open and dense subset of X, then the set of all points of X for which all the properties P_i hold is still dense. For this reason it is common to extend the notion of 'almost all points' (in the topological sense) in the case of complete metric (or complete metrisable) spaces to 'the points of a set containing a countable intersection of open and dense subsets.' In the case where we deal with complete metrisable spaces we call a subset *residual* if it contains a countable intersection of open and dense subsets; a property is called *generic* if the set of all points having that property is residual. In the literature one also uses the following terminology. A subset is of the first *category* if it is the complement of a residual set; otherwise it is of the second *category*.

Subspace. Let $A \subset X$ be a subset of a topological space X. Then A inherits a topology from X: open sets in A are of the form $U \cap A$ where $U \subset X$ is open. In the same way, if A is a subset of a metric space X, with metric ρ, then A inherits a metric by simply putting $\rho_A(p, q) = \rho(p, q)$ for all $p, q \in A$. It is common to also denote the metric on A by ρ.

Compactness. Let X be a topological space. A collection of (open) subsets $\{U_i \subset X\}_{i \in I}$ is called an *(open) cover* of X if $\cup_{i \in I} U_i = X$. A topological space is called *compact* if for each open cover $\{U_i\}_{i \in I}$ there is a finite subset $I' \subset I$ such that $\{U_i\}_{i \in I'}$ is already an open cover of X; the latter cover is called a finite *subcover* of the former.

If X is metrisable then it is compact if and only if each infinite sequence $\{x_i\}$ of points in X has an *accumulation point*, that is, a point $\bar{x}$ such that each neighbourhood of $\bar{x}$ contains points of the sequence.

In a finite-dimensional vector space (with the usual topology and equivalence class of metrics) a subset is compact if and only if it is closed and bounded. In infinite-dimensional vector spaces there is no such thing as 'the usual topology' and definitely not a 'usual equivalence class of metrics' and for most 'common' topologies and metrics it is not true that all closed and 'bounded' subsets are compact.

Product spaces. Let $\{X_i\}_{i \in I}$ be a collection of topological spaces. On the Cartesian product $X = \Pi_i X_i$, with projections $\pi_i : X \to X_i$, the product topology is generated by the subbase containing all sets of the form $\pi_i^{-1}(U_i)$ with $U_i \subset X_i$ open. If each X_i is compact then the product is also compact. If the topologies of X_i are metrisable and if I is countable, then X is also metrisable: uncountable products are in general not metrisable. A metric on X, defining the product topology in the case of a countable product, can be obtained as follows. We assume that on each X_i the topology is given by the metric ρ_i. Without loss of generality we may assume that ρ_i distances are at most one and that the spaces are indexed by $i \in \mathbb{N}$. For $p, q \in X$ with $\pi_i(p) = p_i$ and $\pi_i(q) = q_i$, $i \in \mathbb{N}$ we define the distance as $\sum_i 2^{-i} \rho_i(p_i, q_i)$.

Quotient spaces. Let X be a topological space and $\sim$ an equivalence relation defined on X. Consider the quotient space $X/\sim$ as well as the canonical projection $\pi : X \to X/\sim$. Then on $X/\sim$ one defines the quotient topology by declaring a subset $U \subset X/\sim$ open, if and only if $\pi^{-1}(U) \subset X$ is open. It is easy to verify that this indeed defines a topology. It also is the finest topology, that is, with the largest number of open sets, such that the projection $\pi : X \to X/\sim$ is continuous.

Function spaces. On the (real vector) space $C^k(\mathbb{R}^n)$ of all C^k-functions on $\mathbb{R}^n$, $0 \leq k < \infty$, the *weak topology* is defined by the subbase consisting of the 'g-neighbourhoods' $U_{K,\varepsilon,g,k}$ which are defined as the set of functions $f \in C^k(\mathbb{R}^n)$ such that in each point of K the (partial) derivatives of $f - g$ up to and including order k are, in absolute value, smaller than ε, where $K \subset \mathbb{R}^n$ is compact, $\varepsilon > 0$, and $g \in C^k(\mathbb{R}^n)$. It can be shown that with this topology $C^k(\mathbb{R}^n)$ is complete metrisable, and hence has the Baire property. An important property of this topology is that for any diffeomorphism $\varphi : \mathbb{R}^n \to \mathbb{R}^n$ the associated map $\varphi^* : C^k(\mathbb{R}^n) \to C^k(\mathbb{R}^n)$, given by $\varphi^*(f) = f \circ \varphi$ is a homeomorphism. All these notions extend to the space of C^k-functions on some open subset of $\mathbb{R}^n$.

For C^∞-functions on $\mathbb{R}^n$ we obtain the *weak C^∞-topology* from the subbase $U_{K,\varepsilon,g,k}$ as defined above (but now interpreted as subsets of $C^\infty(\mathbb{R}^n)$) for any compact $K \subset \mathbb{R}^n$, $\varepsilon > 0$, $g \in C^\infty(\mathbb{R}^n)$, and $k \geq 0$. Also the space of C^∞-functions is complete metrisable and the topology does not change under diffeomorphisms in $\mathbb{R}^n$.

A.2 Differentiable manifolds

Topological manifold. A *topological manifold* is a topological space M which is metrisable and which is locally homeomorphic with $\mathbb{R}^n$, for some n, in the sense that each point $p \in M$ has a neighbourhood which is homeomorphic with an open subset of $\mathbb{R}^n$. Usually one assumes that a topological manifold is *connected*, that is, that it is not the disjoint union of two open subsets. The integer n is called the *dimension* of the manifold.

Differentiable manifold. A *differentiable manifold* consists of a topological manifold M, say of dimension n, together with a *differentiable structure*. Such a differentiable structure can be given in various ways. One is by specifying which of the continuous functions are differentiable. We denote the continuous functions on M by $C^0(M)$. Given a positive integer k, a C^k-*structure* on M is given by a subset $C^k(M) \subset C^0(M)$ satisfying:

- For each point $p \in M$ there are a neighbourhood U of p in M and n elements $x_1, \ldots, x_n \in C^k(M)$ such that $x = (x_1|U, \ldots, x_n|U) : U \to \mathbb{R}^n$ is a homeomorphism of U to an open subset of $\mathbb{R}^n$ and such that for each $g \in C^k(M)$, the function $(g|U) \circ x^{-1}$ can be extended as a C^k-function to all of $\mathbb{R}^n$. Such (U, x) is called a *local chart*; $x_1|U, \ldots, x_n|U$ are called *local coordinates*.
- If $g \in C^0(M)$ satisfies: for each $p \in M$ there are a neighbourhood U of p in M and some $g_U \in C^k(M)$ such that $g|U = g_U|U$, then $g \in C^k(M)$.
- For each C^k-function $h : \mathbb{R}^n \to \mathbb{R}$ and $g_1, \ldots, g_m \in C^k(M)$, also $h(g_1, \ldots, g_m) \in C^k(M)$.

Note that, as a consequence of this definition, whenever we have a C^k-structure on a manifold M and $\ell < k$ there is a unique C^ℓ-structure on M such that $C^k(M) \subset C^\ell(M)$. It can also be shown that given a C^k-structure on M, $k \geq 1$ one can always find a C^∞-structure $C^\infty(M) \subset C^k(M)$, which is, however, not unique.

Some simple examples of differentiable manifolds are:

- $\mathbb{R}^n$ with the usual C^k-functions.
- An open subset $U \subset \mathbb{R}^n$; note that in this case we cannot just take as C^k-functions the restrictions of C^k-functions on $\mathbb{R}^n$: if $U \neq \mathbb{R}^n$ there are C^k-functions on U which cannot be extended to a C^k-function on $\mathbb{R}^n$; see also below.
- The unit sphere $\{(x_1, \ldots, x_n) \in \mathbb{R}^n \mid \sum x_i^2 = 1\}$ in $\mathbb{R}^n$; as C^k-functions one can take here the restrictions of the C^k-functions on $\mathbb{R}^n$; this manifold, or any manifold diffeomorphic to it, is usually denoted by $\mathbb{S}^{n-1}$.
- The n-dimensional torus $\mathbb{T}^n$ is defined as the quotient of $\mathbb{R}^n$ divided out by $\mathbb{Z}^n$. If π denotes the projection $\mathbb{R}^n \to \mathbb{T}^n$ then the C^k-functions on $\mathbb{T}^n$ are exactly those functions f for which $f \circ \pi \in C^k(\mathbb{R}^n)$. Note that $\mathbb{T}^1$ and $\mathbb{S}^1$ are diffeomorphic.

The second of the above examples motivates the following definition of the C^k-functions on an arbitrary subset $A \subset M$ of a differentiable manifold M. We say that a function $f : A \to \mathbb{R}$ is C^k if, for each point $p \in A$ there are a neighbourhood U of p in M and a function $f_U \in C^k(M)$ such that $f|U \cap A = f_U|U \cap A$. This construction rests on the work of Whitney [245]. The set of these C^k-functions on A is denoted by $C^k(A)$. If $A \subset M$ is a topological manifold and if $C^k(A) \subset C^0(A)$ defines a C^k-structure on A, then A is called a C^k-*submanifold* of M.

A submanifold of M is called *closed* if it is closed as a subset of M; note, however, that a manifold is often called closed if it is compact (and has no boundary, a notion which we do not discuss here).

Differentiability. Let M and N be C^k-manifolds; then a map $f : M \to N$ is C^k-*differentiable* if and only if for each $h \in C^k(N)$, we have that $h \circ f \in C^k(M)$. The set of C^k-maps from M to N is denoted by $C^k(M, N)$. If moreover such a map f is a bijection and such that also f^{-1} is C^k then f is called a C^k-*diffeomorphism*. Although we have here a simple definition of differentiability, the definition of derivative is more complicated and requires the notion of tangent bundle which we now discuss.

Tangent bundle. Let M be a differentiable manifold and $p \in M$. A smooth curve through p is a map $\ell : (a, b) \to M$, with $0 \in (a, b)$, which is at least C^1 and satisfies $\ell(0) = p$. We call two such curves ℓ_1 and ℓ_2 equivalent if for each differentiable $f : M \to \mathbb{R}$ we have $(f \circ \ell_1)'(0) = (f \circ \ell_2)'(0)$. An equivalence class for this equivalence relation is called a *tangent vector of M at p*. These tangent vectors can be interpreted as directional derivatives acting on smooth functions on M. As such they can be added (provided we consider tangent vectors at the same point of M) and they admit scalar multiplication. This makes $T_p(M)$, the *set of tangent vectors of M at p* a vector space.

Note that whenever M is an open subset of $\mathbb{R}^n$ and $p \in M$, then $T_p(M)$ is canonically isomorphic with $\mathbb{R}^n$: for $x \in \mathbb{R}^n$ the vector in $T_p(M)$ corresponding to x is represented by the curve $\ell(t) = p + tx$. This observation, together with the use of local charts makes it possible to show that $T(M) = \cup_{p \in M} T_p(M)$, the *tangent bundle* of M, is a topological manifold, and even a differentiable manifold: if M is a C^k-manifold, then $T(M)$ is a C^{k-1}-manifold.

Derivative. Let $f : M \to N$ be a map which is at least C^1, M and N being manifolds which are at least C^1; then f induces a map $df : T(M) \to T(N)$, the *derivative* of f which is defined as follows. If a tangent vector v of M at a point p is represented by a curve $\ell : (a, b) \to M$ then $df(v)$ is represented by $f \circ \ell$. Clearly df maps $T_p(M)$ to $T_{f(p)}(N)$. The restriction of df to $T_p(M)$ is denoted by df_p; each df_p is linear. If M and N are open subsets of $\mathbb{R}^m$, respectively, $\mathbb{R}^n$, and $p \in \mathbb{R}^m$ then df_p is given by the matrix of first-order partial derivatives of f at p. From this last observation, and using local charts one sees that if f is C^k then df is C^{k-1}.

Maps of maximal rank. Let $f : M \to N$ be a map which is at least C^1. The *rank* of f at a point $p \in M$ is the dimension of the image of df_p; this rank is maximal if it equals $\min\{\dim(M), \dim(N)\}$. Related to this notion of 'maximal rank' there are two important theorems.

Theorem A.1 (Maximal Rank). *Let* $f : M \to N$ *be a map which is at least* C^1 *having maximal rank in* $p \in M$; *then there are local charts* (U, x) *and* (V, y), *where* U, V *is a neighbourhood of* p, *respectively* $f(p)$, *such that* $y \circ f \circ x^{-1}$, *each of these maps suitably restricted, is the restriction of a linear map. If* f *is* C^k *then the local coordinates* x *and* y *can be taken to be* C^k.

This means that if a map between manifolds of the same dimension has maximal rank at a point p then it is locally invertible. If f maps a 'low'-dimensional manifold

to a higher-dimensional manifold and if it has maximal rank in p then the image of
a neighbourhood of p is a submanifold of N; If it has maximal rank in all points
of M, then f is called an *immersion*. If such an immersion f is injective and if f
induces a homeomorphism from M to $f(M)$, with the topology inherited from N,
then $f(M)$ is a submanifold of N and f is called an *embedding*.

The second theorem we quote here uses a notion 'set of measure zero' from
measure theory which is discussed in the next section.

Theorem A.2 (Sard). *Let $f : M \rightarrow N$ be a C^k-map, M and N manifolds of
dimension m, respectively, n. Let $K \subset M$ be the set of points where f does not
have maximal rank. If $k - 1 \geq \max\{(m - n), 0\}$ then $f(K)$ is a set of measure zero
with respect to the (uniquely defined) Lebesgue measure class on N (intuitively this
means that $f(K)$ is very small).*

The theorem remains true if one takes K as the (often larger) set of points where
the rank of f is smaller than n. This theorem is of fundamental importance in the
proof of the transversality theorem; see below.

Transversality. Let N be a submanifold of M, both at least C^1-manifolds. A map
$f : V \rightarrow M$, V a differentiable manifold and f at least C^1, is called *transversal
with respect to* N if for each point $p \in V$ we have either:

- $f(p) \notin N$, or
- $df_p(T_p(V)) + T_{f(p)}(N) = T_{f(p)}(M)$.

A special case of transversality occurs whenever f in the above definition
is a *submersion*, that is, a map from a 'high'-dimensional manifold to a lower-
dimensional manifold which has in each point maximal rank: such a map is transversal with respect to any submanifold of M.

One can prove that whenever f is transversal with respect to N, then $f^{-1}(N)$ is
a submanifold of V; if f and N are both C^k, this submanifold is C^k, and its dimension is $\dim(V) - \dim(M) + \dim(N)$; if this comes out negative, then $f^{-1}(N) = \emptyset$.

Theorem A.3 (Transversality). *For M, N, and V as above, each having a C^k-
structure, $k \geq 1$, it is a generic property for $f \in C^k(V, M)$ to be transversal with
respect to N; $C^k(V, M)$ denotes the set of C^k maps from V to M: the topology on
this set is defined in the same way as on $C^k(\mathbb{R}^n)$.*

We give here a brief outline of the proof of this theorem for the special case
that $M = \mathbb{R}^n$, N is a closed C^∞-submanifold of M, and V is a compact C^∞-
manifold. In this case it is easy to see that the set of elements of $C^k(V, M)$ which
are transversal with respect to N is open. So we only have to prove that the N-
transversal mappings are dense. Because each C^k-map can be approximated (in
any C^k-neighbourhood) by a C^∞-map, it is enough to show that the N-transversal
elements are dense in $C^\infty(V, M)$. Let $f : V \rightarrow M$ be a C^∞-map; we show how
to approximate it by an N-transversal one. For this we consider $F : V \times \mathbb{R}^n \rightarrow
M = \mathbb{R}^n$, defined by $F(p, x) = f(p) + x$; note that the product $V \times \mathbb{R}^n$ has
in a natural way the structure of a C^∞-manifold. Clearly F is a submersion and

hence transversal with respect to N. Denote $F^{-1}(N)$, which is a C^∞-submanifold of $V \times \mathbb{R}^n$, by W. Finally $g : W \to \mathbb{R}^n$ is the restriction to W of the canonical projection of $V \times \mathbb{R}^n$ to $\mathbb{R}^n$. By the theorem of Sard, the image under g of the set of critical points of g, here to be interpreted as the set of points where g has rank smaller than n, has Lebesgue measure zero in $\mathbb{R}^n$. This means that arbitrarily close to $0 \in \mathbb{R}^n$ there are points which are not in this critical image. If x is such a point, then f_x, defined by $f_x(p) = f(p) + x$ is transversal with respect to N.

Remark. The transversality theorem A.3 has many generalisations which can be proven in more or less the same way. For example, for differentiable maps $f : M \to M$ it is a generic property that the associated map $\tilde{f} : M \to M \times M$, defined by $\tilde{f}(p) = (p, f(p))$ is transversal with respect to $\Delta = \{(p, p) \in M \times M\}$. This means that it is a generic property for maps of M to itself that the fixed points are isolated and that the derivative at each fixed point has no proper value one.

Vector fields and differential equations. For ordinary differential equations of the form $x' = f(x)$, with $x \in \mathbb{R}^n$, f can be interpreted as a *vector field* on $\mathbb{R}^n$: indeed a solution of such an equation is a curve $\ell : (a, b) \to \mathbb{R}^n$ such that for each $t \in (a, b)$ we have that $\ell'(t) = f(\ell(t))$ and this derivative of ℓ can be interpreted as a vector (see our definition of a tangent vector). These notions generalise to manifolds. A *vector field* on a differentiable manifold M is a *cross-section* $X : M \to T(M)$; with 'cross-section' we mean that for each $p \in M$ the vector $X(p)$ is in $T_p(M)$. A solution of the *differential equation* defined by the vector field X is a smooth curve $\ell : (a, b) \to M$ such that for each $t \in (a, b)$ we have that the tangent vector defined by ℓ at t equals $X(\ell(t))$; with this tangent vector defined by ℓ at t we mean the tangent vector at $\ell(t)$ represented by the curve $\ell_t(s) = \ell(t + s)$.

Existence and uniqueness theorems for ordinary differential equations easily generalise to such equations on manifolds.

Orientation and degree. Let M be a smooth manifold, which we assume to be connected. For each $p \in M$ the tangent space can be given an orientation, for example by choosing a base and declaring this base to be positively oriented. Another basis is positively, respectively, negatively, oriented if the matrix transforming one to the other has a positive, respectively, negative, determinant. Clearly there are two orientations possible in such a tangent space. Next one can try to orient all tangent spaces $T_q(M)$ for $q \in M$ in such a way that the orientations depend continuously on the q. With this we mean that if we have a positively oriented base in $T_p(M)$ and if we move it continuously to other points (during this motion it should remain a base) then it remains positively oriented. It is not possible to define such a global orientation on any manifold, for example, consider the manifold obtained from $\mathbb{S}^2$ by identifying antipodal points, the so-called projective plane: such manifolds are called *nonorientable*. If M is *orientable* and connected, it admits two orientations. If one is chosen then we say that the manifold is *oriented*.

If M_1 and M_2 are two compact oriented manifolds of the same dimension and if $f : M_1 \to M_2$ is a smooth map then the degree of f is defined as follows. Take a point $p \in M_2$ such that f has maximal rank in all points of $f^{-1}(p)$; by

Sard's theorem such points exist. Now the degree of f is the number of points of $f^{-1}(p)$ properly counted. Due to the maximal rank condition and the compactness the number of these points is finite. We count a point $q \in f^{-1}(p)$ as plus one, respectively, minus one, if df_q is an orientation- preserving map, respectively, an orientation-reversing map from $T_q(M_1)$ to $T_p(M_2)$. This degree does not depend on the choice of $p \in M_2$ if M_2 is connected.

If we reverse the orientation on M_1 or on M_2 then the degree of f in the above situation is multiplied by -1. This means that if $M_1 = M_2 = M$ then we do not have to specify the orientation as long as we use the same orientation on M as domain and as image; still we need M to be orientable for the degree to be defined.

An important special case is the degree of maps of the circle $\mathbb{S}^1$. For this we interpret $\mathbb{S}^1 = \mathbb{R}/\mathbb{Z} = \mathbb{T}^1$, where the canonical projection from $\mathbb{R}$ to $\mathbb{S}^1$ is denoted by π. If $f : \mathbb{S}^1 \to \mathbb{S}^1$ is a continuous map and $\pi(x) = f(0),$[1] then there is a unique continuous *lift* $\tilde{f} : [0, 1) \to \mathbb{R}$ of f (i.e., $\pi \tilde{f}(x) = f(x)$ for all $x \in [0, 1)$). In general $\lim_{s \to 1} \tilde{f}(s) = \tilde{x}$ exists but is not equal to x; the difference $\tilde{x} - x$ is an integer. In fact it is not hard to show that this integer is the degree of f. For details see Exercise 3.11.

Riemannian metric. A Riemannian metric g on a differentiable manifold M is given by a positive definite inner product g_p on each tangent space $T_p(M)$. Usually one assumes such a metric to be differentiable: it is C^k if for any C^k-vector fields X_1 and X_2, the function $p \mapsto g_p(X_1(p), X_2(p))$ is C^k. Note that a Riemannian metric is not a metric in the sense we defined before, however, it defines such a metric, at least if the manifold M is connected. In order to see this we define the length of a curve $\ell : [a, b] \to M$ which is at least C^1 as

$$L(\ell) = \int_a^b \sqrt{g_{\ell(s)}(\ell'(s), \ell'(s))}\, ds;$$

note the analogy with the length of curves in $\mathbb{R}^n$. For $p, q \in M$ we define the distance from p to q as the infimum of the lengths of all smooth curves from p to q. It is not very hard to prove that in this way we have indeed a metric as defined before.

It can be shown that each manifold admits a smooth Riemannian metric (even many such metrics). Such a Riemannian metric can even be constructed in such a way that the corresponding metric is complete.

---

[1] We denote the elements of $\mathbb{S}^1$ just as reals x instead of by their equivalence classes $[x]$.

A.3 Measure theory

Basic definitions. A *measure* on a set X is given by a nonempty collection $\mathcal{B}$ of subsets of X and a map m from $\mathcal{B}$ to the non negative real numbers, including ∞, such that:

- $\mathcal{B}$ is a *sigma algebra*; that is, $\mathcal{B}$ is closed under formation of countable unions and complements (algebra: addition corresponds to taking the union; multiplication corresponds to taking the intersection).
- m is *sigma additive*; that is, for each countable collection $B_i \in \mathcal{B}$ of pairwise disjoint subsets, $m(\cup_i B_i) = \sum_i m(B_i)$.

The elements of the sigma algebra $\mathcal{B}$ are called *measurable sets*; for $B \in \mathcal{B}$, $m(B)$ is the *measure* of B. A triple $(X, \mathcal{B}, m)$ as above is called a *measure space*.

Given such a sigma algebra $\mathcal{B}$ with measure m, the *completion* $\mathcal{B}'$ of $\mathcal{B}$ is obtained by including all sets $B' \subset X$ which are contained in a set $B \in \mathcal{B}$ with $m(B) = 0$ (and – of course – then also including countable unions and intersections of the new collection); the extension of m to $\mathcal{B}'$ is unique (and obvious). Often taking the completion of a sigma algebra $\mathcal{B}$, with respect to a given measure, goes without saying.

A measure on a topological space is called a *Borel measure* if all open sets are measurable. The smallest sigma algebra containing all open sets of a topological space is called its *Borel sigma algebra*; its elements are called *Borel subsets*. In order to avoid pathologies we assume all Borel measures satisfy the following additional assumptions.

- The topological space on which the measure is defined is locally compact, complete metrisable, and separable (i.e., it contains a countable dense subset).
- The measure is locally finite (i.e., each point has a neighbourhood of finite measure).

These conditions are always satisfied for the Borel measures we consider in the theory of dynamical systems (we note that Borel measures satisfying these extra conditions are also called Radon measures, but several different definitions of Radon measures are used).

A measure is called *nonatomic* if each point has measure zero: an atom is a point with positive measure.

A measure m on X is called a *probability measure* if $m(X) = 1$; for $A \subset X$, $m(A)$ is then interpreted as the *probability* to find a point in A.

If $(X_i, \mathcal{B}_i, m_i)$, $i = 1, 2$, are two measure spaces, then a surjective map $\varphi : X_1 \to X_2$ is called a *morphism*, or a *measure preserving map*, if for each $A \in \mathcal{B}_2$ we have $\varphi^{-1}(A) \in \mathcal{B}_1$ and $m_1(\varphi^{-1}(A)) = m_2(A)$. If moreover $X_1 = X_2$ we call φ an *endomorphism*. If the morphism φ is invertible and induces an isomorphism between $\mathcal{B}_1$ and $\mathcal{B}_2$, then φ is an *isomorphism*.

These three notions also appear in a weaker form: *morphisms mod 0*, *endomorphisms mod 0*, and *isomorphisms mod 0*. A morphism mod 0 is given by a map $\varphi : X_1 \to X_2$, which satisfies:

- φ is defined on all of X_1 except maybe on a set A of m_1-measure zero.
- The complement of $\varphi(X_1 - A)$ in X_2 has m_2-measure zero.
- For each $C \in \mathcal{B}_2$, $\varphi^{-1}(C) \in \mathcal{B}_1$ and $m_1(\varphi^{-1}(C)) = m_2(C)$.

The other two 'mod 0' notions are defined similarly. The latter definitions are motivated by the fact that sets of measure zero are so small that they can be neglected in many situations.

Note that the sets of measure zero form an ideal in the sigma algebra of a measure space. An isomorphism mod 0 induces an isomorphism between the sigma algebras $\mathcal{B}_2$ and $\mathcal{B}_1$ after they have been divided out by their ideal of sets of measure zero.

If X_1 and X_2 are topological spaces, then a map $\varphi : X_1 \to X_2$ is called *measurable* if for each Borel set $A \subseteq X_2$ also $\varphi^{-1}(A) \subseteq X_1$ is a Borel set.

Lebesgue measure on $\mathbb{R}$. As an example we discuss the Lebesgue measure on $\mathbb{R}$. This measure assigns to each open interval (a, b) the measure $b - a$. It turns out that there is a unique sigma additive measure on the Borel sigma algebra of $\mathbb{R}$ which assigns this measure to intervals, called the Lebesgue measure. It is not possible to extend this measure to all subsets of $\mathbb{R}$, so the notion of being measurable is a nontrivial one. Clearly the Lebesgue measure is a Borel measure. Its restriction to the interval $[0, 1]$ is a probability measure.

Product measure. On $\mathbb{R}^n$ the Lebesgue measure can be obtained in the same way. This is a special case of product measures which are defined for Borel measure spaces. If $(X_i, \mathcal{B}_i, m_i)_{i \in I}$ is a finite collection of Borel measure spaces, then the product is defined as follows.

- As a set, the product is $X = \Pi_i X_i$; the canonical projections $X \to X_i$ are denoted by ϖ_i.
- The sigma algebra $\mathcal{B}$ of measurable subsets of X is the smallest sigma algebra containing all sets of the form $\varpi_i^{-1}(B_i)$ with $i \in I$ and $B_i \in \mathcal{B}_i$ (or its completion with respect to the measure defined below).
- For $B_i \in \mathcal{B}_i$ we have $m(\cap_i \varpi_i^{-1}(B_i)) = \Pi_I m_i(B_i)$; if in such a product one of the terms is zero then the result is zero, even if a term ∞ occurs. The extension of this function m to all of $\mathcal{B}$ as a sigma additive function turns out to be unique.

The reason why we had to restrict to *finite* products is that infinite products of the form $\Pi_i m_i(B_i)$, for an infinite index set I need not be defined, even if all terms are finite and ∞ as a result is allowed. There is, however, a very important case where countable infinite products are defined, namely for *Borel probability measures* on compact spaces. In this case also infinite products are well defined because all terms are less than or equal to 1. We require the spaces to be compact, because otherwise the product might not be locally compact; we require the product to be only countably infinite so that the product space is still separable. Indeed countable infinite products of Borel probability measures on compact spaces can be defined as above.

Measure class. Let $(X, \mathcal{B}, m_i)$, $i = 1, 2$, denote two measures. We say that these measures belong to the same *measure class* if they define the same subsets of measure zero. An important measure class is that of the Legesgue measure and an important property of this measure class is that it is invariant under diffeomorphisms in the following sense. If $K \subset \mathbb{R}^n$ is a set of measure zero and if $\varphi : U \to \mathbb{R}^n$ is a diffeomorphism of a neighbourhood of $\overline{K}$ to an open subset of $\mathbb{R}^n$, then $\varphi(K)$ is also a set of measure zero (this holds for diffeomorphisms which are at least C^1 but not for homeomorphisms). This last fact can be easily derived from the following characterisation of sets of Lebesgue measure zero. A subset $K \subset \mathbb{R}^n$ has measure zero if and only if, for any $\varepsilon > 0$, it is contained in a countable union of cubes with sides of length ε_i with $\sum_i \varepsilon_i^n < \varepsilon$.

The fact that the Lebesgue measure class is preserved under diffeomorphisms means that also on smooth manifolds one has the notion of sets of measure zero.

Example A.1 ($\mathbb{Z}_2^{\mathbb{N}}$ *versus* $[0, 1)$). On $\mathbb{Z}_2 = \{0, 1\}$ we consider the measure which assigns to 0 and 1 each the measure $\frac{1}{2}$. On $\mathbb{Z}_2^{\mathbb{N}}$, which is the set of sequences $(s_1, s_2, \ldots)$ with $s_i \in \mathbb{Z}_2$, we have the product measure. In the discussion of the doubling map we used that this measure space is isomorphic (mod 0) with the Lebesgue measure on $[0, 1)$. The isomorphism mod 0 is constructed by considering such a sequence $(s_1, s_2, \ldots)$ as the diadic expression for $\sum_i s_i 2^{-i}$. It is easy to verify that this is transforming the Lebesgue measure to the product measure on $\mathbb{Z}_2^{\mathbb{N}}$. It is only an isomorphism mod 0 because we have to exclude sequences $(s_1, s_2, \ldots)$ for which there is some N with $s_i = 1$ for all $i > N$.

Example A.2 (Measure and category). We continue this discussion with an example showing that $\mathbb{R}$ is the disjoint union of a set of the first category, that is, a set which is negligible in the topological sense, and a set of Lebesgue measure zero, that is, a set which is negligible in the measure-theoretical sense. For this construction we use the fact that the rational numbers are countable and hence can be given as $\{q_i\}_{i \in \mathbb{N}}$. We consider the following neighbourhoods of the rational numbers $U_j = \cup_i D_{2^{-(i+j)}}(q_i)$, where $D_{2^{-(i+j)}}(q_i)$ is the $2^{-(i+j)}$-neighbourhood of q_i. Then we define $U = \cap_j U_j$. It is clear that the measure of U_j is smaller than 2^{1-j}. Hence U has Lebesgue measure zero. On the other hand each U_j is open and dense, so U is residual and hence the complement of U is of the first category. This means that probably a better way to formalise the notion 'almost all', in the case of $\mathbb{R}$ or $\mathbb{R}^n$, would be in terms of sets which are both residual and have a complement of measure zero. This is what one usually gets when applying the transversality theorem; this comes from the role of Sard's theorem A.2 in the proof of the transversality theorem A.3. On the other hand, in the study of resonances and KAM tori, one often has to deal with sets which are residual but have a complement of positive Lebesgue measure. Compare Exercise 5.15.

Example A.3 (A homeomorphic swap of 'measure 0' and 'nonmeasurable'). Observe that the properties of 'having measure 0' and 'being nonmeasurable' are nontopological concepts and therefore do not have to be preserved by a homeomorphism. We illustrate this by presenting a homeomorphism that maps nonmeasurable sets to sets of measure 0. In particular we recall from Chapter 3, in particular from § 3.4.3, the family of modified doubling maps (3.4) $\varphi_p : [0, 1) \to [0, 1)$ defined by

$$\varphi_p(x) = \begin{cases} \frac{1}{p}x & \text{as } 0 \le x < p, \\ \frac{1}{1-p}(x-p) & \text{as } p \le x < 1, \end{cases}$$

with $p \in (0,1)$. Next we recall the homeomorphism $h_p : [0,1) \to [0,1)$ that conjugates the standard doubling map $\varphi_{\frac{1}{2}}$ with φ_p, that is, such that

$$h_p \circ \varphi_{\frac{1}{2}} = \varphi_p \circ h_p.$$

The conjugating homeomorphism h_p is completely determined by requiring that $h_p(0) = 0$ and $h_p(\frac{1}{2}) = p$. Finally we recall the set

$$B_p = \left\{ x \in [0,1) \mid \lim_{n \to \infty} \frac{\#\{j \in \{0,1,2,\dots,n-1\} \mid \varphi_p^j(x) \in [0,p)\}}{n} = p \right\},$$

which for each $p \in (0,1)$ has Lebesgue measure 1. For $p \ne \frac{1}{2}$ the map h_p maps $B_{\frac{1}{2}}$ into the complement of B_p, which has Lebesque measure 0.

We now claim that, for $p \ne \frac{1}{2}$, the homeomorphism h_p maps nonmeasurable sets to sets of measure 0. Indeed, let $A \subset [0,1)$ be any Lebesque nonmeasurable set; compare [139]. Then, becuase $B_{1/2}$ has full measure, also $A \cap B_{1/2}$ is nonmeasurable. On the other hand $h_p(A \cap B_{1/2})$ has measure 0. This proves our claim.

Appendix B
Miscellaneous KAM theory

We further review certain results in KAM theory, in continuation of Chapter 2, §§ 2.2 and 2.5, and of Chapter 5. The style is a little less formal and for most details we refer to [27, 37, 87] and to the many references given there. In §§ 5.2 to 5.4 we described the persistent occurrence of quasi-periodicity in the setting of diffeomorphisms of the circle or the plane. As already indicated in § 5.5, a similar theory exists for systems with continuous time, which we now describe in broad terms, as this fits in our considerations. In particular we deal with the 'classical' KAM theorem regarding Lagrangian invariant tori in nearly integrable Hamiltonian systems, for details largely referring to [58, 194, 195, 253, 254]. Also we discuss the 'dissipative' counterpart of the 'classical' KAM theorem, regarding a family of quasi-periodic attractors [87]. At the level of KAM theory these two results are almost identical and we give certain representative elements of the corresponding proof. Also see [27, 37, 76, 87] and many references therein.

B.1 Introduction

As we saw in Chapter 2, § 2.2 and Chapter 5, quasi-periodicity is defined by a smooth conjugation. We briefly summarise this setup for quasi-periodic n-tori in systems with continuous time. First on the n-torus $\mathbb{T}^n = \mathbb{R}^n/(2\varpi Z)^n$ consider the vector field

$$\mathbb{X}_\omega = \sum_{j=1}^n \omega_j \frac{\partial}{\partial x_j},$$

where $\omega_1, \omega_2, \ldots, \omega_n$ are called frequencies [79, 175]. Now, given a smooth[1] vector field X on a manifold M, with $T \subseteq M$ an invariant n-torus, we say that the restriction $X|_T$ is *multiperiodic* if there exists $\omega \in \mathbb{R}^n$ and a smooth diffeomorphism $\Phi : T \to \mathbb{T}^n$, such that $\Phi_*(X|_T) = \mathbb{X}_\omega$. We say that $X|_T$ is *quasi-periodic* if the frequencies $\omega_1, \omega_2, \ldots, \omega_n$ are independent over $\mathbb{Q}$.

---

[1] Say, of class C^∞.

A quasi-periodic vector field $X|_T$ leads to an *affine* structure on the torus T as follows. In fact, because each orbit is dense, it follows that the self-conjugations of $\mathbb{X}_\omega$ exactly are the translations of $\mathbb{T}^n$, which completely determine the affine structure of $\mathbb{T}^n$. Then, given $\Phi : T \to \mathbb{T}^n$ with $\Phi_*(X|_T) = \mathbb{X}_\omega$, it follows that the self-conjugations of $X|_T$ determine a natural affine structure on the torus T. Note that the conjugation Φ is unique modulo translations in T and $\mathbb{T}^n$.

Remarks.

– In Exercise 5.2 we discussed *integer affine* structures on such quasi-periodic tori, given by an atlas, the transition maps of which are torus automorphisms of the form (5.34)

$$x \mapsto a + Sx,$$

with $a \in \mathbb{R}^n$ and $S \in \mathrm{GL}(n, \mathbb{Z})$. For further details see [162] and references therein.
– The current construction is compatible with the integer affine structure on the the Liouville tori of an integrable Hamiltonian system [21]. Note that in that case the structure extends to all multiperiodic tori.

B.2   Classical (conservative) KAM theory

Classical KAM theory deals with smooth, nearly integrable Hamiltonian systems of the form

$$\dot{x} = \omega(y) + f(y, x)$$
$$\dot{y} = g(y, x), \tag{B.1}$$

where y varies over an open subset of $\mathbb{R}^n$ and x over the standard torus $\mathbb{T}^n$. The phase space thus is an open subset of $\mathbb{R}^n \times \mathbb{T}^n$, endowed with the symplectic form $dy \wedge dx = \sum_{j=1}^n dy_j \wedge dx_j$.[2] The functions f and g are assumed small in the C^∞-topology, or in another appropriate topology, compare §§ 5.2 and 5.5, with Appendix A or [142]. For further reference to Hamiltonian dynamics we refer to [20–24]; also see [44]. For a brief guide to these works also see Appendix E.

Adopting vector field notation, for the unperturbed and perturbed Hamiltonian systems we write

$$X(x, y) = \omega(y)\partial_x \quad \text{and} \quad \tilde{X} = (\omega(y) + f(x, y))\partial_x + g(x, y)\partial_y, \tag{B.2}$$

where $\partial_x = \partial/\partial x$, and so on. Note that y is the action variable and x the angle variable for the unperturbed system X. Thus the phase space is foliated by X-invariant

---

[2] This means that a Hamiltonian function $H : \mathbb{T}^n \times \mathbb{R}^n \to \mathbb{R}$ exists, of the form $H(x, y) = H_0(y) + H_1(x, y)$, such that $\omega = \partial_y H_0$, $f = \partial_y H_1$ and $g = -\partial_x H_1$.

tori, parametrised by y. Each of the tori is parametrised by x and the corresponding motion is multi- (or conditionally) periodic with frequency vector $\omega(y)$.

KAM theory asks for persistence of the X-invariant n-tori and the multiperiodicity of their motion for the perturbation $\tilde{X}$. The answer needs two essential ingredients. The first of these is that of *Kolmogorov nondegeneracy* which states that the map $y \in \mathbb{R}^n \mapsto \omega(y) \in \mathbb{R}^n$ is a (local) diffeomorphism. Compare the twist condition of § 5.3. The second ingredient is formed by the diophantine conditions (5.29) on the frequency vector $\omega \in \mathbb{R}^n$, for $\gamma > 0$ and $\tau > n - 1$ leading to the Cantor bundle $\mathbb{R}^n_{\tau,\gamma} \subset \mathbb{R}^n$ of half-lines, compare Chapter 5, in particular § 5.5; also see Figure 5.5.

Completely in the spirit of Theorem 5.3, the classical KAM theorem roughly states that a Kolmogorov nondegenerate nearly integrable system $\tilde{X}$, for sufficiently small f and g is smoothly conjugated to the unperturbed integrable approximation X (see (B.2)) provided that the 'conjugation condition' is restricted to the set $\omega^{-1}(\mathbb{R}^n_{\tau,\gamma})$. In this formulation smoothness has to be taken in the sense of Whitney [88, 194, 253, 254]; also see [27, 37, 76, 87, 195].

As a consequence we may say that in Hamiltonian systems of n degrees of freedom typically quasi-periodic invariant (Lagrangian) n-tori occur with positive measure in phase space; compare Chapter 3, coined there as 'weak persistence'. It should be mentioned that also an isoenergetic version of this classical result exists, implying a similar conclusion restricted to energy hypersurfaces [21, 37, 59, 85]. Here we have to work with equivalences instead of conjugations; compare §§ 3.4.2, and 5.2.1. The twist map theorem 5.3 is closely related to the isoenergetic KAM theorem in particular (cf. Example 5.5 on coupled pendula).

Remarks.
- We chose the quasi-periodic stability format as in § 5.3. For regularity issues compare the remark following Theorem 5.2 in § 5.2 and the one following Theorem 5.7 in § 5.5.
- One extension of the above deals with global KAM theory [79]. In a Liouville–Arnold integrable system [21] the set of Lagrangian invariant tori often forms a nontrivial bundle [22, 117] and the question is what remains of this bundle structure in the nearly integrable case. The classical KAM theorem only covers local trivialisations of this bundle, at the level of diophantine quasi-periodic tori giving rise to local conjugations of Whitney class C^∞. In [79] these local conjugations are glued together so as to obtain a global Whitney smooth conjugation that preserves the nontriviality of the integrable bundle. This gluing process uses the integer affine structure on the quasi-periodic tori, which enables taking convex combinations of the local conjugations, subjected to a suitable partition of unity [142].

The nontriviality of these bundles is of importance for certain semiclassical quantisations of the classical systems at hand, in particular for certain spectrum defects [107]. One example is the spherical pendulum [22, 79, 117]; for other examples see [107] and references therein.

– As indicated in the Historical Remarks § 2.2.5, the literature on the classical KAM theorem is immense. As an entrance to this, see [37, 87]. Here we just mention a difference between the case $n = 2$ and $n \geq 3$. In the former of these cases the energy surface is 3-dimensional in which the 2-dimensional KAM tori have codimension 1. This means that the perturbed evolutions are trapped forever, which can lead to a form of perpetual stability. In the latter case this no longer holds true and the evolutions can escape. For further details about the stickiness of quasi-periodic tori and on Nekhoroshev stability [178, 179], again see the above references and their bibliographies.

B.3 Dissipative KAM theory

As already noted by Moser [174, 175], KAM theory extends outside the world of Hamiltonian systems, such as to volume-preserving systems, or to equivariant or reversible systems. This also holds for the class of general smooth systems, often called 'dissipative'. In fact, the KAM theorem allows for a Lie algebra proof, that can be used to cover all these special cases [37, 84, 87, 88]. It turns out that in many cases parameters are needed for persistent occurrence of (diophantine) quasi-periodic subsystems; for instance, see §§ 5.2 and 5.5. One notion that has to be generalised from the Hamiltonian case is that of integrable and nearly integrable. Integrability can generally be defined in terms of equivariance with respect to a (locally) free action of the torus group $\mathbb{T}^n$, whereas nearly integrable amounts to small perturbations of integrable as usual.

As another example we now consider the dissipative setting, where we discuss a parametrised system with normally hyperbolic invariant n-tori carrying quasi-periodic motion. From [143], Theorem 4.1, it follows that the existence of invariant tori is persistent and that, up to a smooth[3] diffeomorphism, we can restrict to the case where $\mathbb{T}^n$ is the phase space. Compare a similar discussion in § 5.1. To fix thoughts we consider the smooth system

$$\dot{x} = \omega(\alpha) + f(x, \alpha)$$
$$\dot{\alpha} = 0, \tag{B.3}$$

where $\alpha \in \mathbb{R}^n$ is a multiparameter. The results of the classical KAM theorem regarding system (B.1) largely carry over to system (B.3).

As before we write

$$X(x, \alpha) = \omega(\alpha)\partial_x \quad \text{and} \quad \tilde{X}(x, \alpha) = (\omega(\alpha) + f(x, \alpha))\partial_x \tag{B.4}$$

for the unperturbed and perturbed system. The unperturbed system X is integrable, which in the notation of (B.4) amounts to x-independence. Hence, for the

---

[3] In this case of class C^k for large k.

unperturbed system X the product of phase space and parameter space as an open subset of $\mathbb{T}^n \times \mathbb{R}^n$ is completely foliated by invariant n-tori and for any perturbation f this foliation is persistent. The interest is with the multiperiodic dynamics; compare the setting of Theorem 5.2. As just stated, KAM theory here gives a solution similar to the Hamiltonian case. The analogue of the Kolmogorov nondegeneracy condition here is that the frequency map $\alpha \mapsto \omega(\alpha)$ is a (local) submersion [142]. Then, in the spirit of Theorem 5.2, the perturbed system $\tilde{X}$ is smoothly conjugated to the unperturbed case X (see (B.4) as before), provided that we restrict the 'conjugation condition' to the diophantine set $\mathbb{R}^n_{\tau,\gamma}$. Again the smoothness has to be taken in the sense of Whitney [88, 194, 253, 254]; also see [27, 37, 76, 87].

It follows that the occurrence of normally hyperbolic invariant tori carrying (diophantine) quasi-periodic flow is typical for families of systems with sufficiently many parameters, where this occurrence has positive measure in parameter space. In fact, if the number of parameters equals the dimension of the tori, the geometry as sketched in Figure 5.5 carries over by a diffeomorphism. The same holds for the remarks regarding the format of quasi-periodic stability and regularity following § B.2 and Theorem 5.2.

Remarks.
- In Example 5.2 we considered two coupled Van der Pol type oscillators and were able to study invariant 2-tori and their multiperiodicity by the circle map theorem 5.2. In this respect, the present KAM theory covers the case of n coupled oscillators, for $n \geq 2$.
- Dissipative KAM theory provides families of quasi-periodic attractors in a robust way. As in the conservative case, here there is also an important difference between the cases $n = 2$ and $n \geq 3$, namely that a flow on $\mathbb{T}^3$ can have chaotic attractors, which it cannot on $\mathbb{T}^2$. This is of importance for fluid mechanics [31, 207]; for a discussion see § C.3 below.

B.4   On the KAM proof in the dissipative case

This section contains a sketch of the proof of [87], Theorem 6.1 (p. 142 ff.), which is just about the simplest case of the general Lie algebra proof in [84,88], following the ideas of Moser [175] and techniques of Pöschel [194]. As announced in Chapter 1 at the end of § 1.3.4 and later indicated in Chapter 5, in particular § 5.5, the KAM theory proofs involved here are based on an appropriate, somewhat adapted Newton algorithm, where at each iteration step a linear small divisor problem has to be solved.

We like to illustrate this further regarding the proof in the dissipative setting; compare the setup of Theorem 5.2. Here it should be emphasized that the KAM theoretical aspects of the conservative (classical) and dissipative results are essentially the same, the former only further complication being that one has to keep track of the symplectic form.

B.4.1 Reformulation and some notation

We here derive the naive conjugation equation and its (adapted) linearisation, that points at the linear small divisor problem to be used for the Newtonian iteration mentioned before; compare Chapter 5, in particular § 5.5.1.

So we are looking for a conjugation between the unperturbed and perturbed systems (B.4)

$$X(x,\alpha) = \omega(\alpha)\partial_x, \quad \text{respectively,} \quad \tilde{X}(x,\alpha) = (\omega(\alpha) + f(x,\alpha)\partial_x.$$

As a first simplification, using the inverse function theorem [142], we reparametrise $\alpha \leftrightarrow (\omega, \nu)$, where ν is an external multiparameter that we can suppress during the proof. As in §§ 5.2 and 5.5 the naive search is for a diffeomorphism Φ conjugating X to $\tilde{X}$, that is, such that

$$\Phi_* X = \tilde{X},^4 \tag{B.5}$$

Again searching for Φ of the skew form

$$\Phi(x,\omega) = (x + \tilde{U}(x,\omega), \omega + \tilde{\Lambda}(\omega)), \tag{B.6}$$

the conjugation equation (B.5) rewrites to the nonlinear equation

$$\frac{\partial \tilde{U}(x,\omega)}{\partial x}\omega = \tilde{\Lambda}(\omega) + f(x + \tilde{U}(x,\omega), \omega + \tilde{\Lambda}(\omega)),$$

which is exactly equation (5.20). The corresponding linearisation looks like

$$\frac{\partial \tilde{U}(x,\omega)}{\partial x}\omega = \tilde{\Lambda}(\omega) + f(x,\omega); \tag{B.7}$$

compare equation (5.21) and the linear small divisor problem (5.24), the torus flow case. In this linearised form intuitively we think of f as an error that has to be reduced in size during the iteration, thereby in the limit yielding the expression (B.5). Recall from §§ 5.5.1 and 5.5.2 that necessary for solving the equation (B.7) is that the right-hand side has average zero, which amounts to

$$\tilde{\Lambda}(\omega) = -f_0(.,\omega),$$

leading to a parameter shift.

---

[4] As in § 5.5, equation (5.19), this is tensor notation for $D_{(x,\alpha)}\Phi \cdot X(x,\alpha) \equiv \tilde{X}(\Phi(x,\alpha))$; also see Exercise 3.10 in § 3.5.

B.4.2 On the Newtonian iteration

As said in §5.5 the present proof is based on approximation by real analytic functions and the simplest version of the present KAM theory is in this analytic setting where we use the compact open topology on holomorphic extensions. This is the topology corresponding to uniform convergence on compact complex domains. So we return to a complex domain (5.30)

$$\Gamma + \rho := \bigcup_{\omega \in \Gamma} \{\omega' \in \mathbb{C}^n \mid |\omega' - \omega| < \rho\},$$

$$\mathbb{T}^n + \kappa := \bigcup_{x \in \mathbb{T}^n} \{x' \in (\mathbb{C}/2\varpi\mathbb{Z})^n \mid |x' - x| < \kappa\},$$

for positive constants κ and ρ, considering the product

$$\mathrm{cl}\,((\mathbb{T}^n + \kappa) \times (\Gamma + \rho)) \subset (\mathbb{C}/2\pi\mathbb{Z})^n \times \mathbb{C}^n,$$

where the size of perturbations like f is measured in the supremum norm, thereby generating the topology at hand; see Chapter 5, in particular §5.5.3. For technical reasons setting $\Gamma' = \{\omega \in \Gamma \mid \mathrm{dist}\,(\omega, \partial\Gamma) \geq \gamma\}$, we abbreviate

$$\Gamma'_{\tau,\gamma} = \Gamma' \cap \mathbb{R}^n_{\tau,\gamma},$$

The conjugation Φ is obtained as the Whitney-C^∞ limit of a sequence $\{\Phi_j\}_{j \geq 0}$ of real analytic near-identity diffeomorphisms, with $\Phi_0 = \mathrm{Id}$, all of the skew format (B.6). Here we just sketch the important elements of this, for details referring to [27, 87, 88, 194, 195, 253, 254].

Remarks.
- The fact that $\Phi : \mathbb{T}^n \times \Gamma'_{\tau,\gamma} \to \mathbb{T}^n \times \Gamma$ is of class Whitney-C^∞, means that it can be seen as the restriction of a C^∞-map defined on all of $\mathbb{T}^n \times \Gamma$. Compare the formulations of Theorems 5.2 and 5.3.
- The compact open topology on the holomorphic functions restricted to a compact complex domain and using the supremum norm is well known just to give rise to a Banach space. When passing to a general noncompact domain one takes the inverse limit over Banach spaces, thus ending up with a Fréchet space. Fréchet spaces are complete metrisable, therefore here the Baire property is valid; see Appendix A, [96], and references therein.
- The fact that Φ cannot be real analytic can be guessed from the fact that, for dynamical reasons, generically the conjugation relation (B.5) cannot possibly hold true in the gaps of the Cantor bundle $\Gamma'_{\tau,\gamma}$ of half-lines; also see Exercise B.3. This being said, it may be mentioned here that the conjugation Φ can be proven to be Gevrey regular [240], also compare [37] and its references.

Each of the Φ_j is defined on a complex neighbourhood D_j,

$$\mathbb{T}^n \times \Gamma'_{\tau,\gamma} \subset D_j \subset \mathrm{cl}\,((\mathbb{T}^n + \kappa) \times (\Gamma + \rho)),$$

where $D_{j+1} \subset D_j$ for all $j \geq 0$. In fact, the D_j shrink towards $(\mathbb{T}^n + \frac{1}{2}\kappa) \times \Gamma'_{\tau,\gamma}$ as $j \to \infty$.

Remark. Note that in the $\mathbb{T}^n$-direction the complex domain does not shrink completely, which implies that the Whitney-C^∞ limit Φ even is analytic in the x-variables. A similar regularity can be obtained in the ray direction of the set $\mathbb{R}^n_{\tau,\gamma}$; see Lemma 5.5 and Figure 5.5.

Recalling that $\tilde{X}(x,\omega) = (\omega + f(x,\omega))\,\partial_x = X(x,\omega) + f(x,\omega)\,\partial_x$, we describe the first step of the Newtonian iteration. Instead of solving equation (B.5)

$$\Phi_*(X) = \tilde{X}$$

for the conjugation Φ, we solve a 'linearised' equation (on a suitable domain) giving

$$\Phi_0 : D_1 \to D_0,$$

defining $\tilde{X}_1 = \left(\Phi_0^{-1}\right)(\tilde{X})$. Writing $(\xi,\sigma) = \Phi_0^{-1}(x,y)$, for $\tilde{X}_1$ we get the format

$$\tilde{X}_1(\xi,\sigma) = (\sigma + f_1(\xi,\sigma))\,\partial_\xi = X(\xi,\sigma) + f_1(\xi,\sigma)\,\partial_\xi,$$

with f_1 smaller than f, thus with $\tilde{X}_1$ closer to X than $\tilde{X}$.
To be more precise, we require Φ_0 to have the familiar skew format

$$\Phi_0 : (\xi,\sigma) \mapsto (\xi + U_0(\xi,\sigma), \sigma + \Lambda_0(\sigma)),$$

where U_0 and Λ_0 are determined from a homological equation

$$\frac{\partial U_0(\xi,\sigma)}{\partial \xi}\,\sigma = \Lambda_0(\sigma) + {}^{d_0}(f \circ \Phi_0)(\xi,\sigma); \qquad (\text{B.8})$$

cf. the conjugation equation (B.5). Here

$$^{d_0}(f \circ \Phi_0)(\xi,\sigma) = \sum_{|k| \leq d_0} (f \circ \Phi_0)_k(\sigma)\,e^{i\langle \sigma,k\rangle}$$

is the truncation of the Fourier series at an appropriate order d_0. As before, the term $\Lambda_0(\sigma)$ serves to get the $\mathbb{T}^n$-average of the right-hand side equal to zero. The fact that $^{d_0}(f \circ \Phi_0)$ is a trigonometric polynomial implies that U_0 can be solved from equation (B.8) as another trigonometric polynomial in ξ, for which only finitely many diophantine conditions are needed. This accounts for the fact that Φ_0 is defined as an analytic map on the full neighbourhood D_1 of $\mathbb{T}^n \times \Gamma'$.

The second Newtonian iteration step repeats the above procedure with $\tilde{X}_1$ instead of $\tilde{X}$, thus obtaining $\Psi_1 : D_2 \to D_1$ and giving a vector field $\tilde{X}_2$ on D_2, defining a new 'error' f_2, and so on. Also in $f_1 \circ \Psi_1$ a new truncation error d_1 is introduced. We write $\Phi_1 = \Phi_0 \circ \Psi_1$ for the composed map $D_2 \to D_0$.

Continuation of the iteration leads to a sequence of maps

$$\Phi_j : D_{j+1} \to D_0, \quad \text{where } \Phi_{j+1} = \Phi_j \circ \Psi_{j+1},$$

$j = 0, 1, 2, \ldots$, with corresponding 'errors' f_j and truncation orders d_j. The remainder of the proof is a matter of refined bookkeeping, where the approximation properties of Whitney smooth maps by analytic ones [244, 253, 254] should be kept in mind. To this end, the size of the shrinking domains

$$\mathbb{T}^n \times \Gamma'_{\tau,\gamma} \subset \cdots \subset D_2 \subset D_1 \subset D_0 \subset \mathrm{cl}\left((\mathbb{T}^n + \kappa) \times (\Gamma + \rho)\right)$$

is determined by geometric sequences whereas the truncation d_j grows exponentially with $j \in \mathbb{Z}_+$. Moreover, for the size of the 'error' we have $|f_j|_{D_j} \le \delta_j$, with the exponential sequence $\delta_{j+1} = \delta_j^{1+p}$, $j \in \mathbb{Z}_+$ for a fixed $0 < p < 1$.[5] For details see [87] (pp. 146–154); also see references given before.

Remarks.
- In the solution (B.8) we need only finitely many diophantine conditions of the form $|\langle \sigma, k \rangle| \ge c|k|^{-\tau}$, namely only for $0 < |k| \le d$. A crucial lemma [87] (p. 147) ensures that this holds for all $\sigma \in \Gamma_\gamma + r_j$, where r_j is an appropriate geometric sequence; compare the Exercises B.1 and B.2.
- The analytic bookkeeping mentioned above includes many applications of the mean value theorem and the Cauchy integral formula. The latter serves to express the derivatives of a (real) analytic function in terms of the function itself, leading to useful estimates of the C^1-norm in terms of the C^0-norm.

B.5 Historical remarks

KAM theory in its present form was started by Kolmogorov in 1954 (compare the appendix of [20]). In Arnold [58] a more detailed proof of Kolmogorov's result was given and Moser in [173] proved a first version of the twist map theorem 5.3. Together these constitute the foundation of classical KAM theory regarding conservative systems and form the basis of § B.2. For good accounts of this development see, for example, [44, 59]. A first version of the circle map theorem, which is at the basis of Theorem 5.2, was given in Arnold [57].

---

[5] Compare this with the familiar quadratic convergence of the standard Newtonian iteration process.

One branch of KAM theory runs via Moser [174, 175], where the modified term formalism was introduced. The idea is that for persistence of a (diophantine) quasi-periodic torus in a given system, the system has to be modified by extra terms. In [88] this theory was transformed into parametrised KAM theory, where one starts working with families of systems and where generally parameter shifts are needed for the persistence. This theory also applies outside the context of Hamiltonian systems. In fact if applies for classes of systems that preserve a given structure, such as a symplectic or volume form, a group action (including reversibility), and the dependence on external parameters. The simplest examples of this are the Theorems 5.2 and the results in § B.3, where no structure at all has to be preserved and where external parameters clearly are needed for a sound persistence result. This particular result (see [87]) is of importance for proving the existence of families of quasi-periodic attractors (e.g., see [31]).

The element of Whitney-differentiability [244, 245] in the 'Cantor-directions' was introduced in [194, 253, 254] in the conservative case and later in [28, 37, 77, 78, 80–82, 84, 86–88, 133, 241] for the parametrised KAM theory in the general structure-preserving setting. Recently Gevrey regularity also became of importance here [240, 241].

The Newton-like method as introduced in § B.4 is already present in Kolmogorov's proof and explanation of this was one of the subjects in [177, 208]. Also other texts, such as [27, 76, 159, 195, 250] (partly) give tutorial explanations of aspects of KAM theory.

B.6   Exercises

Exercise B.1 (Homothetic role of γ). By a scaling of the time t and of ω show that the above proof only has to be given for the case $\gamma = 1$. What happens to the bounded domain Γ as γ gets small? (Hint: Set $\bar{t} = \gamma t$ and $\bar{\omega} = \gamma^{-1}\omega$.)

Exercise B.2 ($(*, *)$ Order of truncation and diophantine conditions). Following Exercise B.1 we restrict to the case where $\gamma = 1$, taking the set Γ sufficiently large to contain a nontrivial Cantor half-line bundle of parameter values corresponding to $(\tau, 1)$-diophantine frequencies. Consider a geometric sequence $\{s_j\}_{j=0}^{\infty}$ tending to 0. For $j \geq 0$ consider the complexified domain

$$D_j = (\mathbb{T}^n + \tfrac{1}{2}\kappa + s_j) \times (\Gamma'_{\tau,1} + r_j),$$

where $r_j = \tfrac{1}{2}s_j^{2\tau+2}$. For the order of truncation d_j take

$$d_j = \text{Entire}\left(s_j^{-2}\right).$$

1. Show that for all integer vectors $k \in \mathbb{Z}^n$ with $0 < |k| < d_j$ one has $|k|^{\tau+1} \leq (2r_j)^{-1}$;

2. Next show that for all $\sigma \in \Gamma'_{\tau,1} + r_j$ and all k with $0 < |k| < d_j$ one has $|\langle \sigma, k \rangle| \geq \frac{1}{2}|k|^{-\tau}$;

3. As an example take $s_j = \left(\frac{1}{4}\right)^j$, $j \geq 0$, and express the order of truncation d_j as a function of j.

(Hints: For a while dropping the index j, regarding item 1 note that by definition $d \leq s^{-2} = (2r)^{-1/(\tau+1)}$. Next, to show 2, first observe that for $\sigma \in \Gamma'_{\tau,1} + r$ there exists $\sigma^* \in \Gamma'_{\tau,1}$ such that $|\sigma - \sigma^*| \leq r$. It then follows for all $k \in \mathbb{Z}^n$ with $0 \leq |k| \leq d$ that

$$|\langle \sigma, k \rangle| \geq |\langle \sigma^*, k \rangle| - |\sigma - \sigma^*||k| \geq |k|^{-\tau} - r|k| \geq \frac{1}{2}|k|^{-\tau}.)$$

Exercise B.3 ((∗∗) Real-analytic unicity). Let $\tau > n - 1$ and $\gamma > 0$ and consider the (τ, γ)-diophantine subset $\mathbb{R}^n_{\tau,\gamma} \subset \mathbb{R}^n$. Let $f : \mathbb{R}^n \to \mathbb{R}$ be a real-analytic function, such that $f(\omega) = 0$ for all $\omega \in \mathbb{R}^n_{\tau,\gamma}$. Show that $f(\omega) \equiv 0$ on all of $\mathbb{R}^n$.

Appendix C
Miscellaneous bifurcations

Consider a smooth dynamical system depending smoothly on parameters, where we study the changes in the dynamics due to variation of parameters. When the dynamics changes in a qualitative way, we say that a bifurcation occurs. In the main text we already saw several examples of this. For instance, in Chapter 1 where, studying the Van der Pol system, we encountered the Hopf bifurcation; here a point attractor loses its stability and an attracting periodic orbit comes into existence. Moreover in § 3.2 we met sequences of period doubling bifurcations as these occur in the logistic family, whereas in Example 3.1 in § 3.3 we saw a saddle-node bifurcation.

These and other examples are discussed below in more detail. Speaking in terms of Chapter 3, one important issue is, whether the bifucation itself is persistent under variation of (other) parameters, that is, whether it occurs in a robust way, or – in other words – whether a given bifurcation family is typical as a family. Another point of interest is the possibility that the dynamics changes from regular to complicated, especially when, upon changes of parameters, a cascade of bifurcations takes place; here think of the period doubling bifurcations in the logistic system. In particular we focus on some transitions from orderly to chaotic dynamics.

Local bifurcation theory deals with equilibria of vector fields or fixed points of maps, where bifurcation usually means that hyperbolicity is lost. A similar theory exists for periodic orbits: in the case of flows one reduces to a previous case by turning to a Poincaré map and in the case of maps similarly by choosing an appropriate iterate. Of this local theory a lot is known; for example, compare [2, 3, 5, 180, 219, 220] and with [25–32]. Also in the Hamiltonian case much has been said about local bifurcations; for an overview see [28]. A brief guide to this literature is contained in Appendix E.

Somewhat less generally known are quasi-periodic bifurcations, sometimes also called torus bifurcations, where aspects of local bifurcation theory are intermixed with KAM theory. In fact, versions of the local bifurcation scenarios also show up here, but in a way that is 'Cantorised' due to a dense set of resonances. The Whitney smoothness format of the KAM theory as developed in Chapter 5 and in Appendix B allows for a concise geometrical description; see [27, 28, 37, 77, 80–84, 87, 88, 101].

Below we illustrate the above theories, focusing on the Hopf bifurcation, the Hopf–Neĭmark–Sacker bifurcation and the quasi-periodic Hopf bifurcation. In this way we show that for higher-dimensional modelling quasi-periodic bifurcations are of great importance. As an example we sketch the Hopf–Landau–Lifschitz–Ruelle–

Takens scenario for the onset of turbulence [93, 146, 155, 156, 181, 207]. In this and other bifurcation scenarios, which contain transitions to chaos, also homo- and heteroclinic bifurcations play an important role. For details in this respect we refer to [30], where the Newhouse phenomenon is also treated, and to [39]. For a computer-assisted case study see [91].

Remarks.
– All the bifurcation models dealt with in this appendix can also occur in a higher-dimensional setting, embedded in a normally hyperbolic invariant manifold; compare [143], Theorem 4.1.
– In the complement of a diophantine Cantor bundle in the product of phase space and parameter space, generically there will be coexistence of periodicity, quasi-periodicity, and chaos.
  In the complement of such a Cantor bundle, due to the infinity of resonances, there generally will be an infinite regress of the resonant bifurcation structures; compare [63, 83] and references therein.

C.1 Local bifurcations of low codimension

The theory of bifurcations of equilibrium points of vector fields is best understood, where a similar theory goes for fixed points of maps; a systematic treatment is based on the concept of *codimension*; compare the textbooks [4, 5, 9, 29] and for background on dynamical systems theory also see [2, 3, 8, 17, 45, 46]. Roughly speaking the codimension of a bifurcation is the minimum number $p \in \mathbb{Z}_+$ such that the bifurcation occurs in generic p-parameter families of systems. Here one often speaks of *germs* of such families, so there is no need to specify the relevant neighbourhoods. For a definition of germs in a specific setting see Chapter 5, in particular § 5.4.

Remarks.
– The structure of bifurcation theory can change drastically when the dynamics has to preserve a certain structure, such as, whether a symmetry is present. Below we show certain differences between bifurcations in the general 'dissipative' setting and in the Hamiltonian one.
– The above concept of codimension is rather naive. A more sophisticated notion is based on (trans-) versality; see Appendix A, the references [2, 9, 231], and also see [3, 84, 88]. In this setting it is important to specify the equivalence relation to be used on the spaces of parametrised systems. We already saw that the relation based on smooth conjugations can be too fine to be useful; compare Exercise 3.8. Instead, topological conjugation is often more appropriate; see Definition 3.11 in § 3.4.2. For vector fields we often even resort to topological equivalence; see a remark following Definition 3.11. It should be said that the theory can be quite involved, where, for instance, a difference must be made between the topological equivalences depending continuously on the parameters or not. For an overview see [3, 9, 180].

In this appendix we do not attempt a systematic approach, but we present a few miscellaneous examples, mainly in the general 'dissipative' setting where the local bifurcation theory deals with loss of hyperbolicity. Indeed, by the Hartman–Grobman theorem the dynamics near hyperbolic equilibria and fixed points is persistent under variation of parameters and hence has codimension 0. In both these cases one even has structural stability under topological conjugation [3, 9, 13].

Remark. The Hartman–Grobman theorem expresses *linearisability* of a hyperbolic fixed point under topological conjugation. As indicated in Chapter 3, in particular § 3.4.2 and its exercises, this setup does not hold under smooth conjugations, because then the eigenvalues of the linear parts are invariants. For further discussion compare [3].

Bifurcation occurs when eigenvalues of the linear part are on the imaginary axis (for equilibria) or on the unit circle (for fixed points). The simplest case occurs when one simple eigenvalue crosses the imaginary axis at 0 or the unit circle at 1 and the others remain hyperbolic. Because the eigenvalues depend continuously on parameters, this phenomenon has codimension 1. Under open conditions on higher-order terms in both settings we get a saddle-node bifurcation; see below for more details. For fixed points, another codimension 1 bifurcation is period doubling, occurring when one simple eigenvalue crosses the unit circle at -1. See below for certain details; also see the above references, in particular [4]. The cases when a complex conjugated pair of eigenvalues crosses the imaginary axis or the unit circle leads to the Hopf and the Hopf–Neĭmark–Sacker bifurcation, which we discuss later.

C.1.1 Saddle-node bifurcation

The saddle-node bifurcation for equilibria and fixed points, as already met in Example 3.1 in § 3.3, in its simplest form already takes place in the phase space $\mathbb{R} = \{y\}$, in which case normal forms under topological conjugation are given by

$$\dot{y} = y^2 - \mu \quad \text{and}$$
$$y \mapsto y + y^2 - \mu, \tag{C.1}$$

$y \in \mathbb{R}$, $\mu \in \mathbb{R}$, both near 0. In general each of these bifurcations takes place in a center manifold depending on μ, [143, 180]; see Figure 3.2.

C.1.2 Period doubling bifurcation

Another famous codimension-1 bifurcation is period doubling, which does not occur for equilibria of vector fields. However, it does occur for fixed points of diffeomorphisms, where a topological normal form is given by

$$P_\mu(y) = -(1 + \mu)y \pm y^3, \tag{C.2}$$

where $y \in \mathbb{R}, \mu \in \mathbb{R}$, [180]. It also occurs for periodic orbits of vector fields, in which case the map (C.2) occurs as a Poincaré return map inside a center manifold.

Sequences of period doublings have been detected in the logistic family; for a discussion we refer to § 3.2; in particular see Figure 3.1.

C.1.3   Hopf bifurcation

The other codimension-1 bifurcation for equilibria of vector fields is the Hopf bifurcation, occurring when a simple complex conjugate pair of eigenvalues crosses the imaginary axis. For an example see § 1.3.1. A topological normal form (classifying modulo equivalence) in a center manifold $\mathbb{R}^2 \times \mathbb{R} = \{(y_1.y_2), \mu\}$ is given by

$$\dot{\varphi} = 1,$$
$$\dot{r} = \mu r \pm r^3, \tag{C.3}$$

in corresponding polar coordinates $y_1 = r \cos \varphi, y_2 = r \sin \varphi$, [180]. For phase portraits compare Figure 1.12 in § 1.3.1. Figure C.1 shows a 'bifurcation diagram' in the case '$-r^3$' of equation (C.3). Observe the occurrence of the attracting periodic evolution for $\mu > 0$ of amplitude $\sqrt{\mu}$.

Remark. Note that in our formulation we do not use conjugations but equivalences. This is related to the fact that the periodic orbit has a period that is an invariant under conjugation, which would not lead to a finite classification. For further discussion in this direction see [3, 9, 180].

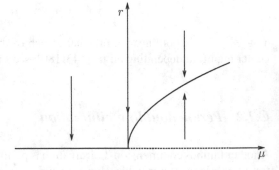

Fig. C.1 'Bifurcation diagram' of the Hopf bifurcation: the amplitude of the oscillation is given as a function of the parameter μ.

C.1.4　Hopf–Neĭmark–Sacker bifurcation

We now turn to the analogue of the Hopf bifurcation for fixed points of
diffeomorphisms, named Hopf–Neĭmark–Sacker bifurcation. Here the eigenval-
ues in a fixed point cross the complex unit circle at $e^{\pm 2\pi i\beta}$, such that β is not
rational with denominator less than 5; see [2, 220]. Again this phenomenon has
codimension-1, in the sense that it occurs persistently in 1-parameter families of
diffeomorphisms.

So let us consider

$$P_{\alpha,\beta}(y) = e^{2\pi(\alpha+i\beta)}y + O(|y|^2), \tag{C.4}$$

$y \in \mathbb{C} \cong \mathbb{R}^2$, near 0, where $O(|y|^2)$ should contain generic third-order terms. For
the linear part this means that the eigenvalues are $e^{2\pi(\alpha\pm i\beta)}$, which for $\alpha = 0$ are
on the complex unit circle. We let $\alpha = \mu$ play the role of bifurcation parameter,
varying near 0. On one side of the bifurcation value $\mu = 0$, this system by normal
hyperbolicity and the normally hyperbolic invariant manifold theorem [143] has an
invariant circle. When trying to classify modulo topological conjugation, observe
that due to the invariance of the rotation numbers of the invariant circles, no topological stability can be obtained [180].

Still, this bifurcation can be characterised by many persistent properties. Indeed,
in a generic 2-parameter family (C.4), say with both α and β as parameters, the
periodicity in the parameter plane is organised in resonance tongues; compare
Figure C.2. For parameter values within a given tongue, periodic orbits occur of a
fixed rotation number. As the parameters change, these periodic orbits get born and
die in saddle-node bifurcations at the tongue boundaries which have been indicated
in Figure C.2.[1] For background also see [2, 89, 91].

If the diffeomorphism is the return map of a periodic orbit of a flow, this bifurcation produces an invariant 2-torus. Usually this counterpart for flows is called
a Neĭmark–Sacker bifurcation. The periodicity is as it occurs in the resonance
tongues, for the vector field is related to phase lock. The tongues are contained in the
complement of a Cantor bundle of quasi-periodic tori with diophantine frequencies.
Compare the discussion in §§ 5.1 and 5.2 and again compare [184].

Remarks.
– A related object is the Arnold family (5.14)

$$P_{\beta,\varepsilon} : x \mapsto x + 2\pi\beta + \varepsilon \sin x$$

of circle maps [2, 4, 91] as described in Example 5.3 at the end of § 5.2 in
Chapter 5; also see the caption of Figure 5.4.[2] As discussed there, quasi-periodic

---

[1] Note that the tongue structure is hardly visible when only one parameter, such as α, is being used.

[2] Again note that for $|\varepsilon| \geq 1$ the map $P_{\beta,\varepsilon}$ is no longer diffeomorphic, in which case the analogy
with the Hopf–Neĭmark–Sacker bifurcations ceases to exist.

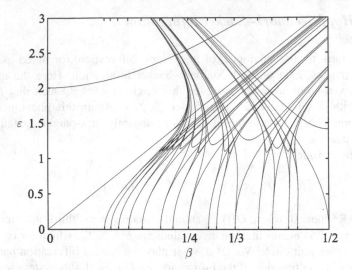

Fig. C.2 Resonance tongues [91] in the Arnold family (5.14) are similar to those in the Hopf–Neĭmark–Sacker bifurcation. Compare Figure 5.4.

and periodic dynamics coexist in the product of phase space and parameter space. Periodicity in the (β, ε)-plane is organised in resonance tongues as shown in Figure 5.4. Also here in the complement of the union of tongues, which is open and dense, there is a Cantor bundle of curves having positive measure, for which the corresponding dynamics is quasi-periodic. Compare Chapter 5, § 5.2.3, Appendix A, and [184].

– From [180] it became clear that, even under topological conjugation, there is no finite classification of the Hopf–Neĭmark–Sacker bifurcation. On the other hand, when using a more appropriate number of parameters, a rather good partial description of the bifurcation can be given, resorting to the persistence notions introduced in Chapter 2 but not using the notion of topological conjugation; also see [91].

C.1.5 The center-saddle bifurcation

As we saw already in Chapters 1 and 2, persistence under variation of parameters depends a lot on the class of systems with which one is dealing. The same holds for the present notions of bifurcation, codimension, and so on. Here we describe the example of the (Hamiltonian) center-saddle bifurcation, which has codimension-1. In fact it is the Hamiltonian counterpart of the saddle-node bifurcation (C.1) as discussed in § C.1. Compare [21, 24, 28].

We consider Hamiltonian systems with 1 degree of freedom, where the local bifurcation theory to a large extent reduces to catastrophe theory [231].

To understand this consider $\mathbb{R}^2 = \{p, q\}$ with the standard symplectic for $\sigma = dp \wedge dq$. For Hamiltonian functions $H, G : \mathbb{R}^2 \to \mathbb{R}$ suppose that, up to an additive constant, $G = H \circ \Phi$ for a (local) orientation-preserving diffeomorphism Φ. At the level of (germs of) functions the corresponding relation is called *right equivalence*. The corresponding Hamiltonian vector fields X_H and X_G then are smoothly equivalent by Φ; see Exercise C.1. Therefore, instead of Hamiltonian vector fields on the plane, we can study Hamiltonian functions under smooth coordinate transformations (or right-equivalences).

Remark. In this case a neat mathematical concept of codimension exists, based on the geometrical notion of transversality [231]; also compare Appendix A. The ensuing families often are called *versal*.

To fix thoughts we now describe a local bifurcation, of the form

$$H_\mu(p, q) = \tfrac{1}{2}p^2 + V_\mu(q), \tag{C.5}$$

where $p, q \in \mathbb{R}$ and where μ is a multiparameter and where p, q, μ all vary near 0.[3] Observe that the Hamiltonian (C.5) is of the natural form 'kinetic' plus 'potential energy'.

The function V_0 has $q = 0$ as a critical point, corresponding to an equilibrium point of H_0. In the case where this critical point has a nondegenerate Hessian, by the Morse lemma, the function V_0 is structurally stable under right-equivalences Φ, so this case has codimension-0. The next case is that $V_0(x) \sim \tfrac{1}{3}q^3$, which has the following versal right-equivalence normal form

$$V_\mu(q) = \tfrac{1}{3}q^3 - \mu q, \text{ with } \mu \in \mathbb{R},$$

and thus is of codimension-1.[4] This is the *fold* catastrophe, the first of the celebrated Thom list [45,46,231]. In the product $\mathbb{R} \times \mathbb{R} = \{q, \mu\}$, the set of critical points is the parabola $\mu = q^2$, for each $\mu \geq 0$ leading to $q = \pm\sqrt{\mu}$: a maximum for $q = -\sqrt{\mu}$ and a minimum for $q = +\sqrt{\mu}$. For the corresponding Hamiltonian systems given by (C.5) these correspond to a center and a saddle-point, respectively, which is why we call this the center-saddle bifurcation. Compare Exercise C.2.

Remarks.
- In a similar way we can get a hold on more degenerate singularities, partly contained in the Thom list [45,46,231], such as the so-called cuspoids, umbilics, and so on,[5] to obtain bifurcation models in Hamiltonian systems using smooth equivalence (e.g., see [28]). Here the Hamiltonian functions no longer have the format 'kinetic' plus 'potential energy'.

---

[3] Formally we should express this in terms of germs.

[4] In the language of catastrophe theory this family is a versal (even universal) unfolding of the singularity $\tfrac{1}{3}q^3$.

[5] The nomenclature in Arnold et al. [45,46] is different and based on group theory.

– For more degrees of freedom there generally is no hope of obtaining structural stability in any sense whatsoever. A quite representative picture is given by Figure C.6 in §C.3.

C.2   Quasi-periodic bifurcations

For the generic bifurcations of equilibria and periodic orbits, the bifurcation sets and diagrams are generally determined by the geometry of singularities of the Thom list [45, 46, 231] in the product of phase space and parameter space, often using singularity theory. Quasi-periodic bifurcation theory concerns the extension of these bifurcations to invariant tori in nearly integrable systems, for example, when the tori lose their normal hyperbolicity or when certain resonances occur. In that case the dense set of resonances, also responsible for the small divisors, leads to a 'Cantorisation' of the classical geometries obtained from singularity theory (see, e.g., [27, 28, 37, 77, 80–82, 84, 87, 88, 101, 133]).

C.2.1   The quasi-periodic center-saddle bifurcation

We start with an example in the Hamiltonian setting, where we consider the phase space $M = \mathbb{R}^2 \times \mathbb{T}^2 \times \mathbb{R}^2 = \{(I_1, I_2), (\varphi_1, \varphi_2), (p, q)\}$ with symplectic form $d I_1 \wedge d \varphi_1 + d I_2 \wedge d \varphi_2 + d p \wedge d q$. Here a robust model for the quasi-periodic center-saddle bifurcation is given by the nearly integrable family

$$H_{\mu,\varepsilon}(I, \varphi, p, q) = \omega_1 I_1 + \omega_2 I_2 + \tfrac{1}{2} p^2 + V_\mu(q) + \varepsilon f(I, \varphi, p, q) \qquad (C.6)$$

with $V_\mu(q) = \tfrac{1}{3} q^3 - \mu q$; see §C.1.5 and [28, 133]. The unperturbed case $\varepsilon = 0$ is *integrable* in the sense that it is $\mathbb{T}^2$-symmetric . That case, by factoring out the $\mathbb{T}^2$-symmetry, boils down to a standard center-saddle bifurcation, involving the fold catastrophe [45, 46, 231] in the potential function $V = V_\mu(q)$; again see §C.1.5. This results in the existence of two invariant 2-tori, one elliptic and the other hyperbolic. For $0 \neq |\varepsilon| \ll 1$ the dense set of resonances complicates this scenario, as sketched in Figure C.3, determined by the diophantine conditions, requiring that for $\tau > 1$ and $\gamma > 0$

$$|\langle k, \omega \rangle| \geq \gamma |k|^{-\tau}, \qquad \text{for } q < 0,$$
$$|\langle k, \omega \rangle + \ell \beta(q)| \geq \gamma |k|^{-\tau}, \qquad \text{for } q > 0 \qquad (C.7)$$

for all $k \in \mathbb{Z}^n \setminus \{0\}$ and for all $\ell \in \mathbb{Z}$ with $|\ell| \leq 2$. Here $\beta(q) = \sqrt{2q}$ is the normal frequency of the elliptic torus given by $q = \sqrt{\mu}$ for $\mu > 0$. As before (cf. Chapter 5 and Appendix B), this gives a Cantor bundle of positive measure as sketched in Figure C.3.

Fig. C.3 Sketch of the
Cantorised fold, as the
bifurcation set of the
(Hamiltonian) quasi-periodic
center-saddle bifurcation for
$n = 2$ [28, 133], where the
horizontal axis indicates the
frequency ratio $\omega_2 : \omega_1$,
cf. (C.6). The lower part of
the figure corresponds to
hyperbolic tori and the upper
part to elliptic ones. See the
text for further
interpretations.

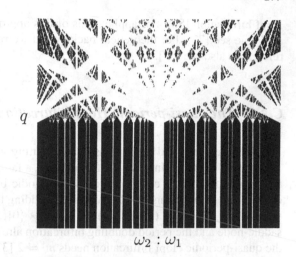

q

$\omega_2 : \omega_1$

To explain Figure C.3 first consider the integrable case $\varepsilon = 0$, in which case for
each value of (ω_1, ω_2) in the 3-dimensional (p, q, μ)-space the curve of equilibria
$p = 0, q^2 = \mu$ is folded in the origin $(p, q, \mu) = (0, 0, 0)$. For simplicity replacing
(ω_1, ω_2) by the ratio $\omega_2 : \omega_1$, we obtain a 1-parameter family of folded curves,
yielding a folded surface. This surface is the graph over the $(q, \omega_1 : \omega_2)$-plane,
where the fold line corresponds with the line $q = 0$. The diophantine conditions
(C.7) give rise to a Cantor bundle[6] that in Figure C.3 is projected on the $(q, \omega_1 : \omega_2)$-plane.

For $|\varepsilon| \ll 1$ this Cantor bundle is distorted somewhat by a near-identity dif-
feomorphism that also conjugates the (diophantine) quasi-periodic dynamics with
the integrable case $\varepsilon = 0$. This is a consequence of parametrised KAM theory. For
references see the beginning of this section and § B.5. Also compare the formulations of Theorems 5.2 and 5.3 and the discussions in Appendix B. Therefore on the
perturbed Cantor bundle the dynamics likewise is quasi-periodic.

We summarise our approach. The model (C.6) was based on the right-
equivalence normal form $\frac{1}{2}p^2 + \frac{1}{3}q^3 - \mu q$, together with the extra terms $\omega_1 I_1 +
\omega_2 I_2$, leading to a $\mathbb{T}^2$-symmetric, or integrable, family. Parametrised KAM theory
ensures persistence of the families of 2-tori carrying (diophantine) quasi-periodic
dynamics, so this part of the model indeed is robust. In this way an open set
is created in the space of all 3-parameter families of Hamiltonian systems on
$M = \mathbb{R}^2 \times \mathbb{T}^2 \times \mathbb{R}^2 = \{I, \varphi, (p, q)\}$, the elements of which are nearly integrable
families, share this property of having a Cantor bundle of positive measure, like
Figure C.3, with associated quasi-periodic dynamics.

Remark. Referring to a remark at the end of § C.1, we note that similar approaches
exist for all cuspoid and umbilic catastrophes [81,82] as well as for the Hamiltonian

---

[6] In fact, for $q > 0$ it is the intersection of five Cantor bundles of lines.

Hopf bifurcation [80]. For applications of this approach see [83]. For a reversible analogue see [77]. For theoretical background we refer to [28, 84, 175]; for more references also see [37].

C.2.2   The quasi-periodic Hopf bifurcation

The approach followed in the conservative setting works equally well for general 'dissipative' systems. In this way, for instance, a family of quasi-periodic attractors as described in § B.3, can undergo quasi-periodic bifurcations that are variations on the local bifurcations of § C1. When embedding the torus-attractor in the phase space $\mathbb{T}^n \times \mathbb{R}^m = \{x \,(\mathrm{mod}\, 2\pi\mathbb{Z}), y\}$ as $\mathbb{T}^n \times \{0\}$, quasi-periodic versions of the saddle-node and the period doubling bifurcation already occur for $m = 1$, whereas the quasi-periodic Hopf bifurcation needs $m = 2$ [37, 87, 88]. Again we start with a $\mathbb{T}^n$-symmetric, or integrable case [88], where we can reduce to $\mathbb{R}^m = \{y\}$ and consider the bifurcations of relative equilibria. Then the concern is with small nearly integrable perturbations of such integrable models. As in the previous section we establish robustness on a Cantor bundle of positive measure of (diophantine) quasi-periodic tori. Indeed, the property of having such a Cantor bundle again will be persistent.

We now turn to the *quasi-periodic Hopf bifurcation* [70, 88]. For simplicity we restrict our description to the setting of the quasi-periodic response problem; compare Examples 5.5 and 5.6 of § 5.5, where we only considered the linear problem. Here, instead of (5.25), we consider a general (nonlinear) quasi-periodically forced oscillator

$$y'' + cy' + ay = g(t, y, y'; a, c),\tag{C.8}$$

again with parameters a and c with $a > 0, c > 0$, and $c^2 < 4a$. The quasi-periodicity of the forcing term g is expressed by requiring that $g(t, y, z; a, c) = G(t\omega_1, t\omega_2, \ldots, t\omega_n, y, z; a, c)$ for some real-analytic function $G : \mathbb{T}^n \times \mathbb{R}^4 \to \mathbb{R}$, $n \geq 2$, and a *fixed* nonresonant frequency vector $\omega \in \mathbb{R}^n$ (i.e., with $\omega_1, \omega_2, \ldots, \omega_n$ independent over $\mathbb{Q}$). The problem is to find a response solution $y = y(t, a, c)$ of (5.25) that is quasi-periodic in t with the same frequencies $\omega_1, \omega_2, \ldots, \omega_n$. We refer to this as the Stoker problem [218].

The corresponding vector field on $\mathbb{T}^n \times \mathbb{R}^2$ in this case is given by the system of differential equations

$$
\begin{aligned}
x' &= \omega \\
y' &= z \\
z' &= -ay - cz + G(x, y, z; a, c),
\end{aligned}\tag{C.9}
$$

where $x = (x_1, x_2, \ldots, x_n)$ modulo 2π. The response problem now translates to the search for an invariant torus which is a graph $(y, z) = (y(x; a, c), z(x; a, c))$ over $\mathbb{T}^n$ with dynamics $x' = \omega$. We call this a Stoker torus.

As in § 5.5, for convenience of notation we complify by

$$\zeta = z - \bar{\lambda} y \quad \text{and} \quad \lambda = -\frac{c}{2} + i \sqrt{a - \frac{c^2}{4}}.$$

Notice that $\lambda^2 + c\lambda + a = 0$, $\Im \lambda > 0$, and that

$$y'' + cy' + ay = \left[\frac{d}{dt} - \lambda\right]\left[\frac{d}{dt} - \bar{\lambda}\right] y.$$

Identifying $\mathbb{R}^2 \cong \mathbb{C}$ and using complex multiplication, the vector field (C.9) on $\mathbb{T}^n \times \mathbb{C} = \{x, \zeta\}$ obtains the form

$$
\begin{aligned}
x' &= \omega \\
\zeta' &= \lambda\zeta + G(x, \zeta; \lambda).
\end{aligned}
\tag{C.10}
$$

The problem of finding an invariant torus of the form $\zeta = \zeta(x; \lambda)$ with dynamics $x' = \omega$ becomes a KAM perturbation problem by assuming that G is small[7] and that the Stoker torus $\zeta(x; \lambda)$ is a perturbation of $\zeta = 0$, that is, of the torus $\mathbb{T}^n \times \{0\} \subset \mathbb{T}^n \times \mathbb{C}$.

Writing $\lambda = \alpha + i\beta$, we introduce diophantine conditions in the pair (ω, λ) of the form (C.7): given $\tau > n - 1$, $\gamma > 0$ and $M \in \mathbb{N}$ we require that

$$|\langle k, \omega \rangle + \ell\beta| \geq \gamma|k|^{-\tau},
\tag{C.11}$$

for all $k \in \mathbb{Z}^n \setminus \{0\}$ and for all for $\ell \in \mathbb{Z}$, with $|\ell| \leq M$. First note that this condition for the case $\ell = 0$, for the fixed nonresonant frequency vector ω implies that $\omega \in \mathbb{R}^n_{\tau,\gamma}$. Second, the conditions (C.11) then give rise to a Cantor bundle of horizontal lines in the (α, β)-parameter plane. As in similar cases met before, this bundle has positive measure for any choice of $M \in \mathbb{N}$. Let Γ be a suitable compact rectangle in the (α, β) plane that contains an open segment of the β-axis; see Figure C.4. By $\Gamma_{\tau,\gamma,M}$ we denote the set of (α, β) satisfying (C.11).

In [70] it was shown that under the diophantine conditions (C.11), up to a near-identity C^∞-diffeomorphism, the vector field (C.10) can be put on a $\mathbb{T}^{n+1}$-symmetric normal form

$$
\begin{aligned}
x' &= \omega \\
\zeta' &= \lambda\zeta + \sum_{j=1}^{N} c_j(\lambda)\zeta|\zeta|^{2j} + r(x, \zeta, \lambda) + p(x, \zeta, \lambda).
\end{aligned}
\tag{C.12}
$$

---

[7] This smallness is to be understood in terms of the compact open topology on holomorphic extensions; compare § 5.5 and B.4.

Fig. C.4 Cantor bundle $\Gamma_{\tau,\gamma,M} \subset \Gamma$ of horizontal lines in the (α, β)-parameter plane as determined by the diophantine conditions (C.11).

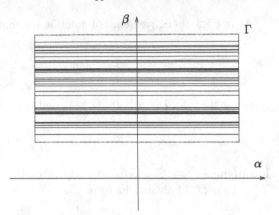

Here $N = [\frac{1}{2}M - 1]$, where [.] takes the integer part. Moreover the functions c_j, $1 \leq j \leq N$, r, and p are of class C^∞, and $r = O(|\zeta|^M)$ and p is infinitely flat on $\Gamma_{\tau,\gamma,M}$.[8] All the involved estimates are uniform on Γ.

A proof of this normal form theorem runs along the same lines as the proof in § B.4; also compare the Exercises 5.8 and 5.9 as well as the references [48, 88]. The smooth functions c_j, $1 \leq j \leq N$, r, and p are obtained by the Whitney extension theorem [245, 253, 254].

Stoker n-tori. From the normal form (C.12) for $M = 2$ it directly follows that for all $(\alpha, \beta) \in \Gamma_{\tau,\gamma,2}$ a Stoker response torus exists, which moreover satisfies the diophantine conditions (C.11) for $M = 2$.

Evidently, for $\alpha \neq 0$ these tori are normally hyperbolic. By the normally hyperbolic invariant manifold theorem [143], the nowhere dense set $\Gamma_{\tau,\gamma,M} \subset \Gamma$ can be fattened to an open parameter domain with Stoker tori. Note that these new tori generally only satisfy the diophantine conditions (C.11) for $\ell = 0$, which just expresses that we fixed $\omega \in \mathbb{R}^n_{\tau,\gamma}$.

Because for $\alpha \approx 0$ the hyperbolicity gets small, an issue is how this fattening can be done optimally near the β-axis. In [70], a careful application of the uniform contraction principle [25] and exploiting the flatness of the function p on $\Gamma_{\tau,\gamma,M}$, led to the following result near the β-axis.

Considering a horizontal line in the set $\Gamma_{\tau,\gamma,2}$, the fattening of the domain shrinks to zero for $\alpha \to 0$, but the domain contains an open disc on each side, the (smooth) boundaries of which are tangent to the β-axis to infinite order. In Figure C.5 we sketched the situation in a number of cases, where the near-identity diffeomorphic distortion was taken into account. The line $\mathcal{H}$ is the diffeomorphic image of the β-axis, where the normal hyperbolicity of (C.10) vanishes, and $\mathcal{H}_c \subset \mathcal{H}$ is the Cantor set corresponding to $\Gamma_{\tau,\gamma,2} \cap \{\alpha = 0\}$. The attached discs are named $\mathcal{A}_\sigma$, attached to $\sigma \in \mathcal{H}_c$, for the attracting n-tori and similarly $\mathcal{R}_\sigma$ for the repelling ones.

---

[8] This means that the derivatives to any order of p vanish on all of $\Gamma_{\tau,\gamma,M}$.

Fig. C.5 Sketch of the fattening by normal hyperbolicity of the nowhere dense parameter domain $\Gamma_{\tau,\gamma,2}$ of Stoker tori. The curve $\mathcal{H}$ interpolates the Cantor set of the persistent n-tori that are not normally hyperbolic and the discs $\mathcal{A}_{\sigma_1,\sigma_2}$ and $\mathcal{R}_{\sigma_1,\sigma_2}$ form open extensions of the nowhere dense parameter domain with normally hyperbolic n-tori. In between a resonance 'bubble' is visible. The Stoker domain corresponds to the n-tori that are 'very' normally hyperbolic and that can be found right away without KAM theory [218].

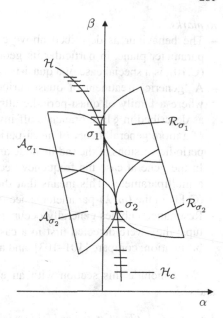

In the nonempty intersections $\mathcal{A}_{\sigma_1} \cap \mathcal{A}_{\sigma_2}$ and $\mathcal{R}_{\sigma_1} \cap \mathcal{R}_{\sigma_2}$, $\sigma_1, \sigma_2 \in \mathcal{H}_c$, the invariant n-tori coincide. In the complement of

$$\bigcup_{\sigma \in \mathcal{H}_c} (\mathcal{A}_\sigma \cup \mathcal{R}_\sigma)$$

there is a countable union of discs, often called resonance 'bubbles', which are centered around the resonant parameter points. Loosely speaking, one might say that the bubbles contain the gaps in the complement $\mathcal{H} \setminus \mathcal{H}_c$ of the Cantor set.

Invariant $(n+1)$-tori. Another application of the $\mathbb{T}^{n+1}$-symmetric normal form (C.12), for $M = 4$, leads to the existence of invariant $(n+1)$-tori of system (C.10). Referring to a slight adaptation of Figure C.5, these tori emanate from points $\sigma \in \Gamma_{\tau,\gamma,4} \cap \{\alpha = 0\}$ to the right,[9] also giving rise to the attachment of an open disc, the boundary of which has flat contact with the β-axis. This part of the existence proof rests on normal hyperbolicity and only uses another careful application of the uniform contraction principle [25, 70]. These invariant $(n+1)$-tori generically will be quasi-periodic on a nowhere dense parameter subset of positive measure.

---

[9] This is in accordance with the choices in (C.3) leading to Figure C.1; in the normal form (C.12) this means that $c_1(0, \beta) < 0$.

Remarks.
- The behaviour as described above concerning the rectangle Γ in the (α, β)-parameter plane, in particular its geometry and the corresponding dynamics of (C.10), is a special case of a quasi-periodic Hopf bifurcation.
- A 'generic' treatment of quasi-periodic Hopf bifurcations was given in [88], where a family of quasi-periodic attractors (corresponding to invariant n-tori) as described in § B.3, branches off invariant $(n + 1)$-tori. This treatment is part of a more general theory of quasi-periodic bifurcations, that also contains quasi-periodic versions of the saddle node and the period doubling bifurcation.

  In the general case the frequency vector ω cannot be kept fixed, but serves as a multiparameter. This means that the geometry of the diophantine conditions (C.11) in the (ω, λ)-parameter space gets more involved.
- Resonance bubbles generally occur in dissipative quasi-periodic bifurcation setups; they were detected first in a case study of the quasi-periodic saddle node bifurcation; compare [101–103], and also see [63].

  We conclude this section with an example that connects Figures C.3, C.4, and C.5.

Example C.1 (The Cantorised fold revisited). Consider a family of maps that undergoes a generic quasi-periodic Hopf bifurcation from circle to 2-torus. It turns out that here the Cantorised fold of Figure C.3 is relevant again,[10] where now the vertical coordinate is the bifurcation parameter. The Cantor bundle contains the quasi-periodic dynamics, whereas in the complement we can have chaos. A fattening process as explained above, also can be carried out here.

C.3 Transition to chaos

One of the interests over the second half of the twentieth century has been the transition between orderly and complicated forms of dynamics upon variation of either initial states or of system parameters. By 'orderly' we here mean equilibrium and periodic dynamics and by 'complicated' quasi-periodic and chaotic dynamics, although we note that only chaotic dynamics is associated with unpredictability; see § 2.2.4. As already discussed in § 5.1, systems such as a forced nonlinear oscillator or the planar 3-body problem exhibit coexistence of periodic, quasi-periodic, and chaotic dynamics; also compare Figure C.6 in Example C.2 below on the parametrically forced pendulum.

Onset of turbulence. Similar remarks go for the onset of turbulence in fluid dynamics. Around 1950 this led to the scenario of Hopf–Landau–Lifschitz [146, 155, 156], which roughly amounts to the following. Stationary fluid motion

---

[10] In fact, this is a slight variation of Figure C.3, because the diophantine conditions (C.7) have to hold for all $\ell \in \mathbb{Z}$ with $|\ell| \leq 7$ instead of only with $|\ell| \leq 2$.

corresponds to an equilibrium point in an ∞-dimensional state space of velocity fields. The first transition is a Hopf bifurcation (see §C.1 and [5, 29, 145]) where a periodic solution branches off. In a second transition of similar nature a quasi-periodic 2-torus branches off, then a quasi-periodic 3-torus, and so on. Also compare Example C.1.

The idea is that the motion picks up more and more frequencies and thus obtains an increasingly complicated power spectrum. In the early 1970s this idea was modified in the Ruelle–Takens route [207] to turbulence, based on the observation that, for flows, a 3-torus can already carry chaotic attractors [11] arbitrarily close to quasi-periodic dynamics, giving rise to a broadband power spectrum; see Newhouse, Ruelle, and Takens [181]. By the quasi-periodic bifurcation theory [37, 87, 88], as sketched in §C.2.2, these two approaches are unified in a persistent way, keeping track of measure-theoretic aspects. Also compare the discussion in Chapter 4, in particular in §4.3.

Remark. In scenarios for the onset of turbulence in fluid or gas flows, one usually varies one parameter, such as the Reynolds number, and studies the transitions from laminar to turbulent. Referring to the beginning of this section, we recall the scenario of Hopf–Landau–Lifschitz–Ruelle–Takens as discussed before.

Any 1-parameter family as considered in such a scenario, should be regarded as a generic subfamily [12] of a multiparameter setting, involving several bifurcations, say a standard Hopf or Hopf–Neĭmark–Sacker bifurcation as described in §C.1 and a quasi-periodic Hopf bifurcation; see §C.2.

Period doubling sequences. Another transition to chaos was detected in the logistic family of interval maps

$$\Phi_\mu(x) = \mu x(1 - x),$$

as met in Chapter 1; compare [4, 12, 167, 169], also for a holomorphic version. This transition consists of an infinite sequence of period doubling bifurcations (see §C.1) ending up in chaos; it was discovered by Feigenbaum–Coullet–Tresser that this family has several universal aspects and occurs generically in 1-parameter families of dynamical systems; compare the above references.

Homoclinic bifurcations. Often also homoclinic bifurcations play a role, where the transition to chaos is immediate when parameters cross a certain boundary and, on the other hand, period doubling and other cascades are possible. For general theory see [3,30,39,66,67]. Also compare the discussion in §4.2.2 on nonhyperbolic attractors. There exist quite a number of case studies where all three of the above scenarios play a role; for example, see [91–93] and many of their references.

---

[11] In [207] the term 'strange' attractor was coined.

[12] Or 'path'.

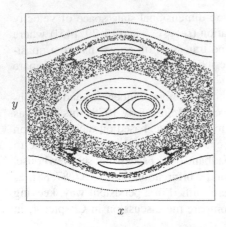 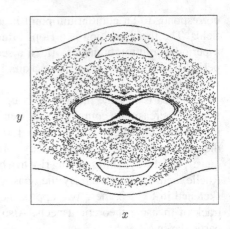

Fig. C.6 Poincaré maps of the undamped swing (C.13) for $\Omega = 2, \omega = 1$. In the left picture we took $\varepsilon = 0.25$ and in the right picture $\varepsilon = .40$. Several evolutions can be distinguished.

The swing. We conclude by revisiting the pendulum with periodic forcing, in a variation called swing, presenting a few diagrams of its dynamics, obtained from its stroboscopic (or Poincaré map). We discuss several aspects both in the conservative and in the dissipative case, indicating a few open problems with respect to chaos. This discussion is related to §§ 4.2.2 and 4.3.

Example C.2 (Chaos in the swing). Consider the parametrically forced pendulum, also called the *swing*, given by the equation of motion

$$x'' = -cx' - (\omega^2 + \varepsilon \cos(\Omega t))\sin x, \tag{C.13}$$

which is a slight variation on (1.5) of § 1.1. In the Figures C.6 and C.7 we depict a few significant orbit segments

$$\left\{ \left(x\left(\frac{2}{\Omega}\pi n \right), y\left(\frac{2}{\Omega}\pi n \right) \right) \mid 0 \leq n \leq N \right\},$$

where $y = x'$, of the Poincaré or stroboscopic map. Figure C.6 corresponds to the undamped case of (C.13) (i.e., with $c = 0$.) Here we see several 'islands' in a 'chaotic sea'. Also we see a number of dotted closed curves. A probable interpretation runs as follows. The islands are domains where the Moser twist map Theorem 5.3 applies, giving a (topological) disc that contains a union of quasi-periodic invariant circles constituting a set of positive measure. In the center of the discs there is a periodic orbit (not shown here), corresponding to a (subharmonic) periodic evolution of (C.13).[13] In both the left and the right diagram we witness four larger islands, which contain two periodic sets of period 2 (each with a mirror symmetry under reflections in the origin), corresponding to subharmonic periodic

---

[13] The central periodic point probably is a Lebesgue density point of quasi-periodicity [173, 194].

Fig. C.7 Poincaré map of the damped swing, again for $\omega = 1, \Omega = 2$. We took $\varepsilon = 0.5$ and $c = 0.05$. The cloud of points depicted belongs to one evolution. The geometric shape becomes clearer when more iterates are added, and it attracts all nearby evolutions. This may well be a Hénon-like attractor as described in § 4.2.

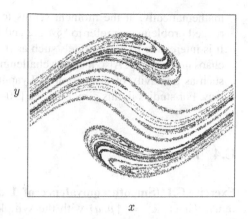

evolutions of period $4/\Omega\pi$. These solutions are entrained by the fact that $\Omega \approx 2\omega$: this phenomenon is also called *resonance*. This $1 : 2$ resonance is related to an area-preserving version of the period doubling bifurcation, where the central fixed point at $(x, y) = (0, 0)$ turns from stable into unstable, and a stable period 2 orbit emerges. This kind of resonance often can be used for explaining natural phenomena (compare, e.g., [21, 73, 170]).

Remarks.
- The chaotic evolutions are not yet well understood. For the conservative cases of Figure C.6 it was conjectured in 1967 [59] that each orbit that is neither periodic nor quasi-periodic densely fills a set of positive area; compare Chapter 4, in particular § 4.3.
  This conjecture applies generally for similar area-preserving maps with transversal homo- or heteroclinic intersections, for instance, also for certain (isoenergetic) Poincaré maps in the 3-body problem. Up to now such conjectures only have been proven in certain toy models.
  Note that the occurrence of transversal homoclinic points (see Chapter 4, in particular § 4.4.3) implies the existence of horseshoes; for example, see Figure 4.10, which means that there is positive topological entropy [18, 59]. An open question is whether the dispersion exponent is also positive on a set of initial states with positive measure.
- In Figure C.7 only one orbit of the Poincaré map is shown in the damped, that is, dissipative case, again close to $1 : 2$ resonance, that is, with $\Omega = 2\omega$. It turns out that almost any initial state after a transient is attracted towards this 'figure', which probably is a chaotic, Hénon-like attractor; compare §§ 4.2.2 and 4.3 of Chapter 4 and Exercise 1.26. In accordance with (4.4), the present conjecture is that this attractor coincides with the closure

$$\overline{W^u((0,0))}$$

of the unstable manifold to the saddle point $(x, y) = (0, 0)$. Also questions related to chaoticity play here. Again, to settle this and similar problems

mathematically, at the moment seems to be out of reach. For a discussion of related problems, we refer to §§ 4.2.2 and 4.3.
– It is interesting to scan models such as (C.13) for bifurcations and transitions to chaos numerically. Also it seems challenging to prove certain of the above points, such as positivity of dispersion or Lyapunov exponents, in a computer-assisted way. For studies in this direction, see [90–94].

C.4 Exercises

Exercise C.1 (Smooth equivalence of 1 degree of freedom Hamiltonian systems). Given $\mathbb{R}^2 = \{p, q\}$ with the symplectic form $\sigma = dp \wedge dq$. Consider a Hamilton function $H : \mathbb{R}^2 \to \mathbb{R}$ and a smooth orientation-preserving diffeomorphism $\Phi : \mathbb{R}^2 \to \mathbb{R}^2$. Define $G = H \circ \Phi^{-1}$ and consider the Hamiltonian vector fields X_H and X_G. Show that Φ is a smooth equivalence from X_H to X_G.
(Hint: Show that $\Phi_*(X_H)(p, q) = (\det d_{(p,q)}\Phi) \cdot X_G(\Phi(p, q))$.)

Exercise C.2 (Phase portraits of the Hamiltonian center-saddle bifurcation). Draw graphs of the potential functions $V_\mu(q) = \frac{1}{3}q^3 - \mu q$ for $\mu < 0$, $\mu = 0$ and $\mu > 0$. Sketch the corresponding phase portraits of the vector field X_{H_μ}, where $H_\mu(p, q) = \frac{1}{2}p^2 + V_\mu(q)$.

Appendix D
Derivation of the Lorenz equations

The derivation of the Lorenz equations is quite involved and at certain moments facts known from fluid dynamics have to be used. We proceed as follows. First a precise description of the geometry of the problem is given. Next an analysis is given of the equations of motion of incompressible fluids of constant density within this geometry. Finally we introduce the correction in these equations of motion as a consequence of temperature differences. In this way a system of partial differential equations is obtained. These have to be reduced to a system of three ordinary differential equations. To this end we choose an appropriate 3-dimensional subspace of our function space, such that the projection on this subspace contains as much as possible of the original dynamics. To arrive at a good choice, we need the help of a stability analysis that Rayleigh [198] carried out for this kind of system. After this a projection on a 3-dimensional subspace turns out to be feasible. For this derivation we gratefully make use of [238].

D.1 Geometry and flow of an incompressible fluid

Our starting position is a 2-dimensional layer, in which there is a temperature difference between the lower and upper boundary. In the horizontal direction it extends infinitely. For the horizontal coordinate we take x and for the vertical one y. Our layer can be described as $\{y_1 \leq y \leq y_2\}$. Without loss of generality we may assume that $y_1 = 0$. For the height we can also take whatever we want, which can be arranged by choosing appropriate units. It turns out that $y_2 = \pi$ is convenient and so we arrive at

$$\{0 \leq y \leq \pi\}.$$

In reality 3-dimensional layers are even more relevant, however, in a 3-dimensional layer 2-dimensional flows can also occur. For such flows, there is no component in the second horizontal direction and also all velocities are independent of the second horizontal coordinate.

A 2-dimensional fluid flow is described by a velocity field $v = (v_x, v_y)$, where both components are functions of x, y, and the time t. The equation of motion is deduced from Newton's law $F = m\,a$. The acceleration of a 'fluid particle' is given by

$$(\partial_t + \partial_v)v,$$

where we abbreviate $\partial_t = \partial/\partial t$ and $\partial_v = v_x\partial_x + v_y\partial_y$. We assume that the density of our fluid equals 1. The force effected per unit of volume, and hence per unit mass, is constituted by three contributions. The first is the negative gradient of the pressure

$$-\text{grad } p = (-\partial_x p, -\partial_y p).$$

Secondly there is a force due to friction in the fluid and that will have the form

$$\nu\Delta v,$$

where $\Delta = \partial_{xx} + \partial_{yy}$ is the Laplace operator and where ν is a constant determined by the viscosity of the fluid (i.e., the degree in which the fluid is experiencing internal friction). Thirdly, we keep the possibility open of an 'external force'

$$F_{\text{ext}}.$$

In this way we find the Navier–Stokes equation

$$\partial_t v = -\partial_v v - \text{grad } p + \nu\Delta v + F_{\text{ext}}. \tag{D.1}$$

A more extensive explanation of this equation can be found in any good book on fluid dynamics (e.g., see [65]).

Remarks.
- We imposed the requirement that the fluid is incompressible and has constant density. This means that the velocity field is *divergence-free*, meaning that

$$\text{div } v = \partial_x v_x + \partial_y v_y = 0;$$

for this and various other facts from vector analysis that we use here, we refer to [74]. This implies that the right-hand side of the Navier–Stokes equation (D.1) also has to be divergence-free. It can be shown that, for given v and F_{ext}, this requirement, up to an additive constant, uniquely determines the pressure p.
- For any divergence-free vector field v in dimension 2 there exists a *stream function* (i.e., a function Ψ), such that

$$v_x = -\partial_y\Psi, \quad v_y = \partial_x\Psi.$$

Because in our case the velocity of the fluid along the boundaries of the strip must be horizontal, any associated stream function has to be constant on these boundary components. If there is a difference between these constants for the upper and lower boundary, there will be a net flow in the horizontal direction. We exclude this and therefore assume that our stream functions are zero on the boundaries of the strip. Moreover we note that, contrary to what happens often, we do not require that the velocity along the boundaries vanishes, but just that this velocity is parallel to these.

— For 2-dimensional vector fields $v = (v_x, v_y)$ we define the *curl* by

$$\text{curl}\, v = \partial_x v_y - \partial_y v_x.$$

In the Navier–Stokes equation (D.1) both the left- and the right-hand side are divergence–free and the vector fields have to be tangent to the boundaries, therefore we don't lose information if we apply the curl operator on all terms in the equation. We here note that if Ψ is the stream function of v then

$$\text{curl}\, v = \Delta \Psi.$$

Thus we transform the Navier–Stokes equation (D.1) into its *vorticity form*

$$\partial_t \Delta \Psi = -\partial_v \Delta \Psi + v \Delta^2 \Psi + \text{curl}\, F_{\text{ext}}. \tag{D.2}$$

D.2   Heat transport and the influence of temperature

As noted before, we assume that the lower boundary of the strip is warmer than the upper boundary. We set the temperature of the lower boundary at the value T and keep the temperature of the upper boundary at 0. This gives rise to a stationary temperature distribution

$$\tau_{\text{stat}}(x, y, t) = T - yT/\pi,$$

recalling that the height of the strip is π. This is indeed the temperature distribution when the fluid is at rest and if the only heat transfer is due to heat conduction. We now describe a temperature distribution $\tau(x, y, t)$ in terms of the deviation from the stationary distribution:

$$\Theta(x, y, t) = \tau(x, y, t) - \tau_{\text{stat}}(x, y, t).$$

From this definition it follows that Θ, as well as our stream function, has to vanish at the strip boundaries. If in the fluid there is no heat conduction, and hence the temperature moves with the fluid, then necessarily

$$\partial_t \tau = -\partial_v \tau.$$

In the case that heat conduction does take place in the fluid, the right-hand side needs an extra term $\kappa \Delta \tau$, where κ depends on the way in which the fluid conducts the heat. For the function Θ this means

$$\partial_t \Theta = -\partial_v \Theta - \partial_v(-yT/\pi) + \kappa \Delta \Theta. \tag{D.3}$$

Now we still have to describe the influence of the temperature distribution on the motion of the fluid. This influence is a consequence of the fact that warm fluid is less dense than cold fluid. This leads to a problem because we have assumed that the fluid is incompressible and has everywhere the same density. In the textbook [65], as already mentioned, arguments are given that under reasonably general circumstances, the conclusions of the following 'reasoning' hold to good approximation. The density, for small differences in temperature, can be assumed equal to

$$\rho = \rho_0 - c\tau,$$

for positive constants c and ρ_0. This means that we have a gravitational force per unit of volume, as a function of time and position, which is equal to a constant plus $c\tau$.[1] In summary, in the Navier–Stokes equation (D.1), (D.2), the external force F_{ext}, up to an additive constant, is set equal to $(0, c\tau)$, which casts the vorticity form (D.2) of the Navier–Stokes equations into the form

$$\partial_t \Delta \Psi = -\partial_v \Delta \Psi + \nu \Delta^2 \Psi + c\partial_x \Theta. \tag{D.4}$$

Now, finally, to get the whole system of equations (D.3) and (D.4) in a simpler form, we note that, if Ψ is the stream function of the velocity field v, then for an arbitrary function f the directional derivative can be expressed as follows.

$$\partial_v f = \partial_x \Psi \, \partial_y f - \partial_y \Psi \, \partial_x f = \frac{\partial(\Psi, f)}{\partial(x, y)},$$

where in the latter expression we use notation familiar from transforming integrals. A direct computation confirms this. We now express the total set of equations as

$$\partial_t \Delta \Psi = -\frac{\partial(\Psi, \Delta \Psi)}{\partial(x, y)} + \nu \Delta^2 \Psi + c\partial_x \Theta$$

$$\partial_t \Theta = -\frac{\partial(\Psi, \Theta)}{\partial(x, y)} + \frac{1}{\pi} T \partial_x \Psi + \kappa \Delta \Theta. \tag{D.5}$$

D.3 Rayleigh stability analysis

The equations (D.5) have a trivial solution, namely where both Ψ and Θ are identically zero. This solution corresponds to the case where the fluid is at rest and heat conduction occurs exclusively by diffusion. Now Rayleigh attempted to answer the following question. For which differences of temperature T is this zero solution

---

[1] NB: Here two minus signs have cancelled as follows. Gravitation points downward, whereas the positive y-directions points upward: this accounts for one minus sign. The formula for the density contains the second minus sign.

stable? A feasible method for this is to linearise at this zero solution. (Recall how we obtained a good idea of the dynamics near the saddle point of the Hénon map by linearisation.) Presently we linearise by omitting the terms

$$-\frac{\partial(\Psi, \Delta\Psi)}{\partial(x, y)} \quad \text{and} \quad -\frac{\partial(\Psi, \Theta)}{\partial(x, y)}$$

from (D.5), evidently leading to

$$\partial_t \Delta\Psi = \nu\Delta^2\Psi + c\partial_x\Theta$$
$$\partial_t\Theta = \tfrac{1}{\pi}T\partial_x\Psi + \kappa\Delta\Theta. \tag{D.6}$$

We first determine a great number of solutions of the linearised system (D.6), later explaining the meaning of these for the stability problem. We try solutions of the form

$$\Psi = X(t)\Psi_{n,a}, \quad \Theta = Y(t)\Theta_{n,a},$$

where

$$\Psi_{n,a} = \sin(ax)\sin(ny), \quad \Theta_{n,a} = \cos(ax)\sin(ny), \tag{D.7}$$

with $a \in \mathbb{R}$ and $n \in \mathbb{Z}$ both positive. Substituting these in the linearised system (D.6) gives the following equations for the scalar functions $X(t)$ and $Y(t)$:

$$\partial_t X = -\nu(a^2 + n^2)X + \frac{ca}{a^2 + n^2}Y$$
$$\partial_t Y = \tfrac{1}{\pi}Ta X - \kappa(a^2 + n^2)Y.$$

This is a linear system with matrix

$$A = \begin{pmatrix} -\nu(a^2 + n^2) & ca\,(a^2 + n^2)^{-1} \\ \tfrac{1}{\pi}Ta & -\kappa(a^2 + n^2) \end{pmatrix}.$$

Because all our constants are positive, the trace of A is negative. Moreover, for sufficiently small values of T both eigenvalues are negative. For increasing T this remains true, at least as long as $\det A > 0$. The turning point occurs for the temperature $T_{n,a}$, where we get $\det A = 0$; that is, where

$$\nu\kappa(a^2 + n^2)^2 = \frac{T_{n,a}a^2 c}{\pi(a^2 + n^2)}$$

or

$$T_{n,a} = \frac{\nu\kappa\pi(a^2 + n^2)^3}{a^2 c}.$$

If we view the solutions of (D.6) as very small perturbations of the zero solution of (D.5), then we see that for $T > T_{n,a}$ these small solutions can grow exponentially. For such values of T we don't expect stability. Our interest is now with values of n and a for which $T_{n,a}$ is minimal. Clearly, to minimise $T_{n,a}$, we have to take

$n = 1$. Next, to determine the value of a where $T_{n,a}$ is minimal, we have to find the minimum of $(a^2 + 1)^3/a^2$. This occurs for $a^2 = \frac{1}{2}$. The corresponding 'critical' temperature then is

$$T_c = T_{1,\sqrt{1/2}} = \frac{27 \nu \kappa \pi}{4c}. \tag{D.8}$$

We now return to the stability problem. According to Fourier theory, any initial state (Ψ, Θ) of the equation of motion (D.5) can be expressed as an infinite sum (or integral) of contributions in terms of $\Psi_{n,a}$ and $\Theta_{n,a}$. We used this before when discussing the heat equation in § 1.3.5. Again we refer to [200] for background information.

Remarks.
- Apart from these, contributions also occur as indicated earlier in this section, namely when in the definitions of $\Psi_{n,a}$ and $\Theta_{n,a}$ of (D.7), we just replace $\sin(ax)$ by $\cos(ax)$ or $\cos(ax)$ by $-\sin(ax)$, respectively. This does not essentially change the argument.
- The functions Ψ and Θ both vanish at the boundaries $\{y = 0, \pi\}$ of the domain of definition. This is the reason why in the y-directions only contributions show up of the form $\sin(ny)$.
- In the x-direction the domain is unbounded, whence the parameter a can attain all (positive) values. This means that the Fourier expansion in the x-direction gives rise to an integral. This might be different if we would bound the domain in the x-direction as well, or by requiring so-called 'periodic boundary conditions'. We won't go into this matter here.
- We won't consider any aspects regarding the convergence of these Fourier expansions.

When taking an arbitrary initial state (Ψ, Θ) of (D.5), expressing it as a sum (or an integral) of contributions as (D.7) and so on, then the corresponding solution again can be written as such a sum (or integral) of the solutions corresponding to the various (n, a). For $T < T_c$ (see (D.8)) we expect the solution to tend to the zero-solution (inasmuch as all contributions tend to 0). In this case the zero-solution will be stable. We already said that for $T > T_c$ the zero-solution is not stable, in the sense that solutions of the linearised system can grow exponentially.

Remark. We must point out that our argumentation, in order to obtain mathematical rigour, needs completion. In this discussion we systematically have omitted all considerations dealing with function spaces, (i.e., *infinite-dimensional vector spaces*). Nevertheless we consider T_c as a formal stability boundary.

D.4 Restriction to a 3-dimensional state space

After the above stability analysis it may be feasible to look for approximating solutions of the equations of motion (D.5), starting from solutions $\Psi_{n,a}$ and $\Theta_{n,a}$, (see (D.7)) of the linearised system (D.6) found so far. Here we have to keep in mind that such a low-dimensional approximation can only have metaphoric value as a model. It then also seems reasonable to start with the solution that is 'the first' to be unstable, so for $a^2 = 1/2$ and $n = 1$. This has to to do with the theory of center manifolds [143], discussed elsewhere in the text. Presently, however, the parameter a is still kept free, although we fix $n = 1$. We first check to what extend the solution of the linearised system (D.6) satisfies the complete nonlinear equations (D.5). To this end we compute

$$\frac{\partial(\Psi_{1,a}, \Delta\Psi_{1,a})}{\partial(x, y)} = 0$$

$$\frac{\partial(\Psi_{1,a}, \Theta_{1,a})}{\partial(x, y)} = \tfrac{1}{2}a \sin(2y).$$

We see a 'new' function $\sin(2y)$ show up. Note, however, that neither the Laplace operator Δ nor the partial derivative ∂_x, applied to this function, yields any new functions. So if we want to restrict to a 3-dimensional state space, it seems feasible to consider approximating solutions of the form

$$\Psi = X(t)\Psi_{1,a};$$
$$\Theta = Y(t)\Theta_{1,a} - Z(t) \sin(2y).$$

When substituting in the equation of motion (D.5) the only essentially new term is

$$X(t)Z(t)\frac{\partial(\Psi_{1,a}, \sin(2y))}{\partial(x, y)} = X(t)Z(t)2a \cos(ax) \sin y \cos(2y)$$

$$= X(t)Z(t)(-a \cos(ax)(\sin y - \sin(3y))$$

$$= X(t)Z(t)(-a\Theta_{1,a} + a \cos(ax) \sin(3y)).$$

So, the only term outside the basis specified so far, is

$$a X(t)Z(t) \cos(ax) \sin(3y) \tag{D.9}$$

and it is this term that we delete in order to get a projection on a 3-dimensional state space. We have to admit that here there is some arbitrariness: we also could have deleted the whole new term. What we did here is carry out an orthogonal projection with respect to the inner product in which the basis functions for the Fourier decomposition are orthogonal. And this choice is inspired by the fact that these basis functions of the Fourier decomposition also are the eigenfunctions of the linearised system (D.6). After carrying out the substitution in (D.5) (including this deletion) we get:

$$\partial_t(-(a^2 + 1)X(t)\Psi_{1,a}) = \nu(a^2 + 1)^2 X(t)\Psi_{1,a} - acY(t)\Psi_{1,a}$$

$$\partial_t(Y(t)\Theta_{1,a} - Z(t) \sin(2y)) = -\frac{1}{2}a X(t)Y(t) \sin(2y) - a X(t)Z(t)\Theta_{1,a}$$

$$+ \frac{1}{\pi}aT X(t)\Theta_{1,a} - \kappa(a^2 + 1)Y(t)\Theta_{1,a}$$

$$+ 4\kappa Z(t) \sin(2y),$$

which, by taking together the terms with $\Psi_{1,a}$, $\Theta_{1,a}$, and $-\sin(2y)$, yields the following equations for X, Y, and Z.

$$\partial_t X = -\nu(a^2 + 1)X + ac\,(a^2 + 1)^{-1}Y$$
$$\partial_t Y = -aXZ + \tfrac{1}{\pi}aT\,X - \kappa(a^2 + 1)Y$$
$$\partial_t Z = \tfrac{1}{2}a\,XY - 4\kappa Z.$$

This is the 3-dimensional system of differential equations as announced. They don't yet have the format in which the Lorenz equations are usually shown. We now carry out a few changes to obtain the standard Lorenz format, displayed below as (D.11).

As a first step we divide the right-hand sides of all three equations by $\kappa(a^2 + 1)$, which ensures that the coefficient of Y in the second equation equals -1. Such a multiplication only makes the system faster or slower, but all other properties of the solutions, such as being (quasi-) periodic or chaotic, and the like, will not be changed by this. This gives:

$$X' = -\frac{\nu}{\kappa}X + \frac{ac}{\kappa(a^2 + 1)^2}Y$$
$$Y' = -\frac{a}{\kappa(a^2 + 1)}XZ + \frac{aT}{\kappa\pi(a^2 + 1)}X - Y$$
$$Z' = \frac{a}{2\kappa(a^2 + 1)}XY - \frac{4}{a^2 + 1}Z. \tag{D.10}$$

Next we carry out scaling substitutions of the form $X = Ax$, $Y = By$ and $Z = Cz$ to get the system on the standard form

$$x' = \sigma y - \sigma x$$
$$y' = rx - y - xz$$
$$z' = -bz + xy; \tag{D.11}$$

see (1.28) in §1.3. It turns out that for all positive values of the parameters s ν, κ, a, c, and T, there exists a unique choice for corresponding values of A, B, C, σ, r, and b. In this way the physical meaning of the coefficients in the Lorenz equations (D.11) can also be determined.

To show how A, B, C, σ, r, and b should be chosen, we first carry out the scaling substitutions $X = Ax$, and soon, in (D.10). This gives

$$x' = -\frac{\nu}{\kappa}x + \frac{acB}{\kappa(a^2 + 1)A}y$$
$$y' = -\frac{aAC}{\kappa(a^2 + 1)B}xz + \frac{aTA}{\kappa\pi(a^2 + 1)B}x - y$$
$$z' = \frac{aAB}{2\kappa(a^2 + 1)C}xy - \frac{4}{a^2 + 1}z.$$

Fig. D.1 Typical evolution of the Lorenz system (D.11), projected on the (x, z)-plane; compare Figure 1.23.

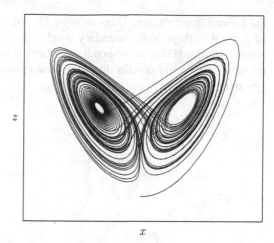

Thus we find as necessary conditions to get the systems (D.10) and (D.11) the same for given v, κ, a, c, and T:

$$\sigma = \frac{v}{\kappa};$$

$$\frac{v}{\kappa} = \frac{ac}{\kappa(a^2 + 1)} \frac{B}{A} \Rightarrow \frac{B}{A} = \frac{v(a^2 + 1)^2}{ac};$$

$$r = \frac{aT}{\kappa\pi(a^2 + 1)} \frac{A}{B} = \frac{Ta^2c}{\kappa v\pi(a^2 + 1)^3} = \frac{T}{T_{1,a}},$$

where $T_{1,a}$ is the critical temperature (D.8). Moreover,

$$1 = \frac{a}{\kappa(a^2 + 1)} \frac{AC}{B} \Rightarrow C = \frac{\kappa v(a^2 + 1)^3}{a^2c};$$

$$1 = \frac{a}{2\kappa(a^2 + 1)} \frac{AB}{C} \Rightarrow AB = \frac{2\kappa^2 v(a^2 + 1)^4}{a^3c}.$$

Inasmuch as both B/A and AB are determined (and positive), this means that the (positive) values of A and B are uniquely determined. Next

$$b = \frac{4}{a^2 + 1}.$$

Concerning the latter equation we recall that the most feasible value is $a^2 = \frac{1}{2}$, which leads to $b = 8/3$, which is the value mostly given to this coefficient.

Although for the value of b in the standard Lorenz equations (D.11) there is a natural choice, this is not the case for r. For r, at least in the case that leads to the Lorenz attractor, one chooses $r = 28$. We note, however, that for this value of r, even with periodic boundary conditions, several basis solutions of the linearised system (D.6), corresponding to higher values of n, have already become unstable. This casts certain doubts on the choice $r = 28$ for the 3-dimensional approximation.

Appendix E
Guide to the literature

In this appendix we present a brief guide to the literature.

E.1 General references

The items [2–5, 7–13, 15–17] of the bibliography form general contributions to the theory of nonlinear dynamical systems at a textbook level.

Classics. The references Arnold [2] and Guckenheimer and Holmes [5] are classics in the dynamical systems field, which both give a thorough connection between the theories of ordinary differential equations and of dynamical systems, with many applications. The background mathematics necessarily contains quite a lot of geometry. These works surely have inspired a wealth of research in the dynamical systems area and both still well deserve reading.

Palis and De Melo [9] is somewhat more mathematically oriented and stresses the geometrical point of view even more, focusing on the concept of genericity in connection to transversality. One of its highlights is a treatment of the Kupka–Smale theorem that describes generic properties of dynamical systems in terms of its critical elements (equilibria, periodic evolutions, transversality of stable and unstable manifolds). Using the Palis inclination lemma, a proof is given of the Hartmann–Grobmann linearisation theorem.

The references [4] by Devaney, [7] by Hirsch, Smale, and Devaney, and [12] by Peitgen, Jürgens, and Saupe are rather introductory to the field. The first of these mainly deals with discrete time dynamics, that is, with maps, focusing on the logistic family. The second starts from ordinary differential equations, also stressing the difference between linear and nonlinear dynamics and treating the Poincaré–Bendickson theorem, among other things. An introduction to various areas of applications (predator–prey systems, electrical networks, simple mechanical systems, etc.) also is contained. The third reference more directly addresses fractal sets and their various notions of dimension, also discussing how fractals occur in chaos theory, in theoretically more difficult (and often undecided) situations restricting to numerical illustration.

State of the art. The references [3, 8, 11, 13, 14, 16, 17] together give a broad overview of the state of the art in the early 1990s. Katok and Hasselblatt [8] deal with many fundamental aspects of the field, also introducing elements of ergodic theory, both for discrete and continuous time systems (for instance, dealing with Anosov and Axiom A systems) often pointing in directions left open in Chapter 4 and Appendix C. Robinson [13] has a similar depth and thoroughness, more directed towards ordinary differential equations and with an interest for the closing lemma, the existence of invariant manifolds, and the Hartman–Grobman linearisation theorem.

Ott et al. [10, 11] contain a general and broad introduction to dynamics and in particular chaos oriented towards physics. Strogatz [15] presents the theory with a focus on synchronisation, which is one of the resonance phenomena met in the main text. A link between resonance in chaos is presented in [6].

Verhulst [16] and Wiggins [17] mainly deal with the dynamics of ordinary differential equations and are more directed towards applications, where both contain a thorough mathematical background. Moreover, the former is interested in averaging and perturbation theory, whereas the latter is more focused on detecting transversal homoclinic points with the Melnikov criterion.

A broad variety of material is given by a number of sources. we like to mention Broer, Dumortier, Van Strien, and Takens [3], which contains a variety of material used in PhD courses on dynamical systems over a number of years, and which deals with quite a few concepts also treated in the present book. Similarly Afraimovich and Hsu [1] contains a number of lectures in broads perspective. Finallly we like to mention Shilnikov, Turaev, and Chua [14], which contains a broad entrance to the entire theory as well, where Part II pays a lot of attention to bifurcations.

Handbooks and Encyclopædia

For further inspection we include a few handbooks on dynamical systems: [33–36], by Hasselblatt and Katok, Fiedler, and by Broer, Hasselblatt, and Takens. In the latter reference we like to point at the introductory chapter [38] by Broer and Takens. We also mention the *Russian Encyclopædia* [42] edited by Gamkrelidze, as well as the *Encyclopædia on Complexity* [47] edited by Meyers.

E.2   On ergodic theory

In several chapters of the present book we encountered measure-theoretical aspects. One important issue is the existence of measures that are invariant under the dynamics and of ensuing properties, such as ergodic and other mixing properties. Background material at a graduate textbook level can be found in Mañé [18] and in Walters [19].

Handbooks and Encyclopædia

At the level of handbooks we refer to several contributions in Hasselblatt and Katok [33, 34]. In the *Russian Encyclopædia* [42] we refer to Volume 2 edited by Sinai [43].

E.3   On Hamiltonian dynamics

An important special class of dynamical system is formed by Hamiltonian systems, used for modelling the dynamics of frictionless mechanics. Although we pay some attention to this broad field, for a more systematic study, we refer to [20–24].

Abraham and Marsden [20] and Arnold [21] form classical references that contain early accounts of the mathematics behind classical mechanics. The style of the former of these focuses on the mathematical formalism, whereas the latter is more directed towards examples and applications. Both aim at the interesting research problems of integrable and nearly integrable Hamiltonian systems, for example, regarding the dynamics of the rigid body or of the N-body problem.

Cushman and Bates, [22] continue this line of thought, dealing with the classically known examples of integrable Hamiltonian systems and describing the global geometrial structure of the state space. One of their interests concerns examples such as the spherical pendulum as met in the present book and the nontriviality of the ensuing bundles of Lagrangian tori.

The textbooks of Marsden and Ratiu [23] and Meyer and Hall [24] are more introductory in related directions.

Handbooks and Encyclopædia

Concerning handbooks we further refer to Hasselblatt and Katok [33,34]. We would also like to mention the *Russian Encyclopædia* [42], in particular Volume 3, edited by Arnold, Kozlov and Neishtadt [44] as well as the *Encyclopædia on Complexity* [47], in particular the contributions by Broer and Hanßmann [49], Chierchia [50], and Hanßmann [51].

E.4   On normal forms and bifurcations

The present book emphasises certain conceptual aspects of dynamical systems theory. However, when classifying bifurcations or studying applications, more computational methods are needed, based on algorithmic and often formal techniques. In the direction of normal form and bifurcation theory we like to point at [25–32]. Regarding bifurcation theory the present book contains a brief sketch in Appendix C.

Local bifurcation theory of stationary evolutions (equilibria, fixed points) and of periodic points of maps, as well as their invariant manifolds, form the subject matter of Chow and Hale [25] and Chow, Li, and Wang [26]. The former of these references focuses on the uniform contraction principle and the ensuing existence theory for stable and unstable manifolds.[1] The latter covers a great many results regarding equilibria of planar vector fields and maps, in particular their bifurcations. Periodic orbits in vector fields theoretically are covered by considering their Poincaré maps.

A more applied and computational reference is given by Kuznetsov [29]. The same holds for [32] by Sanders, Verhulst, and Murdock, who exploit the relationship between normal forms and averaging and also cover periodic orbits in vector fields in concrete systems. The latter of the two references also deals with the averaging theorem, estimating approximation properties of specific evolutions in terms of perturbation parameters.

Ciocci, Litvak-Hinenzon and Broer [27] extend the bifurcation theory of stationary and periodic evolutions to the quasi-periodic case; compare Appendix C. Palis and Takens [30] consider aspects of homoclinic bifurcations, investigating the dynamics near homoclinic tangency, in certain cases discovering 'small' Hénon-like attractors. Both references contain advanced course material.

In Hanßmann [28] a wide overview is given in which a major part of the above programme concerning local and global bifurcations, is described for the Hamiltonian case. Ruelle [31] presents a bird's-eye view of a mathematical physicist over the entire area of dynamical systems and bifurcation theory.

Handbooks and Encyclopædia

We further refer to the handbook Broer, Hasselblatt, and Takens [36], in particular to the chapters by Takens and Vanderbauwhede [41] and by Homburg and Sandstede [39]. In the *Russian Encyclopædia* [42] we recommend Volumes 5 and 6, both edited and (co-) authored by Arnold [45, 46] and in the *Encyclopædia on Complexity* [47] the contributions by Broer [48], by Broer and Hanßmann [49], by Hanßmann [51], Niederman [52], Verhulst [53], and Walcher [54] as well.

---

[1] Compare the normally hyperbolic invariant manifold theory [143] mentioned several times in the present book.

Bibliography

General References

1. V. Afraimovich and S.-B. Hsu, *Lectures on chaotic Dynamical Systems*, AMS/IP Studies in Advanced Mathematics, International Press, 2003.
2. V.I. Arnold, *Geometrical Methods in the Theory of Ordinary Differential Equations*, Springer-Verlag 1983.
3. H.W. Broer, F. Dumortier, S.J. van Strien, and F. Takens, Structures in dynamics, finite dimensional deterministic studies, In E.M. de Jager and E.W.C. van Groesen (eds.) *Studies in Mathematical Physics II*, North-Holland 1991.
4. R.L. Devaney, *An Introduction to Chaotic Dynamical Systems*, Benjamin, 1986; reprint Addison-Wesley 1989.
5. J. Guckenheimer and P. Holmes, *Nonlinear Oscillations, Dynamical Systems, and Bifurcations of Vector Fields*, Appl. Math. Sciences, Vol. **42**, Springer-Verlag 1983.
6. G. Haller, *chaos near Resonance*, Springer-Verlag 1999.
7. M.W. Hirsch, S. Smale, and R.L. Devaney, *Differential Equations, Dynamical Systems & an Introduction to Chaos*, Second Edition, Academic Press 2004.
8. A. Katok and B. Hasselblatt, *Introduction to the Modern Theory of Dynamical Systems*, Cambridge University Press 1995.
9. J. Palis and W.C. de Melo, *Geometric Theory of Dynamical Systems*, Springer-Verlag 1982.
10. E. Ott, *Chaos in Dynamical Systems*, Cambridge University Press 1993.
11. E. Ott, T. Sauer, and J.A. Yorke (eds.), *Coping with Chaos*, Wiley Series in Nonlinear Science, Wiley 1994.
12. H.-O. Peitgen, H. Jürgens, and D. Saupe, *Chaos and Fractals*, Springer-Verlag 1992.
13. R.C. Robinson, *Dynamical Systems*, CRC Press 1995.
14. L. P. Shilnikov, A. Shilinkov, D. Turaev, and L. Chua, *Methods of Qualitative Theory in Nonlinear Dynamics* Parts I and II, World Scientific 1998 and 2001.
15. S.H. Strogatz, *Nonlinear Dynamics and Chaos*, Addison-Wesley 1994.
16. F. Verhulst, *Nonlinear Differential Equations and Dynamical Systems*, Second Edition, Springer-Verlag 1996.
17. S. Wiggins, *Introduction to Applied Nonlinear Dynamical Systems and Chaos*, Springer-Verlag 1990.

On Ergodic Theory

18. R. Mañé, *Ergodic Theory of Differentiable Dynamics*, Springer-Verlag 1983.
19. P. Walters, *An Introduction to Ergodic Theory*, Springer-Verlag 1981.

On Hamiltonian Dynamics

20. R.H. Abraham and J.E. Marsden, *Foundations of Mechanics*, Second Edition, with an appendix by A.N. Kolmogorov, Benjamin/Cummings 1978.
21. V.I. Arnold, *Mathematical Methods of Classical Mechanics*, GTM **60** Springer-Verlag 1978; Second Edition, Springer-Verlag 1989.
22. R.H. Cushman and L.M. Bates, *Global Aspects of Classical Integrable Systems*, Birkhäuser 1997.
23. J.E. Marsden and T.S. Ratiu, *Introduction to Mechanics and Symmetry*, TAM **17**, Springer-Verlag 1994.
24. K.R. Meyer and G.R. Hall, *Introduction to Hamiltonian Dynamical Systems and the N-Body Problem*, Appl Math Sciences Vol. **90**, Springer-Verlag 1992.

On Bifurcation Theory and Normal Form Theory

25. S.-N. Chow and J.K. Hale, *Methods of Bifurcation Theory*, Springer-Verlag 1982.
26. S-N. Chow, C. Li, and D. Wang, *Normal Forms and Bifurcation of Planar Vector Fields*, Cambridge University Press 1994.
27. M.C. Ciocci, A. Litvak-Hinenzon, and H.W. Broer, Survey on dissipative KAM theory including quasi-periodic bifurcation theory based on lectures by Henk Broer. In: J. Montaldi and T. Ratiu (eds.): *Geometric Mechanics and Symmetry: the Peyresq Lectures*, LMS Lecture Notes Series, **306**. Cambridge University Press 2005, 303–355.
28. H. Hanßmann, *Local and Semi-Local Bifurcations in Hamiltonian Dynamical Systems. Results and Examples*, Lecture Notes in Mathematics **1893**, Springer-Verlag 2007.
29. Yu.A. Kuznetsov, *Elements of Applied Bifurcation Theory*, Third Edition, Appl. Math. Sciences, Vol. **112**, Springer-Verlag 2004.
30. J. Palis and F. Takens, *Hyperbolicity & Sensitive Chaotic Dynamics at Homoclinic Bifurcations*, Cambridge University Press 1993.
31. D. Ruelle, *Elements of Differentiable Dynamics and Bifurcation Theory*. Academic Press 1989.
32. J.A. Sanders and F. Verhulst, *Averaging Methods in Nonlinear Dynamical Systems*. Appl Math Sciences **59**, Springer-Verlag 1985.
    J.A. Sanders, F. Verhulst, and J. Murdock, *Averaging Methods in Nonlinear Dynamical Systems*, Revised Second Edition, Appl Math Sciences **59**, Springer-Verlag 2007.

Handbooks and Encyclopædia

33. B. Hasselblatt and A. Katok (eds.), *Handbook of Dynamical Systems* Vol. **1A**, North-Holland 2002.
34. B. Hasselblatt and A.I. Katok (eds.), *Handbook of Dynamical Systems* Vol. **1B**, North-Holland 2006.
35. B. Fiedler (ed.), *Handbook of Dynamical Systems* Vol. **2**, North-Holland 2002.
36. H.W. Broer, B. Hasselblatt, and F. Takens (eds.), *Handbook of Dynamical Systems* Vol. **3**, North-Holland 2010.
37. H.W. Broer and M.B. Sevryuk, KAM theory: Quasi-periodicity in dynamical systems. In: H.W. Broer, B. Hasselblatt, F. Takens (eds.), *Handbook of Dynamical Systems* Vol. **3**, North-Holland 2010.

38. H.W. Broer and F. Takens, Preliminaries of dynamical systems. In: H.W. Broer, B. Hasselblatt, F. Takens (eds.), *Handbook of Dynamical Systems* Vol. **3**, North-Holland 2010.
39. A.J. Homburg and B. Sandstede, Homoclinic and heteroclinic bifurcations in vector fields. In: H.W. Broer, B. Hasselblatt, F. Takens (eds.), *Handbook of Dynamical Systems* Vol. **3**, North-Holland 2010.
40. F. Takens, Reconstruction theory and nonlinear time series analysis. In: H.W. Broer, B. Hasselblatt, F. Takens (eds.), *Handbook of Dynamical Systems* Vol. **3**, North-Holland 2010.
41. F. Takens and A. Vanderbauwhede, Local invariant manifolds and normal forms. In: H.W. Broer, B. Hasselblatt, F. Takens (eds.), *Handbook of Dynamical Systems* Vol. **3**, North-Holland 2010.
42. R.V. Gamkrelidze (ed.), Dynamical systems I – IX, *Encyclopædia of Mathematical Sciences*, **1, 2, 3, 4, 5, 6, 16, 39, 66**, Springer-Verlag 1988–1995.
43. Ya.G. Sinai (ed.), Dynamical systems II. Ergodic theory with applications to dynamical systems and statistical mechanics. *Encylopædia of Mathematical Sciences*, Vol. **2**, Springer-Verlag 1989.
44. V.I. Arnold, V.V. Kozlov, and A.I. Neishtadt, Mathematical aspects of classical and celestial mechanics. In: V.I. Arnold (ed.), Dynamical Systems III, *Encylopædia of Mathematical Sciences*, Vol.**3**, Springer-Verlag 1988; Third Edition, Springer-Verlag 2006.
45. V.I. Arnold (ed.), Dynamical systems v. bifurcation theory and catastrophe theory, *Encyclopædia of Mathematical Sciences*, Vol. **5**, Springer-Verlag 1994.
46. V.I. Arnol'd, V.A. Vasil'ev, V.V.Goryunov, and O.V. Lyashko, Singularity theory I : Singularities, local and global theory. In: V.I. Arnold (ed.), Dynamical Systems VI, *Encylopædia of Mathematical Sciences*, Vol. **6**, Springer-Verlag 1993.
47. R. Meyers (ed.), *Encyclopædia of Complexity & System Science*, Springer-Verlag 2009, LXXX, 10,370 pp.
48. H.W. Broer, Normal forms in perturbation theory. In: R. Meyers (ed.), *Encyclopædia of Complexity & System Science*, Springer-Verlag 2009, LXXX, 10,370 pp.
49. H.W. Broer and H. Hanßmann, Hamiltonian perturbation theory (and transition to chaos). In: R. Meyers (ed.), *Encyclopædia of Complexity & System Science*, Springer-Verlag 2009, LXXX, 10,370 pp.
50. L. Chierchia, Kolmogorov-Arnold-Moser (KAM) Theory. In: R. Meyers (ed.), *Encyclopædia of Complexity & System Science*, Springer-Verlag 2009, LXXX, 10,370 pp.
51. H. Hanßmann, Dynamics of Hamiltonian systems. In: R. Meyers (ed.), *Encyclopædia of Complexity & System Science*, Springer-Verlag 2009, LXXX, 10,370 pp.
52. L. Niederman, Nekhoroshev Theory. In: R. Meyers (ed.), *Encyclopædia of Complexity & System Science*, Springer-Verlag 2009, LXXX, 10,370 pp.
53. F. Verhulst, Perturbation analysis of parametric resonance. In: R. Meyers (ed.), *Encyclopædia of Complexity & System Science*, Springer-Verlag 2009, LXXX, 10,370 pp.
54. S. Walcher, Convergence of Perturbative Expansions. In: R. Meyers (ed.), *Encyclopædia of Complexity & System Science*, Springer-Verlag 2009, LXXX, 10,370 pp.

Further references

55. D. Aeyels, Generic observability of differentiable systems, *SIAM J. Control Optim* **19** (1981), 595–603.
56. V.I. Arnold, On the classical perturbation theory and the stability problem of the planetary system, *Dokl. Akad. Nauk SSSR* **145** (1962), 487–490.
57. V.I. Arnold, Small divisors I: On mappings of the circle onto itself, *Izvestiya Akad. Nauk SSSR, Ser. Mat.* **25** (1961), 21–86 (in Russian); English translation: *Amer. Math. Soc. Transl., Ser.* 2 **46** (1965), 213–284; Erratum: *Izvestiya Akad. Nauk SSSR, Ser. Mat.* **28** (1964), 479–480 (in Russian).

58. V.I. Arnold, Proof of a theorem by A.N. Kolmogorov on the persistence of conditionally periodic motions under a small change of the Hamilton function, *Russian Math. Surveys* **18** (5) (1963), 9–36 (English; Russian original).
59. V.I. Arnold and A. Avez, *Probèmes Ergodiques de la Mécanique classique*, Gauthier-Villars, 1967; *Ergodic problems of classical mechanics*, Benjamin 1968.
60. V.I. Arnold and B.A. Khesin, *Topological Methods in Hydrodynamics*, Springer-Verlag 1998.
61. R. Badii and A. Politi, Hausdorff dimension and uniformity factor of strange attractors, *Phys. Rev. Letters* **52** (1984), 1661–1664.
62. R. Badii and A. Politi, Statistical description of chaotic attractors: The dimension function, *J. Stat. Phys.* **40** (1985), 725–750.
63. C. Baesens, J. Guckenheimer, S. Kim, and R.S. MacKay, Three coupled oscillators: Mode-locking, global bifurcation and toroidal chaos, *Physica D* **49**(3) (1991), 387–475.
64. J. Barrow-Green, *Poincaré and the Three Body Problem*, History of Mathematics, Vol. 11, American Mathematical Society; London Mathematical Society 1997.
65. G.K. Batchelor, *An introduction to Fluid Dynamics*, Cambridge University Press 1967.
66. M. Benedicks and L. Carleson, On iterations of $1 - ax^2$ on $(-1, 1)$, *Ann. Math.* **122** (1985), 1–24.
67. M. Benedicks and L. Carleson, The dynamics of the Hénon map, *Ann. Math.* **133** (1991), 73–169.
68. M. Bennett, M.F. Schatz, H. Rockwood, and K. Wiesenfeld, Huygens's clocks, *Proc. R. Soc. Lond. A* **458** (2002), 563–579.
69. P. Blanchard, R.L. Devaney, and G.R. Hall, *Differential Equations*, Third Edition, Thomson Brooks/Cole 2006.
70. B.L.J. Braaksma and H.W. Broer, On a quasi-periodic Hopf bifurcation, *Ann. Institut Henri Poincaré, Analyse non linéaire* **4**(2) (1987), 115–168.
71. D.R. Brillinger, *Time series*, McGraw-Hill 1981.
72. M. Brin and A. Katok, On local entropy. In: J. Palis (ed.), *Geometric Dynamics*, Lecture Notes in Mathematics **1007** (1983), 30–38.
73. H.W. Broer, De chaotische schommel, *Pythagoras* **35**(5) (1997), 11–15.
74. H.W. Broer, *Meetkunde en Fysica*, Epsilon Uitgaven **44** 1999.
75. H.W. Broer, Coupled Hopf-bifurcations: Persistent examples of n-quasiperiodicity determined by families of 3-jets, *Astérisque* **286** (2003), 223–229.
76. H.W. Broer, KAM theory: The legacy of Kolmogorov's 1954 paper, *Bull. Amer. Math. Soc. (New Series)* **41**(4) (2004), 507–521.
77. H.W. Broer, M.-C. Ciocci, and H. Hanßmann, The quasi-periodic reversible Hopf bifurcation, *Intern. J. Bifurcation Chaos*, **17**(8) (2007), 2605–2623.
78. H.W. Broer, M.-C. Ciocci, H. Hanßmann, and A. Vanderbauwhede, Quasi-periodic stability of normally resonant tori, *Physica D* **238** (2009), 309–318.
79. H.W. Broer, R.H. Cushman, F. Fassò, and F. Takens, Geometry of KAM tori for nearly integrable Hamiltonian systems. *Ergod. Th. & Dynam. Sys.* **27** (2007), 725–741.
80. H.W. Broer, H. Hanßmann, and J. Hoo, The quasi-periodic Hamiltonian Hopf bifurcation, *Nonlinearity* **20** (2007), 417–460.
81. H.W. Broer, H. Hanßmann, and J. You, Bifurcations of normally parabolic tori in Hamiltonian systems, *Nonlinearity* **18** (2005), 1735–1769.
82. H.W. Broer, H. Hanßmann, and J. You, Umbilical torus bifurcations in Hamiltonian systems, *J. Differential Equations* **222** (2006), 233–262.
83. H.W. Broer, H. Hanßmann, À. Jorba, J. Villanueva, and F.O.O. Wagener, Normal-internal resonances in quasi-periodically forced oscillators: a conservative approach, *Nonlinearity* **16** (2003), 1751–1791.
84. H.W. Broer, J. Hoo, and V. Naudot, Normal linear stability of quasi-periodic tori, *J. Differential Equations* **232** (2007), 355–418.
85. H.W. Broer and G.B. Huitema, A proof of the isoenergetic KAM-theorem from the "ordinary" one, *J. Differential Equations* **90** (1991), 52–60.
86. H.W. Broer and G.B. Huitema, Unfoldings of quasi-periodic tori in reversible systems, *J. Dynamics Differential Equations* **7** (1995), 191–212.

87. H.W. Broer, G.B. Huitema, and M.B. Sevryuk, *Quasi-periodic motions in Families of Dynamical Systems*, Lecture Notes in Mathematics **1645**, Springer-Verlag 1996.

88. H.W. Broer, G.B. Huitema, F. Takens, and B.L.J. Braaksma, Unfoldings and bifurcations of quasi-periodic tori, *Memoirs Amer. Math. Soc.* **421**, 1990.

89. H.W. Broer, M. Golubitsky, and G. Vegter, The geometry of resonance tongues: a singularity theory approach, *Nonlinearity* **16** (2003), 1511–1538.

90. H.W. Broer and B. Krauskopf, Chaos in periodically driven systems. In B. Krauskopf and D. Lenstra (eds.), *Fundamental Issues of Nonlinear Laser Dynamics,* American Institute of Physics Conference Proceedings **548** (2000), 31–53.

91. H.W. Broer, C. Simó, and J.C. Tatjer, Towards global models near homoclinic tangencies of dissipative diffeomorphisms, *Nonlinearity* **11**(3) (1998), 667–770.

92. H.W. Broer, C. Simó, and R. Vitolo, Bifurcations and strange attractors in the Lorenz-84 climate model with seasonal forcing, *Nonlinearity* **15**(4) (2002), 1205–1267.

93. H.W. Broer, C. Simó, and R. Vitolo, The Hopf-saddle-node bifurcation for fixed points of 3D diffeomorphisms, the Arnold resonance web, *Bull. Belgian Math. Soc. Simon Stevin* **15** (2008), 769–787.

94. H.W. Broer, C. Simó, and R. Vitolo, The Hopf-saddle-node bifurcation for fixed points of 3D diffeomorphisms, analysis of a resonance 'bubble', *Physica D* **237** (2008), 1773–1799.

95. H.W. Broer and F. Takens, Unicity of KAM tori, *Ergod. Th. & Dynam. Sys.* **27** (2007), 713–724.

96. H.W. Broer and F.M. Tangerman, From a differentiable to a real analytic perturbation theory, applications to the Kupka Smale theorems, *Ergod. Th. & Dynam. Sys.* **6** (1986), 345–362.

97. L.E.J. Brouwer, Zur Analysis Situs, *Math. Ann.* **68** (1910), 422–434.

98. M. Casdagli, Chaos and deterministic *versus* stochastic non-linear modelling, *J. R. Statist. Soc.* **54 B** (1992), 303–328.

99. M. Casdagli and S. Eubank (eds.), *Nonlinear Modelling and Forecasting*, Addison-Wesley 1992.

100. A. Cayley, The Newton-Fourier imaginary problem, *Amer. J. Math.* **2** (1879), p. 97.

101. A. Chenciner, Bifurcations de points fixes elliptiques. I. Courbes invariantes, *Publ. Math. IHÉS* **61** (1985), 67–127.

102. A. Chenciner, Bifurcations de points fixes elliptiques. II. Orbites périodiques et ensembles de Cantor invariants, *Invent. Math.* **80** (1985), 81–106.

103. A. Chenciner, Bifurcations de points fixes elliptiques. III. Orbites périodiques de "petites" périodes et élimination résonnante des couples de courbes invariantes, *Publ. Math. IHÉS* **66** (1988), 5–91.

104. P. Collet and J.-P. Eckmann, *Iterated Maps on the Interval as Dynamical Systems*, Birkhäuser 1980.

105. R. Courant, *Differential and Integral Calculus*, 1937; reprinted Wiley 1988.

106. J.P. Crutchfield, J.D. Farmer, N.H. Packerd, and R.S. Shaw, Chaos, *Sci Amer* Dec. 1986, 38–49.

107. R.H. Cushman, H.R. Dullin, A. Giacobbe, D.D. Holm, M. Joyeux, P. Lynch, D.A. Sadovskíí, and B.I. Zhilinskíí, CO_2 molecule as a quantum realization of the $1 : 1 : 2$ resonant swing-spring with monodromy, *Phys. Rev. Lett.* **93** (2004), 024302.

108. C.D. Cutler, Some results on the behaviour and estimation of the fractal dimension of distributions on attractors, *J. Stat. Phys.* **62** (1991), 651–708.

109. C.D. Cutler and D.T. Kaplan (eds.), Nonlinear dynamics and time series, *Fields Inst. Communications* **11**, Amer. Math. Soc., 1997.

110. D. van Dantzig, *Studiën over topologische algebra*, dissertatie Rijksuniversiteit Groningen, 1931.

111. M. Denker and G. Keller, On U-statistics and v. Mises' statistics for weakly dependent processes, *Z. Wahrscheinlichkeitstheorie verw. Gebiete* **64** (1983), 505–522.

112. M. Denker and G. Keller, Rigorous statistical procedures for data from dynamical systems, *J. Stat. Phys.* **44** (1986), 67–93.

113. F. Diacu and P. Holmes, *Celestial Encounters. The Origins of Chaos and Stability*, Princeton University Press 1996.

114. L.J. Díaz, I.L. Rios, and M. Viana, The intermittency route to chaotic dynamics. In: H.W. Broer, B. Krauskopf, G. Vegter (eds.), *Global Analysis of Dynamical Systems* IoP (2001), 309–328.
115. C. Diks, *Nonlinear Time Series Analysis*, World Scientific 1999.
116. C. Diks, W.R. van Zwet, F. Takens and J. DeGoede, Detecting differences between delay vector distributions, *Phys. Rev.* **53 E** (1996), 2169–2176.
117. J.J. Duistermaat, On global action-angle coordinates, *Comm. Pure Appl. Math.* **33** (1980), 687–706.
118. J.J. Duistermaat and W. Eckhaus, *Analyse van Gewone Differentiaalvergelijkingen*, Epsilon Uitgaven **33** 1995.
119. J.P. Eckmann, S. Oliffson Kamphorst, D. Ruelle, and S. Ciliberto, Liapunov exponents from time series. *Phys. Rev.* **34 A** (1986), 4971–4979.
120. J.P. Eckmann and D. Ruelle, Ergodic theory of chaos and strange attractors, *Rev. Mod. Phys.* **57** (1985), 617–656.
121. R. Engelking, *General Topology*, seecond edition, Heldermann Verlag 1989.
122. K.J. Falconer, *The Geometry of Fractal Sets*, Cambridge University Press 1985.
123. K.J. Falconer, *Fractal Geometry*, Wiley 1990.
124. R.P. Feynman, R.B. Leighton, and M. Sands, *The Feynman Lectures on Physics* Vols. **1**, **2**, and **3**, Addison-Wesley 1963, 64 and 65.
125. A. Galka, *Topics in Nonlinear Time Series Analysis, with Implications for EEG Analysis*, World Scientific 2000.
126. P. Grassberger and I. Procaccia, Characterisation of strange attractors, *Phys. Rev. Lett.* **50** (1983), 346–349.
127. P. Grassberger and I. Procaccia, Estimation of Kolmogorov entropy from a chaotic signal, *Phys. Rev.* **28 A** (1983), 2591–2593.
128. P. Grassberger and I. Procaccia, Dimensions and entropies of strange attractors from a fluctuating dynamics approach,*Physica* **13 D** (1984), 34–54.
129. J. Guckenheimer and R.F. Williams, Structural stability of Lorenz attractors, *Publ. Math. IHÉS*, **50** (1979), 59–72.
130. J.K. Hale, *Ordinary Differential Equations*, Second Edition, Krieger 1980.
131. P.R. Halmos, *Measure Theory*, Van Nostrand 1950.
132. T.C. Halsey, M.H. Jensen, L.P. Kadanoff, I. Procaccia, and B.I. Shraiman, Fractal measures and their singularities: the characterisation of strange sets, *Phys. Rev.* **33 A** (1986), 1141–1151.
133. H. Hanßmann, The quasi-periodic centre-saddle bifurcation, *J. Diff Eq* **142** (1998), 305–370.
134. K.P. Hart, J. Nagata, and J.E. Vaughan (eds.), *Encyclopædia of General Topology*, North-Holland 2004.
135. F. Hausdorff, *Set Theory*, Second Edition, Chelsea 1962. [German original: Veit (Leipzig) 1914.]
136. A. van Helden, *Measuring the Universe, Cosmic Dimensions from Aristarchus to Halley*, The University of Chicago Press 1985.
137. M. Hénon, A two-dimensional map with a strange attractor, *Comm. Math. Phys.* **50** (1976), 69–77.
138. H.G.E. Hentschel and I. Procaccia, The infinite number of generalised dimensions of fractals and strange attractors, *Physica* **8 D** (1983), 435–444.
139. H. Herrlich, *Axiom of Choice*, Springer-Verlag 2006.
140. M.R. Herman, Mesure de Lebesgue et nombre de rotation. In: J. Palis, M do Carmo (eds.), *Geometry and Topology*, Lecture Notes in Mathematics **597**, Springer-Verlag 1977, 271–293.
141. M.R. Herman, Sur la conjugaison différentiable des difféomorphismes du cercle à des rotations, *Publ. Math. IHÉS* **49** (1979), 5–233.
142. M.W. Hirsch, *Differential Topology*, Springer-Verlag 1976.
143. M.W. Hirsch, C.C. Pugh, and M. Shub, *Invariant Manifolds*, Lecture Notes in Mathematics **583**, Springer-Verlag 1977.
144. M.W. Hirsch and S. Smale, *Differential Equations, Dynamical Systems, and Linear Algebra*, Academic Press 1974.

145. E. Hopf, Abzweigung einer periodischen Lösung von einer stationären Lösung eines Differentialsystems, *Ber. Math.-Phys. Kl Sächs. Akad. Wiss. Leipzig* **94** (1942), 1–22.

146. E. Hopf, A mathematical example displaying features of turbulence. *Commun. Appl. Math.* **1** (1948), 303–322.

147. C. Huygens, *Œuvres Complètes de Christiaan Huygens, publiées par la Société Hollandaise des Sciences* **16**, Martinus Nijhoff, The Hague 1929, Vol. 5, 241–262; Vol. 17, 156–189.

148. M.V. Jakobson, Absolutely continuous invariant measures for one-parameter families of one-dimensional maps, *Commun. Math. Phys.* **81** (1981), 39–88.

149. H. Kantz, Quantifying the closeness of fractal measures, *Phys. Rev.* **49 E** (1994), 5091–5097.

150. H. Kantz and T. Schreiber, *Nonlinear Time Series Analysis*, Cambridge University Press 1997.

151. J.L. Kaplan and J.A. Yorke, Chaotic behaviour of multidimensional difference equations. In H.-O. Peitgen, H.O. Walter (eds.), *Functional Differential Equations and Approximations of Fixed Points*. Lecture Notes in Mathematics **730**, Springer-Verlag, 1979.

152. J.L. Kelley, *General Topology*, Van Nostrand, 1955; Reprinted GTM **27**, Springer-Verlag 1975.

153. F. Klein, *Vorlesungen über die Entwicklung der Mathematik im 19. Jahrhundert*, Springer-Verlag 1926-7; Reprinted 1979.

154. S.E. Kuksin, *Nearly integrable infinite-dimensional Hamiltonian dynamics*, Lecture Notes in Mathematics **1556** Springer-Verlag 1993.

155. L.D. Landau, On the problem of turbulence, *Akad. Nauk.* **44** (1944), 339.

156. L.D. Landau and E.M. Lifschitz, *Fluid Mechanics*. Pergamon 1959.

157. F. Ledrappier and L.S. Young, The metric entropy of diffeomorphisms, Part I and II, *Ann. Math.* **122** (1985), 509–539 and 540–574.

158. C.M. Linton, *From Eudoxus to Einstein, a History of Mathematical Astronomy* Cambridge University Press 2004.

159. R. de la Llave, A tutorial on KAM Theory. *Proceedings of Symposia in Pure Mathematics* **69** (2001), 175–292.

160. E.N. Lorenz, Deterministic nonperiodic flow, *J. Atmosph. Sci.* **20** (1963), 130–141.

161. E.N. Lorenz, *The Essence of Chaos*, University of Washington Press 1993.

162. O.V. Lukina, F. Takens, and H.W. Broer, Global properties of integrable Hamiltonian systems, *Regular and Chaotic Dynamics* **13**(6) (2008), 588–630.

163. B.B. Mandelbrot, *The Fractal Geometry of Nature*, Freeman 1982.

164. R. Mañé, A proof of the C^1 stability conjecture, *Publ. Math. IHÉS* **66** (1988), 161–210.

165. W.S. Massey, *Algebraic Topology: An Introduction*, GTM **56**, Springer-Verlag 1967.

166. R.M. May, Simple mathematical models with very complicated dynamics, *Nature* **261** (1976), 459–466.

167. W.C. de Melo and S.J. van Strien, *One-Dimensional Dynamics*, Springer-Verlag 1991.

168. J.W. Milnor, On the concept of attractor, *Comm. Math. Phys.* **99** (1985), 177–195.

169. J.W. Milnor, Dynamics in one complex variable, Third Edition, *Ann. Math. Studies* **160** Princeton University Press 2006.

170. M.G.J. Minnaert, *De Natuurkunde van 't Vrije Veld* Vols. **1**, **2** and **3**, Third Edition, ThiemeMeulenhoff 1971.

171. C.W. Misner, K.S. Thorne, and J.A. Wheeler, *Gravitation*, Freeman 1970.

172. L. Mora, and M. Viana, Abundance of strange attractors, *Acta Math.* **171** (1993), 1–71.

173. J.K. Moser, On invariant curves of area-preserving mappings of an annulus, *Nachr. Akad. Wiss. Göttingen, Math.-Phys. Kl. II.* **1** (1962), 1–20.

174. J.K. Moser, On the theory of quasiperiodic motions, *SIAM Rev* **8**(2) (1966), 145–172.

175. J.K. Moser, Convergent series expansions for quasi-periodic motions, *Math. Ann.* **169** (1967), 136–176.

176. J.K. Moser, Lectures on Hamiltonian systems, *Memoirs Amer. Math. Soc.* **81** (1968), 1–60.

177. J.K. Moser, Stable and random motions in dynamical systems, with special emphasis to celestial mechanics, *Ann. Math. Studies* **77** Princeton University Press 1973.

178. N.N. Nekhoroshev, An exponential estimate of the time of stability of nearly-integrable Hamiltonian systems, *Russ. Math. Surv.* **32** (1977), 1–65.

179. N.N. Nekhoroshev, An exponential estimate of the time of stability of nearly integrable Hamiltonian systems. II. In: O.A. Oleinik (ed.): *Topics in Modern Mathematics, Petrovskii Seminar No.5*, Consultants Bureau 1985, 1–58.

180. S.E. Newhouse, J. Palis, and F. Takens, Bifurcations and stability of families of diffeomorphisms. *Publ. Math. IHÉS* **57** (1983), 5–71.

181. S.E. Newhouse, D. Ruelle, and F. Takens, Occurrence of strange Axiom A attractors near quasi-periodic flows on $\mathbb{T}^m$, $m \leq 3$, *Commun. Math. Phys.* **64** (1978), 35–40.

182. J.R. van Ommen, M.-O. Coppens, J.C. Schouten, and C.M. van den Bleek, Early warnings of agglomeration in fluidised beds by attractor comparison, *A.I.Ch.E. Journal* **46** (2000), 2183–2197.

183. V.I. Oseledets, A multiplicative ergodic theorem, *Trans. of the Moscow Math. Soc.* **19** (1968), 197–221.

184. J. Oxtoby, *Measure and Category.* Springer-Verlag 1971.

185. N.H. Packard, J.P. Crutchfield, J.D. Farmer, and R.S. Shaw, Geometry from time series, *Phys. Rev. Letters* **45** (1980), 712–716.

186. J. Palis, On the C^1 Ω-stability conjecture, *Publ. Math. IHÉS* **66** (1988), 211–215.

187. H.-O. Peitgen (ed.), *Newton's Method and Dynamical Systems*, Kluwer Academic. 1989.

188. Ya.B. Pesin, *Dimension Theory in Dynamical Systems*, University of Chicago Press 1997.

189. Ya.B. Pesin and H. Weiss, The multi-fractal analysis of Gibbs measures: Motivation, mathematical foundation, and examples, *Chaos* **7** (1997), 89–106.

190. M. St. Pierre, Topological and measurable dynamics of Lorenz maps, *Dissertationes Math.* **382** (1999), 56 pp.

191. H. Poincaré, Sur le problème des trois corps et les équations de la dynamique, *Acta Math.* **13** (1980), 1–270.

192. B. van der Pol, De amplitude van vrije en gedwongen triode-trillingen, *Tijdschr. Ned. Radiogenoot.* **1** (1920), 3–31.

193. B. van der Pol, The nonlinear theory of electric oscillations, *Proc. of the Inst. of Radio Eng.* **22** (1934), 1051–1086; Reprinted in: *Selected Scientific Papers*, North-Holland 1960.

194. J. Pöschel, Integrability of Hamiltonian systems on Cantor sets. *Comm. Pure Appl. Math.* **35**(5) (1982), 653–696.

195. J. Pöschel. A lecture on the classical KAM theorem, *Proceedings of Symposia in Pure Mathematics* **69** (2001), 707–732.

196. M.B. Priestley, *Spectral Analysis and Time Series*, Academic Press, 1981.

197. A. Quarteroni, R. Sacco, and F. Saleri, *Numerical Mathematics*, Second Edition, Springer-Verlag 2007.

198. Lord J.W.S. Rayleigh, On convective currents in a horizontal layer of fluid when the higher temperature is on the under side, *Phil. Mag.* **32** (1916), 529–546.

199. A. Rényi, *Probability Theory*, North-Holland 1970.

200. A. van Rooij, *Fouriertheorie*, Epsilon Uitgaven **10** 1988.

201. O.E. Rössler, Continuous chaos – Four prototype equations. In: O. Gurel and O.E. Rössler (eds.), *Bifurcation Theory and Applications in Scientific Disciplines*, The New York Academy of Sciences 1979.

202. W. Rudin, *Real and Complex Analysis*, McGraw-Hill 1970.

203. D. Ruelle, Microscopic fluctuations and turbulence, *Phys. Lett.* **72A** (1979), 81–82.

204. D. Ruelle, *Chaotic Evolution and Strange Attractors*, Cambridge University Press 1989.

205. D. Ruelle, Deterministic chaos: The science and the fiction, *Proc. R. Soc. London* **A 427** (1990), 241–248.

206. D. Ruelle, Historical behaviour in smooth dynamical systems. In: H.W. Broer, B. Krauskopf, G. Vegter (eds.), *Global Analysis of Dynamical Systems,* IoP (2001), 63–66.

207. D. Ruelle and F. Takens, On the nature of turbulence, *Comm. Math. Phys.* **20** (1971), 167–192; **23** (1971), 343–344.

208. H. Rüssmann, Konvergente Reihenentwicklungen in der Störungstheorie der Himmelsmechanik. In: K. Jacobs (ed.), *Selecta Mathematica* **V**, Springer-Verlag 1979, 93–260.

209. T. Sauer, J.A. Yorke, and M. Casdagli, Embedology, *J. Stat. Phys.* **65** (1991), 579–616.

210. M. Shub, Stabilité globale des systèmes dynamiques, *Astérisque* **56** (1978) 1–211.
211. C.L. Siegel, Iteration of analytic functions, *Ann. Math.* (2) **43** (1942), 607–612.
212. S. Smale, Diffeomorphisms with many periodic points. In: S. Cairns (ed.), *Differential and Combinatorial Topology,* Princeton University Press 1965.
213. S. Smale, Differentiable dynamical systems, *Bull. Amer. Math. Soc.* **73** (1967), 747–817.
214. C. Sparrow, *The Lorenz Equations: Bifurcations, Chaos and Strange Attractors*, Applied Mathematical Sciences, **41** Springer-Verlag 1982.
215. M. Spivak, *Differential Geometry* Vol. **I**, Publish or Perish 1970.
216. S. Sternberg, On the structure of local homeomorphisms of Euclidean n-space, II, *Amer. J. Math.* **80** (1958), 623–631.
217. S. Sternberg, *Celestial Mechanics* part **II**, Benjamin 1969.
218. J.J. Stoker, *Nonlinear Vibrations in Mechanical and Electrical Systems*, Interscience 1950.
219. F. Takens, Singularities of Vector Fields, *Publ. Math. IHÉS* **43** (1974), 47–100.
220. F. Takens, Forced oscillations and bifurcations. In: Applications of Global Analysis I, *Comm. of the Math. Inst. Rijksuniversiteit Utrecht* (1974). In: H.W. Broer, B. Krauskopf and G. Vegter (eds.), *Global Analysis of Dynamical Systems*, IoP (2001), 1–62.
221. F. Takens, Detecting strange attractors in turbulence, In: D. Rand, L.-S. Young (eds.), *Dynamical Systems and Turbulence, Warwick* 1980, Lecture Notes in Mathematics **898** (1981), Springer-Verlag.
222. F. Takens, Invariants related to dimension and entropy, in: *Atas do 13º colóquio Bras. de Mat.* 1981, IMPA, Rio de Janeiro, 1983.
223. F. Takens, Heteroclinic attractors: Time averages and moduli of topological conjugacy, *Bol. Soc. Bras. Mat.* **25** (1994), 107–120.
224. F. Takens, The reconstruction theorem for endomorphisms, *Bol. Soc. Bras. Mat.* **33** (2002), 231–262.
225. F. Takens, Linear versus nonlinear time series analysis — Smoothed correlation integrals, *Chem Eng J* **96** (2003), 99–104.
226. F. Takens, Orbits with historic behaviour, or non-existence of averages, *Nonlinearity* **21** (2008), T33–T36.
227. F. Takens and E. Verbitski, Generalised entropies: Rényi and correlation integral approach, *Nonlinearity* **11** (1998), 771–782.
228. F. Takens and E. Verbitski, General multi-fractal analysis of local entropies, *Fund. Math.* **165** (2000), 203–237.
229. J.-C. Tatjer, On the strongly dissipative Hénon map, *ECIT 87*, Singapore World Scientific, 1989, 331–337.
230. J. Theiler, Estimating fractal dimensions, *J. Opt. Soc. Am.* **7** (1990), 1055–1073.
231. R. Thom, *Stabilité Structurelle et Morphogénèse*, Benjamin 1972.
232. H. Tong (ed.), *Chaos and Forecasting*, World Scientific 1995.
233. H. Tong and R.L. Smith (eds.) Royal statistical society meeting on chaos, *J. of the R. Stat. Soc.* **B 54** (1992), 301–474.
234. W. Tucker, The Lorenz attractor exists, *C.R. Acad. Sci. Paris* Sér. **I** Math. **328** (1999) 1197–1202.
235. E.A. Verbitski, Generalised entropies and dynamical systems, *PhD Thesis*, Rijksuniversiteit Groningen, 2000.
236. M. Viana, Strange attractors in higher dimensions, *Bol. Soc. Bras. Mat.* **24** (1993), 13–62.
237. M. Viana, Global attractors and bifurcations. In: H.W. Broer, S.A. van Gils, I. Hoveijn, F. Takens (eds.), *Nonlinear Dynamical Systems and Chaos*, Birkhäuser 1996, 299–324.
238. M. Viana, What's new on Lorenz strange attractors? *Math Intelli* **22** (3) (2000), 6–19.
239. B.L. van der Waerden, *Moderne Algebra* I and II, Springer-Verlag, 1931.
240. F.O.O. Wagener, A note on Gevrey regular KAM theory and the inverse approximation lemma, *Dynam Syst* **18** (2003), 159–163.
241. F.O.O. Wagener, A parametrized version of Moser's modifying terms theory, *Discrete and Continuous Dynamical Systems – Series S* **3**(4) (2010), 719–768.
242. Q. Wang, L.-S. Young, Strange attractors with one direction of instability, *Comm. Math. Phys.* **218** (2001), 1–97.

243. A.S. Weigend and N.A. Gershenfeld (eds.), *Time Series Prediction*, Addison-Wesley 1994.
244. H. Whitney, Analytic extensions of differentiable functions defined in closed sets. *Trans. Amer. Math. Soc* **36**(1) (1934), 63–89.
245. H. Whitney, Differentiable functions defined in closed sets. *Trans. Amer. Math. Soc.* **36**(2) (1934), 369–387.
246. H. Whitney, Differentiable manifolds, *Ann. Math.* **37** (1936), 645–680.
247. R.F. Williams, Expanding attractors, *Publ. Math. IHÉS* **50** (1974) 169–203.
248. R.F. Williams, The structure of the Lorenz attractors, *Publ. Math. IHÉS* **50** (1979), 73–99.
249. J.-C. Yoccoz, C^1-conjugaisons des difféomorphismes du cercle. In: J. Palis (ed), *Geometric Dynamics,* Proceedings, Rio de Janeiro 1981, Lecture Notes in Mathematics **1007** Springer (1983), 814–827.
250. J.-C. Yoccoz, Travaux de Herman sur les tores invariants. In: Séminaire Bourbaki Vol 754, 1991–1992, *Astérisque* **206** (1992), 311–344
251. J.-C. Yoccoz, Théorème de Siegel, nombres de Bruno et polynômes quadratiques, *Astérisque* **231** (1995), 3–88.
252. J.-C. Yoccoz, Analytic linearization of circle diffeomorphisms. In: S. Marmi, J.-C. Yoccoz (eds.) *Dynamical Systems and Small Divisors,* Lecture Notes in Mathematics **1784** (2002), 125–174, Springer-Verlag.
253. E. Zehnder. An implicit function theorem for small divisor problems, *Bull. Amer. Math. Soc.* **80**(1) (1974), 174–179.
254. E. Zehnder. Generalized implicit function theorems with applications to some small divisor problems, I and II, *Comm. Pure Appl. Math.*, **28**(1) (1975), 91–140; **29**(1) (1976), 49–111.

Index